The Superpower Odyssey:
A Russian Perspective on
Space Cooperation

Date Due

The Superpower Odyssey:
A Russian Perspective on
Space Cooperation

Yuri Y. Karash

American Institute of Aeronautics and Astronautics
1801 Alexander Bell Drive
Reston, VA 20191–4344

Publishers since 1930

American Institute of Aeronautics and Astronautics, Inc., Reston, Virginia

1   2   3   4   5

Library of Congress Cataloging-in-Publication Data

Karash, Yuri Y.
The superpower odyssey : a Russian perspective on space cooperation / Yuri Y. Karash.
   p. cm.
   Includes bibliographical references and index.
   ISBN 1-56347-319-4 (alk. paper)
   1. Astronautics—Russia (Federation)—International cooperation.
   I. Title.
   TL788.4.K295   1999   387.8'0947—dc21   99-18100

Cover design by Sara Bluestone. The drawing on the front cover is of cosmonaut Yuri Romanenko and astronaut Donald Slayton. It is taken from a sketch by American artist Paul Calle who documented the astronaut–cosmonaut joint crew training exercises in Star City, July 1974. (Courtesy of NASA.) Other elements are Apollo and Soyuz spacecraft and a Shuttle–Mir docking.

To the memory of Soviet cosmonaut Vladimir Komarov and U.S. astronauts Gus Grissom, Ed White, and Roger Chaffee who could have formed a single crew to the moon but who instead died in a space race on opposite sides of the Iron Curtain.

FOREWORD

It was President John F. Kennedy who first indentified outer space as an "area of common interest" between the United States and the Soviet Union, one in which it might be possible to lessen Cold War rivalries by working together on common projects. In his 20 January 1961 Inaugural Address, Kennedy suggested, "Let both sides seek to invoke the wonders of science instead of its terrors. Together let us explore the stars." Almost 40 years later, the United States and Russia are struggling to translate Kennedy's vision into a lasting reality.

The politics of early 1961 led Kennedy to decide that the United States should compete with the Soviet Union in space, rather than seek areas for cooperation. The result, of course, was Project Apollo and the race to the moon. But Kennedy never abandoned the hope that space could become an arena for superpower cooperation. Addressing the General Assembly of the United Nations on 20 September 1963, Kennedy asked, "Why, therefore, should man's first flight to the moon be a matter of national competition?... Surely we should explore whether the scientists and astronauts of our two countries—indeed of all the world—cannot work together in the conquest of space." Just 10 days before he was assassinated, Kennedy directed his top space officials to see if there was a way to turn this rhetoric into reality. One can only speculate what might have followed, had Kennedy served two full terms in the White House.

In this study, author Yuri Karash traces the long process through which President Kennedy's vision is finally being turned into reality. The launches of the first two elements of the International Space Station (ISS), in which the United States and Russia are central partners, have already been accomplished. After more than three decades of an ambivalent space relationship, well described in this book, the United States invited Russia to become a partner in the space station in 1993; this brought together the only two countries capable of taking humans into Earth's orbit and sustaining them there. While the United States and the Soviet Union (and now Russia) have been cooperating in space on an episodic basis since the mid-1960s, the International Space Station marks the first time they have both committed themselves, together with other partner countries, to a long-term space undertaking. If the multination partnership in the ISS works out well, it certainly will be the precedent for organizing future large-scale government ventures in space.

As Russia struggles to stabilize its economic and political systems during the difficult transition away from more than 70 years of Communist rule to a still-uncertain future, the short-term viability of U.S.–Russian space cooperation is being held hostage to broader currents in Russian society. But as President Kennedy recognized nearly four decades ago, the underlying logic of the partnership—that all will be better off if these two space powers cooperate rather than compete—is so strong that the partnership is likely to survive its current problems. At least I hope that is the case.

The author of this study is perhaps uniquely qualified to carry out such a work. I first met Yuri Karash in the fall of 1992, when he arrived, full of energy and enthusiasm, to take my course in space policy at George Washington University. He recently had come from the Soviet Union, after being one of the finalists in the Soviet "journalist-in-space" competition and approaching preliminary cosmonaut training. He already had decided to dedicate himself to getting the academic background needed to be a scholar of space affairs as well as a practitioner, and in the past six years he has sought my guidance as he pursued his doctoral studies at American University. This book grew out of his doctoral dissertation and thus is the culmination of his desire to understand the roots of this most significant area of space collaboration. Meanwhile, in his professional life, he has been working to link those most closely involved in the United States and Russia in crafting the arrangements for their collaborative activities in human space flight.

It has been gratifying to watch Yuri Karash's academic and professional career develop. He has already made many contributions to helping his homeland and his adopted country work together to open up the space frontier.

John M. Logsdon
Space Policy Institute
George Washington University
March 1999

TABLE OF CONTENTS

ACKNOWLEDGMENTS

This book is based on a Ph.D. dissertation that I defended at American University in 1997. For this reason I would like to thank first and foremost my dissertation advisers—associate dean of the School of International Service (SIS) Nanette Levinson, professor in the School of Public Affairs Howard McCurdy; and Director of the George Washington University Space Policy Institute John Logsdon—for their continuous guidance, assistance, and encouragement during the work on this dissertation. Other academicians who provided support in this endeavor include dean of the SIS Louis Goodman, director of the SIS Ph.D. program John Richardson, and SIS Professor of Foreign Policy William Kincade.

I also owe thanks to the Russian and American space policy makers, space professionals, Russian politicians, and cosmonauts/astronauts who agreed to be interviewed for this research and who provided very important insights into the processes that shape space policy in the two countries.

On the Russian side they include Major General Yuri Glazkov, deputy chief of Gagarin Cosmonaut Training Center (Star City); Anatoly Kiselyov, director of Khrunichev Space Center; Yuri Semyonov, general designer of RKK Energia; Alexander Medvedchikov, Russian Space Agency (RSA) deputy general director; Alexander Botvinko, deputy chairman of RSA manned space flights department; Mikhail Karnaukhov, RSA senior specialist; Alexei Krasnov, RSA deputy director of the Department of International Cooperation; Oleg Shishkin, former Soviet minister of General Machine Building; Inessa Kozlovskaya, Institute of Biomedical Problems (IBMP) department director; Mark Belakovsky, IBMP contract manager; Sergey Zhukov, president of the Moscow Space Club and director of the Center for Technology Transfer; Vitaly Sevastyanov and German Titov, Duma deputies; Gennady Zyuganov and Alexander Lebed, presidential candidates in the 1996 elections; Valery Zorkin, justice of the Constitutional Court of the Russian Federation; Anatoly Dobrynin, former ambassador of the Soviet Union to the United States; Sergei Zhiltsov, Khrunichev Space Center public relations department director; Konstantin Lantratov, Khrunichev Space Center public relations department senior expert; and Vladimir Dezhurov, Sergey Krikalev, Gennady Strekalov, and Vladimir Titov, cosmonauts.

On the U.S. side my thanks go to Richard Truly, former NASA administrator; Arnold Aldrich, former NASA associate administrator; Robert Clark, former NASA associate administrator; John Fabian, former president of ANSER, former vice president of the International Astronautics Federation, and former president of the Association of Space Explorers; Dick Kohrs, former manager of system engineering of the Space Shuttle and former director of the Space Station Freedom program; Thomas Moser, former director of the Space Station Freedom program; John Shumacher, NASA associate administrator for external relations; Lynn Cline, NASA deputy associate administrator for external relations; Richard Beatty, former employee of Booz, Allen and Hamilton; and David Stall, former Nassau Bay city manager.

I also would like to express my appreciation to Shiuchi Miura, the executive director of the Japanese National Aerospace Development Agency for offering insight into the status of Japanese contributions to the International Space Station.

My thanks also to the NASA History Office for providing me with numerous documents, materials, and photographs without which I could not have completed this project. The Media Resource Center of the NASA Johnson Space Center supplied me with unique pictures relevant to the contemporary period of Russian–American cooperation in space.

I also would like to express my appreciation to Marina Grigorovich-Barskaya for letting me use photographs taken by her late husband, Konstantin Grigorovich-Barsky. Special thanks to Bill Ingalls of NASA and Thomas Cones of Nassau Bay for their contributions to the photo section.

I owe great thanks to my Soviet/Russian flight instructors without whose help I would have never been able to apply for cosmonaut training and, ultimately, would have never written a book on space policy issues. I would like to extend these thanks to my U.S. instructors for helping me stay "in shape" and retain the hope of flying one day as a cosmonaut/astronaut—one of the motivations behind my study of U.S.–Soviet/Russian cooperation in space.

My thanks also to ANSER, an independent public service research institute, and particularly to the leadership of the Center for International Aerospace Cooperation (CIAC), where I was employed. Former president and CEO of ANSER and former U.S. Astronaut John Fabian, former CIAC Director Richard Kohrs, former Director Dick Kline, and former Deputy Director Steve Hopkins were all instrumental in allowing me to combine my professional and academic duties.

My thanks to Kirk Douglas for sharing with me his memories of hosting Russian cosmonauts George Beregovoi and Konstantin Feoktistov in his house.

My special gratitude to Irina Kaznacheeva, an anchor of the Moscow TV business news program *Delovaya Moskva*, for providing me with timely information about developments regarding the Mir space station.

I would like to express my deepest apprection to my parents Irina and Yuri M. Karash for helping me to collect in Russia information necessary for this book while I was in the United States.

My ultimate thanks go to Tatiana Dominick (Dzeneeff), a Russian woman born in the United States, for the support, encouragement, and care she provided me during all five years of my Ph.D. training.

Finally, I would like to acknowledge the crucial role that Heather Brennan of the AIAA book program played in the birth of this book. Her outstanding professional skills helped the book meet the highest publication standards.

Yuri Karash
March 1999

Chapter 1

> We are bringing our spacecraft closer together.
> We are bringing our nations closer together.
>
> Discovery *Commander James Wetherbee,*
> *during Shuttle–Mir rendezvous,*
> *6 February 1995*

INTRODUCTION

OVERVIEW

This book examines the space era from 1957 to 1998, from the first satellite launch through the joint U.S.–Russian work on the International Space Station (ISS) project. It focuses on the analysis of three distinct periods of possible and real cooperation between the United States and the Soviet Union/Russia. The goal of this study is to answer two questions: 1) what are the necessary conditions for the emergence of meaningful space cooperation between Russia and the United States, and 2) might this cooperation continue developing on its own merit, even if the conditions that originated the cooperation were to change?

The first possibility for limited Soviet–American cooperation in space emerged in the late 1950s, together with the space age, and continued until the mid-1960s. The major potential joint project of this period was a human expedition to the moon. The global competition/confrontation between the two countries and aspirations of U.S. and Soviet space communities to beat each other in space prevented actual cooperation.

The second period was from the late 1960s until 1985. During this period experimental docking missions were considered, involving Soyuz, Apollo, Salyut, and Skylab spacecraft and space stations, and later a reusable U.S. Space Shuttle and a Soviet Salyut-type station. The global U.S.–Soviet competition still continued, but the confrontation was replaced by détente for a brief period of time lasting from the end of the 1960s until the mid-1970s. Détente gave the first example of U.S.–Soviet cooperation in space—the Apollo–Soyuz Test Project (ASTP), which took place in 1975. The project also became possible because of the interest of U.S. and Soviet space communities in limited cooperation with each other. However, the lack of willingness of the political leaders to continue broadscale cooperation between the two countries, the end of détente, and the certain disappointment of U.S. and Soviet space professionals about joint work on a first significant cooperative project eliminated the possibility for ASTP-like endeavors at least until the late 1980s.

The third period started with Mikhail Gorbachev's *perestroika* in 1985 and was ongoing at the time of publication of this book, in early 1999. It involved almost 100 joint space projects at both the governmental and private sector levels. The main focus of the joint activities became U.S.–Russian work on the ISS, including Phase 1 of the program—Shuttle–Mir flights. The intention of the Kremlin and White House to make space an area of common interest for the two countries, bringing an end to the space race—one of the major driving forces behind U.S. and Soviet space programs—helped spur a U.S.–Russian partnership in probably the most technologically challenging cooperative program of the 20th century.

The realization of the program encountered a number of difficulties, caused in part by Russia's inability to meet its ISS obligations on time. However, the commitment of the U.S. and Russian political leaderships and space communities to both the ISS and cooperation within this program, and the interdependence of the two countries in building ISS, provide good reasons to believe that the U.S.–Russian broadscale space partnership will finally develop a long-term character.

BACKGROUND

To date there are about 20 spacefaring countries. However, there are only two nations in the world that have more experience than others in human space flight. Thus, the near-term future of mankind in space, especially in the area of human space exploration, depends to a considerable extent on the continuing success of these two space programs.

The necessary conditions for this success are 1) economic power, 2) technological power, and 3) societal willingness to venture into space, which is reflected in space budgets. In the wake of the end of the Cold War, bipolar global competition, and the subsequent disintegration of the Soviet Union, some of these conditions have been endangered—the first and third one in Russia, and the third in the United States. One of the most promising ways to overcome these problems is to replace competition with cooperation between Russia and the United States.

There are three primary rationales for undertaking joint space activities. Cooperation in space, as it is tied to national security, may help Russia and the United States develop and maintain mutual trust that may prevent, or at least slow down, the process of the two nations going back to the Cold War in their relationship. Unification of the technological potential of the two countries can contribute to the solution of some of the Earth's problems from space and can also intensify the study of the universe.* Cooperation can potentially help Russia and the United States handle budget constraints by sharing some of each other's technologies and hardware instead of duplicating them.

The best approach to studying Soviet/Russian–American cooperation in space is to analyze it in the context of the structure of the international political system, bilateral U.S.–Soviet/Russian relations; the attitudes of the U.S. and Soviet/Russian leaderships and the space communities of the two countries toward cooperation with each other; and the payoffs that this cooperation brought in the past, is bringing now, and may bring in the future.

One of the clues for understanding the international political role of space exploration is based on the assumption that the state seeks to maximize power. Strength is an inseparable element of power and is expressed through prestige, among other means. The emergence of the Russian and American space programs was, to a large extent, because of global competition between the two superpowers for the dominant position in the hierarchy of prestige in the international system.

This global competition is now over, but Russia and the United States are still independent countries concerned about their interests. The concepts of strength and power continue to play an important role in their bilateral relations as well as in their relations with the rest of the world. Thus the space program, which is still one of the most obvious manifestations of the economic and technological strength of a nation, is still involved in political calculations, to a greater extent in the United States and to a lesser extent nowadays in Russia.

Space activities have a far more complicated character than just a demonstration of power, however. The launch of Sputnik marked the beginning of a new struggle for the balance of world power—a struggle that involved competition for the loyalty and trust of friends and allies. For this reason, in the 1960s the United States helped Europe and Canada begin their own space programs, launching the satellites they had developed. In the 1970s, the United States broadened the basis of its cooperation with Europe and Canada by engaging them in its major post-Apollo development program, the space transportation system (the Space Shuttle). The U.S. government also enabled the licensing of U.S. technology, so that Japan could quickly develop its own

---

* Among the benefits that could be brought to Earth by the exploration of space and, ultimately, its colonization are ecological monitoring, an increase of agricultural effectiveness, the manufacturing of ultra-pure pharmaceuticals, search and rescue, the prevention of collisions between Earth and other celestial bodies, and, in the long-term perspective, solutions to the problem of overpopulation. See also Ref. 1.

space program. In the 1980s, the United States took the lead in forging a multidecade experiment in intimate international cooperation with its closest friends and allies—Space Station Freedom.[2] The Soviet Union also used space to strengthen relations with its friends and allies. From the 1970s to the 1980s, the Soviet Union developed and carried out the Intercosmos program under which it launched a number of cosmonauts from the Socialist camp and Third World countries.

Cooperation in space was used by the United States and the Soviet Union not only as a means of rapprochement with their traditional friends and allies but also with each other. According to the members of the NASA Task Force on International Relations in Space,

> Historically, space relations with the Soviet Union have been controlled by overall U.S. policy toward that nation. When antagonisms lessen, space relations improve. Indeed, they are one of the means by which the superpowers signal to each other and to the world at large changes in the political relationship.[3]

In the 1970s, the Apollo–Soyuz "handshake in space" symbolized a period of U.S.–Soviet détente. Marcia Smith, a specialist in space policy at the Congressional Research Service, believed that the whole idea of the disintegration of the former Soviet Union, the emergence of Russia, and the prospects for the post-Cold War world are all embraced by the concept of Russia participating in the space station program.[4]

Profound changes in U.S.–Soviet relations in the late 1980s and a gradual transformation of a bipolar international system had devalued to some extent an approach that analyzed the space relationship between the Soviet Union and the United States in terms of maximizing prestige. The need for the new approach was outlined in *International Space Policy for the 1990s and Beyond*, which particularly stressed that

> [w]hile both countries will insist on being in full control of the destiny of their own space programs, reflecting their own requirements and priorities, space relations between the two super-space powers have the potential to be more than a simple reflection of their current political relations. Their common interest in the productive use of civil space for science and applications could become a bridge similar to their common interest in arms control, controlling nuclear proliferation, protecting the environment, and the sale and purchase of grains.[3]

THEORETICAL FRAMEWORK

The realist theory of international relations, which is based on the assumption that states seek to maximize power and prestige, explains the political significance attached to the space programs by the Soviet Union and the United States. The theory of international functionalism helps explain the role that the Kremlin and the White House expect space cooperation to play in the overall relationship between Russia and the United States. According to this theory, individuals and states alike could have such a good experience cooperating in one particular field that they would like to try cooperating in other fields too, thus, overall, getting closer to each other.[5]

However, what if Russia and the United States do not quite meet each other's expectations while cooperating in space? What if the costs of this cooperation exceed the benefits? Will this cooperation continue even when the political and economic conditions that spurred it change?

Such questions are not groundless or rhetorical. U.S.–Russian cooperation in space was initiated during the "honeymoon" that dominated Russian–American relations for a few years after the disintegration of the Soviet Union and was aimed particularly at building the ISS faster, better, and cheaper. However, the honeymoon is over and Russia is causing considerable problems to the realization of the ISS program. Will the U.S.–Russian partnership in the space station program survive?

The answer will be "yes" rather than "no" if, in addition to the commitment of U.S. and Russian top political leaders and space communities to this partnership, the two countries develop a cooperative regime within their space station alliance.

Stephen Krasner defined a regime as "implicit or explicit principles, norms, rules and decision-making procedures around which actors' expectations converge in a given area of international relations."[6] Regime also implies an unlimited number of partners, which is particularly important in the case of the ISS—a multinational project. It also stresses the importance of reputation and the "promise of the future" and shifts attention from the immediate costs and benefits to the relevance of agreements that establish principles, procedures, and institutions—both for regulating competition and for facilitating cooperation.[7,8] Besides, if a regime really increases the payoffs of cooperation, it will reinforce the regime itself (Ref. 8, p. 506). Thus, if Russian and U.S. space professionals develop effective business ties that will facilitate cooperation, these ties will become one of the factors strengthening the U.S.–Russian partnership in the ISS program and may help maintain cooperation even if the original reasons for its creation have weakened.

A regime also enhances the importance of reputation. When the threat of station cancellation in the United States became strong enough, Japan's Foreign Minister Taro Nakayama wrote Secretary of State James Baker to say that if the space station were canceled, "major joint efforts which the international partners have made so far would be nullified, Japanese Space Development Programs would be significantly impaired, and furthermore, I fear that the credibility of the United States as a partner in any major big science effort would inevitably be damaged."[9] In 1994, the *Appropriations Bill 1995* "included more than $2,100,000,000 for the station, despite the severe outlay constraints it faces, because of this project's preeminent role in U.S. foreign policy.... To recommend termination of the project at this point would seriously undermine our diplomatic relationship with the European Community, Japan, Canada, and the Republic of Russia."[10]

According to the Russian Space Agency Director General Yuri Koptev, if Russia does not manage to meet its committments to the ISS:

> all the other participants in the project will have to correct their expenditures upward. The station itself will actually undergo substantial structural changes. The time for actually commissioning it will be postponed by several years. Who will like that? Do I really need to argue that after all this the international space community will radically alter its attitude to Russia?[11]

The interest group theory helps to understand the role played by space communities in U.S.–Soviet/Russian cooperation in space. According to this theory, the social reality of mankind could neither be expressed simply in class terms, nor in isolated individual terms, but rather in the complex groupings in which men and women are associated—families, communities, regions, professions and occupations, and the like. The state must accommodate the legitimacy and autonomy of these groupings, and they must have a share in public decision making. Some Western political scientists, such as Gordon Skilling, believe that interest groups are compatible with an authoritarian system, although they are subject to restrictions imposed by the state; this makes this theory particularly valuable for understanding the role played by the Soviet space organizations in U.S.–Soviet cooperation in space.[12,13]

METHODOLOGY

The most appropriate methodology for the study of Soviet/Russian–American space cooperation is a combination of historical–comparative analysis and explanatory case study analysis. The reason for selecting this approach is the need to study Soviet/Russian–American cooperation in its historical perspective to make conclusions about the prevailing trends that could shape cooperation in the future. The best way to do this is to compare different periods in space cooperation and the varied Russian and American approaches in the context of the two nations' bilateral relationship and the global political situation.

Despite similarities, such as the strong political drive behind space efforts and the emphasis on a human space flight, Russian (especially during the

Soviet period) and American space programs have unique features. The record of Soviet behavior, as well as the explanations of and justifications for that Soviet behavior provided by the Soviets themselves, made clear that the factors determining Soviet policies and actions were the product not of conditions and practices similar to American ones but of an environment and view of the world and of a decision-making process very different from those in the United States. The Soviet decision-making process was marked by the absolute authority of the senior Communist Party hierarchy, with no mechanisms for dissent or public influence, and with built-in imperatives to preserve established patterns of thought and rule.[14] The uniquely different features of the Soviet and American approach encourage the adoption of an analysis that will retain "the holistic and meaningful characteristics of real-life events."[15] This kind of analysis was provided by an explanatory case methodological approach.*

REFERENCES

[1]Sagan, C., *Pale Blue Dot: A Vision of the Human Future in Space*, Random House, New York, 1994.

[2]Space Policy Inst., Association of Space Explorers, "International Cooperation in Space—New Opportunities, New Approaches," *Space Policy*, Vol. 8, No. 3, Aug. 1992.

[3]NASA Advisory Council, Task Force on International Relations in Space, *International Space Policy for the 1990s and Beyond*, 12 Oct. 1987, p. 35.

[4]Smith, M., Congressional Research Service, Speech at the National Space Outlook Conference, 22 June 1994.

[5]Mitrany, D., *The Functional Theory of Politics*, St. Martin's Press, New York, 1975.

[6]Krasner, S., (ed.), *International Regimes*, Cornell Univ. Press, Ithaca, NY, 1983, pp. 1–21.

[7]Keohane, R., *After Hegemony*, Princeton Univ. Press, Princeton, NJ, 1984.

8Haggard, S., and Simmons, B. A., "Theories of International Regimes," International Organization, Vol. 41, Summer 1987, pp. 505–506.

[9]Logsdon, J. M., "U.S.–Japanese Relations at a Crossroads," *Science*, Vol. 255, 17 Jan. 1992.

---

* See Ref. 15. Yin outlined three distinct approaches to conducting case study research: exploratory, descriptive, and explanatory. An exploratory case study seeks to delineate the boundaries of a particular case. Conversely, descriptive case studies highlight, in detail, the unique characteristics of the chosen cases. Finally, explanatory case studies attempt to account for the underlying dynamics of the factors that constitute the basis for both exploratory and descriptive case study analysis. As such, explanatory case studies incorporate aspects of exploratory and descriptive approaches and are thus the most comprehensive.

[10]U.S. Congress, Senate, Departments of Veteran Affairs and Housing and Urban Development, and Independent Agencies, *Appropriation Bill 1995*, Rept. prepared by Barbara Mikulski to accompany H.R. 4624, 103d Cong., 2d sess., 1994, p. 123.

[11]Interview with Yuri Koptev, *Rossiyskaya Gazeta* [Russia's Newspaper], 25 Feb. 1997.

[12]Fleron, F. J., Jr., and Hoffmann, E. P., (eds.), *Post-Communist Studies & Political Science: Methodology and Empirical Theory in Sovietology*, Westview Press, Boulder, CO, 1993, p. 46.

[13]Skilling, G. H., and Griffiths, F., (eds.), *Interest Groups in Soviet Politics*, Princeton Univ. Press, Princeton, NJ, 1971.

[14]Kohler, F. D., "An Overview of U.S.-Soviet Space Relations," *U.S.-Soviet Cooperation in Space*, edited by D. Harvey and L. Ciccoritti, Center for Advanced International Studies, Univ. of Miami, Miami, FL, 1974, p. xxvi.

[15]Yin, R., *Case Study Research: Design and Methods*, Sage, Thousand Oaks, CA, 1994, p. 14.

Chapter 2

COOPERATION/COMPETITION DILEMMA

THE BEGINNING: DWIGHT EISENHOWER–NIKITA KHRUSHCHEV PERIOD

To understand fully the history of Soviet–U.S. cooperation in space, aspects of the Khrushchev–Eisenhower period must first be examined. Although that particular period failed to produce any tangible cooperation, it was still a very important formative period. On one hand, both the United States and the Soviet Union were in the process of forming their space policies, and on the other hand, they were trying out their respective attitudes toward space cooperation with each other. As each country tested its respective approach toward bilateral cooperation, there was a certain amount of interaction that, if failing to bring any tangible results, nevertheless helped to determine certain factors that ultimately influenced Soviet/Russian–American cooperation in space.

By the mid-1950s, the race for the weapons of mass destruction was accelerating between the Soviet Union and the United States. Eisenhower had to face the threat presented by Soviet H-bombs, new long-range bombers, and later missiles. Although he wanted to reduce military spending, he did not want to risk falling behind the Soviet Union in nuclear arms.

> Eisenhower felt desperately the need for accurate intelligence about Soviet progress. The U-2 spyplane was a stopgap, space based reconnaissance the solution.... So Eisenhower's administration came to have two priorities in missile and space policy in the mid-1950s. The first was to make up for lost time in R&D leading to the development of American missiles; the second was to ease into the space age in such a way as to preserve American hopes for penetrating the Iron Curtain once and for all. Each consideration contributed, in its own way, to the tardy timing of the first American satellite. (Ref. 1, p. 113).

Although the launch of an artificial satellite was supposed to become a major breakthrough in human knowledge about the Earth and the universe, President Eisenhower's satellite decision was not driven only by scientific considerations. *The Los Angeles Examiner*'s Washington Bureau Chief noticed that "Ike scored again as a psychological warrior in announcing to the world that America is planning to launch a man-made satellite.... Conceivably, such a device may evolve into a floating sentrypost to look down the throat of Red Russia for any symptoms of acute aggressive automatitis."[2] Although President Eisenhower stressed that this was not a defense weapon and that the scientific secrets uncovered would be shared with all nations, including the Soviet Union, he was obviously leading from strength. It was no coincidence that the announcement of the satellite decision was made only a few hours after the official revelation that a stockpile of hydrogen bombs had been built up in the last six months on President Eisenhower's orders.[2]

Moreover, according to a spokesman for the National Science Foundation, one of the reasons for the satellite announcement was "to offset concern over the strong possibility that Russia is ahead of the U.S. in the race to launch a satellite"[3] because the Soviets had already announced their plans to launch the first artificial satellite (although small compared to Wernher von Braun's plans for the American satellite). Von Braun envisioned a military space station equipped not only with television cameras, but also winged A-bombs or even conventional weapons that could be dropped on any place on Earth. Such statements clearly indicated the considerable involvement of space exploration in the balance-of-power struggle between the Soviet Union and the United States.

The political preparation needed to use a reconnaissance satellite "to look down the throat of Red Russia" was as important as the technical preparation. After the blunt rejection by the Soviets of his Open Skies idea that suggested mutual aerial surveillance of all Soviet and American military installations, President Eisenhower had to find a political solution for the overfly of Soviet territory by American reconnaissance satellites. One of the solutions was to create a precedent for a free and legal overfly of different countries by an American satellite. The International Geophysical Year with the participation of 66 nations provided such an opportunity. Similar hopes were shared by the participants in the International Astronautics Congress in Rome in September 1956. Some speakers pointed out that the progress of technology in the fields of rocketry and artificial satellites was so rapid that it possessed a serious danger unless the legal problems that arose were settled beforehand by international agreement. "Fortunately, the first artificial Earth satellites are to be sent up by the United States and, probably, the Soviet Union as part of a vast cooperative international scientific endeavor."[4] The launch of the first American satellite was presented as a contribution to this scientific endeavor and made it easier to justify an overfly of different countries, including the Soviet Union, by an American satellite. A few days after the launch of Sputnik I, President Eisenhower and Deputy Secretary of Defense Donald Quarles discussed this issue; Quarles observed, "... the Russians have...done us a good turn, unintentionally, in establishing the concept of freedom of international space.... The President then looked ahead...and asked about a reconnaissance (satellite) vehicle."[5]

One more probable reason to make a satellite decision announcement at the end of July 1955 occurred just a few weeks after the end of a summit among the leaders of Great Britain, France, the United States, and the Soviet Union. This meeting did not bring any tangible results in terms of the solution of current international problems but created "the spirit of Geneva." The summit established "an outspoken assumption that would underline much subsequent summit diplomacy: the nations whose leaders had dealt with each other face to face would find it difficult to be enemies" (Ref. 6, p. 223). Thus, there was good reason to expect that the satellite decision announcement, although a clear sign of American strength, would not be taken by the Soviets as a direct threat and would not hinder even limited interaction in space between the Soviet Union and the United States.

The initial reaction of the Soviets to possible cooperation with other countries, including the United States, in launching satellites seemed to be positive. When Premier Khrushchev was asked at a Moscow diplomatic reception whether the Soviet Union would agree if the United States asked for cooperation in the satellite scheme, he said, "Yes...if it is in the interests of mankind," but at the same time he added, "I have heard something about it but have not read it carefully so I cannot say anything definite."[7] The next day Leonid L. Sedov, president of the Soviet Commission on Interplanetary Communication considered it "quite possible" for the United States and the Soviet Union to cooperate in sending a globe-circling satellite into space, although he said at the same time that Soviet Union might beat the United States and launch the first man-made satellite, which would be bigger than the basketball-sized American satellite within two years.[8] Khrushchev's confession that he did not understand space issues and Sedov's statement proving the independent thinking of the Soviet scientific elite on space exploration are very important for understanding the Soviet attitude toward cooperation in space with the United States.

USSR ACADEMY OF SCIENCES

The USSR Academy of Sciences was composed in large part of people whose careers had often involved high administrative posts in the military and industrial bureaucracies, and as such they were not natural spokesmen for science in the same sense as most members of the American National Academy of Science. The USSR Academy was itself part of the governing structure of the Soviet Union and part of the military-industrial-party-governmental complex. As such it served as a tool for the advancement of the scientific community's point of view rather than as a mouthpiece for science.

Because no such tool existed in the highest echelons of the U.S. government, it was easier for the president to effectively shut science out of policy implementation and space decision-making than for the Soviet leadership to do so. By co-opting leading scientists into the party, government, and defense elites, the Soviet system not only permitted greater political control of science and scientists but also increased the possibility for science to influence politics.[9] Moreover, "with the combination of prestige and respectability, election to the academy became a real magnet. Stalin was decorated with the title Honorary Academician" (Ref. 10, p. 146). According to Academician Roald Sagdeev, a prominent Soviet scientist who was President Mikhail Gorbachev's Science Adviser, the USSR Academy of Sciences was almost "a state within a state." It was given "responsibility for final judgement on any nationwide initiative, whether it was investment in a new branch of industry or a construction project that might have environmental repercussions" (Ref. 10, p. 148).

KOROLEV'S INFLUENCE

According to Khrushchev's son Sergei Khrushchev, who was a Soviet aerospace engineer, Premier Khrushchev had been strongly influenced by the Soviet Rocket Chief Designer Sergei Korolev when he visited Korolev's design bureau in 1956. Before this visit, the premier did not even think about beating the Americans in space—this thought had never crossed his mind.[11] Korolev convinced him of the possibility of gaining tremendous political prestige by winning a satellite race. Korolev assured Khrushchev that his design bureau would be able to launch a first satellite many times the weight of an American one and do it before the Americans. Soon after Premier Khrushchev's visit to Korolev's bureau, the Central Committee of the USSR's Communist party and the USSR Council of Ministers issued a decree on space exploration on 30 January 1956. The launch of an artificial satellite was described in one of the sections of the decree. Moreover, meeting with Korolev helped Khrushchev to realize the military and political significance of missiles to the extent that this meeting could probably be considered as a starting point in the Soviet missile race with the United States (Ref. 11, pp. 111–112).

Taking into account the influence Korolev had on Khrushchev, it will not be an exaggeration to consider the chief designer and his bureau an institution that shaped Soviet space policy in the beginning of the space age. This observation was confirmed by Sagdeev, who described Korolev as "a mighty feudal lord, capable, it was said, of 'opening the door of Khrushchev's office with his foot.' If this were not the case, then there probably would not have been the first Sputnik" (Ref. 10, p. 201). Yaroslav Golovanov, one of the most distinguished and experienced Soviet/Russian space journalists, who was close to Korolev, covered Soviet activities in space starting in the late 1950s.

According to Golovanov, who was preparing to fly in space as a journalist in the 1960s, Korolev became a politician himself. It happened because the chief designer was a leader in the field of Soviet science and technology, which was shaping, to a significant extent, Soviet domestic and foreign politics (Ref. 12, p. 572). Korolev was thinking of the Soviet–U.S. space relationship only in terms of a race and was instigating it (Ref. 12, p. 530). In a way, he was concerned about beating Americans in space more than the Soviet political leadership itself.* The Soviet government issued a special decree freeing Korolev of any responsibility if he failed to launch the first artificial satellite before the United States. The reason for the decree was that Sputnik was being developed by another famous Soviet scientist, Mstislav Keldysh (who later became a president of the USSR Academy of Sciences), and Korolev had no formal control over Keldysh's work on the satellite. However, beating the Americans in space was Korolev's personal ambition and challenge.

Initially the launch of the first Sputnik was scheduled for 6 October 1957. However, a conference within the International Geophysical Year (IGY) framework was scheduled for early October in Washington, D.C., and the topic was the use of rockets and satellites for space exploration. Korolev learned from the press release of the conference that the U.S. representative was supposed to make a report called "A Satellite Above the Planet." Korolev was very alarmed by this news; he thought that this report could be a reflection that the Americans would have already launched their own satellite. To make sure that he would not be surpassed by the Americans, Korolev rescheduled the launch of the satellite for two days earlier, on 4 October (Ref. 12, p. 537).

Overall, Golovanov believed that if it had not been for Korolev's initiative, Sputnik would have never been designed, manufactured, and integrated with a launch vehicle early enough to be launched before Explorer I (Ref. 12, p. 532). All of these facts taken together give a good reason to conclude that Korolev's attitude was one of the factors that hindered Soviet–U.S. cooperation in space in the late 1950s through early the 1960s.

---

*Korolev's nationalistic approach to space exploration was confirmed by one of his close coworkers Boris Chertok more than 30 years after the general designer's death. In late 1998 the Russian aerospace community debated whether to deorbit the Russian space station Mir, so that Russia could concentrate all its efforts on the ISS, or to continue Mir's operation, possibly delaying even further the fulfillment of Russia's commitments to the ISS. At that time Chertok said that for Korolev the decision would not have been a dilemma at all. He most likely would have dropped from the partnership with the Americans and begun building the Mir-2 station. Korolev, according to Chertok, would never let bureaucrats sell the Russian scientific potential at the dumping price. Korolev could straightly say to the leader of the Soviet Union Leonid Brezhnev, who became general secretary after Khrushchev, "you messed up," and walk out of the leader's office. Korolev would have understood that Mir's deorbiting is not just the destruction of one spaceship, it is the destruction of a national manned space program. Chertok also expressed confidence that one of Korolev's successors in the position of general designer, Valentin Glushko, would have never given the international space program national priority.[67]

The strengthening of Soviet rocketry in technological, economic, and political terms continued influencing the USSR Academy of Sciences in the direction of its transformation into a "spokesman for rockets." The rocket branch of the Academy had more members (102 academicians and correspondent members) and was better sponsored by the state than any other branch of the Academy. Almost all of the representatives of the Soviet rocket elite became members of this branch, including Korolev and Valentin Glushko (the chief designer of rocket engines). This branch demanded that the Academy leadership reshape the activities of the whole Academy, so that it would better serve the interest of Soviet rocketry. When the president of the Academy, Academician Alexander Nesmeyanov, who occupied this post in the late 1950s and early 1960s, tried to resist this pressure to avoid deemphasizing fundamental sciences, he was replaced by Keldysh. The way it was done shows one more example of the relative independence of the Soviet scientific elite from the party leadership. Although forced to resign, Nesmeyanov was replaced by Keldysh via elections in the Academy. Formally, even Khrushchev could not fire a president of the Academy of Sciences (Ref. 12, pp. 705–706).

Another example of the relative independence of the Soviet scientific elite from the party leadership could be the conflict that developed between Korolev and Glushko. Both of them were involved in a serious personal and organizational conflict, first, over the reason for initial failures of the early launch vehicles (Glushko was blaming malfunctions on Korolev's rockets; See Ref. 12, pp. 506–509), and then in the early 1960s, over the appropriate design of new rocket motors for the next generation of the Soviet space launcher. This conflict was aggravated by the struggle of both for undisputed leadership in the field of rocket science and technology. Even Khrushchev's personal intervention could not resolve the conflict between Korolev and Glushko. The two general designers each had enough independent strength that the Kremlin could not simply order them to cooperate (Ref. 11, pp. 449–454). Ten years after Korolev's death, when Glushko became head of Korolev's Design Bureau called NPO Energia, he was promoted to the ruling body of the Communist Party of the Soviet Union Central Committee. From this moment on, "Glushko concentrated in his hands not only the power of an enormous space empire, but also the political power of a commissar, capable of overwhelming anyone in the space establishment" (Ref. 10, p. 209).

The reason why Korolev had to fascinate Premier Khrushchev with the possibility of beating the Americans in space is clear—this was the only way Korolev could convince a Party leadership obsessed with the idea of catching up with the United States, particularly in the field of technology, to give money to the satellite project. The need for a strong political rationale for continuing space exploration was later confirmed by Academician Sedov while attending the 10th International Congress of Applied Mechanics at Stressa, Italy, in September 1960. Hugh Dryden, deputy administrator of NASA, reported that "Sedov discussed cooperation but without very definite results.... He suggested that if we really cooperated on man in space, neither country would have a program because of the competitive element and for political reasons."[13]

REACTION IN THE UNITED STATES

In the United States, scientists did not influence politics to the extent their Soviet colleagues did. Initially, the satellite project was not part of U.S. participation in the IGY program, and a number of U.S. scientists believed that it was a mistake because of the many indications that the Soviet Union was working on its independent satellite project. This matter had been discussed among U.S. scientists during the second meeting of the Comité Special de L'Annee Geophisique Internationale, held in Rome in September 1954. Having agreed that U.S. prestige would be damaged if the Soviet Union were the first to put a satellite into orbit, the group decided to propose Comité Special's endorsement of a satellite effort for the IGY (Ref. 13, pp. 2–3). The proposal was accepted, but it was only on 27 July 1955 that President Eisenhower announced his decision to proceed with an artificial satellite. "Again, the prime argument for action was that the Soviet Union might announce its own satellite plans first" (Ref. 13, p. 3). According to Hugh Odishaw, executive secretary of the United States Committee for the Geophysical Year, after the satellite decision was made in the United States, "there had been no discussion among scientists of trying to get a satellite space-bound just to beat the Russians."[14] Moreover, one of Eisenhower's most important reactions to Sputnik I was to grant U.S. scientists increased access to the highest echelons of national policymaking, which resulted in establishing the position of Special Assistant to the President for Science and Technology by transfering the President's Science Advisory Committee from a relatively limited and low-level position in the Office of Defense Mobilization to the White House (Ref. 15, p. 17). But even after that, scientists feared that a politically motivated space program would require more resources than would have been necessary at that stage of space exploration and that it could result in a lack of attention to other important scientific projects. "To beat the Russians" would be a political goal, which in the United States was traditionally the privilege and concern of politicians, not scientists.

INITIAL U.S. OPPOSITION TO COOPERATIVE EFFORTS

Although U.S. scientists did not have as much say as their Soviet colleagues in the shaping of space policy, there was a powerful space-related institution in the United States that, like Korolev's bureau, did not favor cooperation with the Soviet Union during the period when the first satellites were launched. The National Aeronautics and Space Act of 1958 contained a few provisions regarding international aspects of U.S. activities in space; in particular,

> SEC. 102. The aeronautical and space activities of the United States shall be conducted so as to contribute materially...to cooperation by the United States with other nations and groups of nations in work done pursuant to this Act and in the peaceful applications of the results thereof.

> SEC. 205. The Administration, under the foreign policy guidance of the President, may engage in a program of international cooperation in work done pursuant to this Act, and in the peaceful application of the results thereof, pursuant to agreements made by the President with the advice and consent of the Senate.[16]

However, regardless of the fact that international cooperation in space was one of NASA's goals, the agency's growing budget was dependent on the assumption of competition with the Soviet Union. Another reason why NASA did not favor cooperation was the culture of the first generation of the agency's employees. For most of them, World War II was still clear in their memories and many of them had served in the armed forces. Two decades of Cold War anxiety, the Soviet Union's domination of Eastern Europe, and Soviet thermonuclear arms filled the American public with the general sense that a real war could start at any time. NASA employees of the first generation, who believed that freedom and maybe even the survival of the Western world were at stake, described the early years of the space race in warlike terms.[17] "NASA employees were called upon to wage a major battle in the Cold War, and to do so with all of the technical skills at their command."[17] Clearly such attitudes toward space exploration stood in the way of even theoretical consideration by NASA staff of any large-scale cooperation in space with the Soviet Union. NASA Administrator T. Keith Glennan urged Eisenhower, before his planned trip to the Soviet Union in 1960, to refrain from proposing cooperation in space beyond the safe and useful sharing of meteorological data (Ref. 1, p. 206).

DIFFERENCES BETWEEN THE U.S. AND SOVIET SATELLITE PROJECTS

A military and strategic competition between the Soviet Union and the United States, as well as the different political and economic systems of the two countries, resulted in one more difference between Soviet and American satellite programs. This difference had a considerable influence on Soviet–American cooperation in space. Both President Eisenhower and Premier Khrushchev were concerned about the possible interference of the satellite project with respective Intercontinental Ballistic Missile (ICBM) projects

(Ref. 18, p. 7). From a technical standpoint there was reason for concern: The Soviet and American satellite programs grew out of developing ICBM technologies. (Although the American Jupiter-C booster used for launching the first U.S. satellite Explorer I was a significantly modified Redstone rocket with scaled-down Sergeants for the three upper stages,[19] in Sputnik just a warhead was replaced in the R-7 ICBM or semyorka, which is still in service as a Soyuz spacecraft main launch vehicle.) However, being technically rooted in ICBMs, Soviet and American satellite programs started developing along different political lines. Premier Khrushchev did not want to fritter away resources necessary for strengthening the defense and attack potential of the country. He allowed Korolev to proceed with Sputnik only after he was assured that the only modification a missile would need was to replace its warhead with a satellite (Ref. 11, p. 111). Thus the Soviet satellite program started growing not only technically but also politically out of the ICBM program. This feature of the Sputnik project, together with the fact that "all of the country's launching sites and space proving grounds had been placed, from the very early days of the Soviet Space program, under the supervision of the military" (Ref. 10, p. 166), had covered all Soviet space activities with secrecy and had made the Soviets especially reluctant to share any of the satellite-related data with the United States even under the auspices of the IGY.

However, besides the need to protect new advanced weapons from adversaries' curiosity, there was also one more reason for secrecy, which was deeply rooted in the social and psychological features of the Soviet society. People working in the area of space exploration were quite satisfied with an atmosphere of secrecy. Because nothing was made public, there was always the chance to hide the actual reason for failure; this enabled them to present their interpretation of a failure as a final statement not only for general consumption but also for government officials (Ref. 10, pp. 234–235).

Unlike the Soviet satellite project, the American satellite project had been deliberately separated from the missile program (Ref. 15, p. 13) for basically two reasons. The first was the aforementioned concern of President Eisenhower about the satellite program possibly standing in the way of the missile program. The second reason was the Defense Department's attempt to divert attention from the highly classified WS-117L project by proceeding with the satellite as a scientific program in cooperation with the IGY (Ref. 18, p. 11). Although these reasons had a military-related character, the result was a facilitated integration of the satellite project in an international scientific program.

U.S. RATIONALE FOR COOPERATION

In terms of a rationale for cooperation, the U.S. attitude toward international cooperation in space, including with the Soviet Union, during the

Eisenhower period can be divided into two periods. From the mid-1950s through the end of 1956, the United States was concerned mostly with the creation of an international legal framework for the free operation of reconnaissance satellites. The second period started at the beginning of 1957 and was motivated by three things. The first motivating factor was that the United States became increasingly worried about the acceleration of the ballistic missile race with the Soviet Union, which could have been even further accelerated by a new one: to achieve capabilities to operate through artificial Earth satellites or space platforms in outer space itself. This is why the United States started seeking a twofold goal—to reach an agreement to ensure the use of outer space solely for peaceful purposes, together with efforts to promote international cooperation in the peaceful uses of outer space (Ref. 20, p. 304).

The importance of the policy underlying the second period in the U.S. approach toward international cooperation in space could be clearly demonstrated by the position of Congress. The decision to share data obtained by satellites, made together with the satellite decision, was not necessarily welcomed by Congress. There was no voluntary congressional statement endorsing the principle of sharing scientific information with other nations, including the Soviets.[21] However, on 17 November 1958, Senator Lyndon B. Johnson addressed the United Nations, urging adoption of the resolution to establish an Ad Hoc Committee on the Peaceful Uses of Outer Space:

> On the goal of dedicating outer space to peaceful purposes for the benefit of all mankind there are no differences within our Government, between our parties, or among our people. The executive and the legislative branches of our Government are together....
>
> We know the gains of cooperation. We know the losses of failure to cooperate. If we fail now to apply the lessons we have learned or even if we delay their application, we know that the advances into space may only mean adding a new dimension to warfare. If, however, we proceed along the orderly course of full cooperation, we shall by the fact of cooperation make the most substantial contribution yet made toward perfecting peace. Men who have worked together to reach the stars are not likely to descend together into the depths of war and desolation.[22]

The second motivating factor was linked to the growing concern in the Democratic Congress about the United States losing the space race to the Soviet Union. To prevent this from happening, Congress offered a twofold approach.

1) The United States should become a leader in terms of concrete space achievements, which would imply American technological and political leadership.

2) The United States should lead in organizing international cooperation in space.[23]

To win the space race in terms of concrete space achievements would require a significant increase in budgetary spending, something that Eisenhower had been opposed to. Leadership through cooperation could help find a reasonable compromise between positions of the president and Congress.

The third motivating factor was President Eisenhower's growing concern about the increasing influence of the military–industrial complex and technical–scientific elite on public life and politics as a result of the space race between the Soviet Union and the United States. Sputnik was a blow to U.S. credibility, and it demanded a similar response.

> As a technocratic accomplishment, involving the integration of new science and engineering under the aegis of the state, it called into question the assumptions behind U.S. military, economic, and educational policy—every means by which the mobilization of brainpower is achieved. As an arcane technical feat, it suggested new dependence on a clique of experts, whom the people's representatives had no choice but to trust. All told, Sputnik threatened to undercut Eisenhower's efforts to usher in the missile age without succumbing to centralized mobilization and planning (Ref. 1, p. 139).

One way to soften the kind of impact that the space race could have had on American society was to moderate the race by cooperation.

The starting point of the second period in the U.S. approach toward international cooperation in space was a memorandum submitted by the United States to the first Committee of the U.N. General Assembly on 12 January 1957, which noted that

> scientists in many nations are now proceeding with efforts to propel objects through outer space and to travel in the distant areas beyond the earth's atmospheric envelope. The scope of programs is variously indicated in the terms "earth satellites," "inter-continental missiles," "long-range unmanned weapons," and "space platforms." No one can predict with certainty what will develop from man's excursion into this new field. But it is clear that if this advance into the unknown is to be a blessing rather than a curse the efforts of all nations in this field need to be brought within the purview of a reliable armaments control system. The United States proposes that the first step toward the objective of assuring that future developments in outer space would be devoted exclusively to peaceful and scientific purposes would be to bring the testing of such objects under international inspection and participation.

In this matter, as in other matters, the United States is ready to participate in fair balanced, reliable systems of control.[24]

SOVIET UNION'S EARLY RESPONSE TO U.S. VIEW
OF INTERNATIONAL COOPERATION

In March 1958, in response to the 1957 memorandum and an earlier U.S.
proposal to apply international control to outer space activities, the Soviet
Union issued a proposed agenda item for the upcoming 13th session of the
U.N. General Assembly entitled "The Banning of the Use of Cosmic Space
for Military Purposes, the Elimination of Foreign Military Bases on the
Territories of Other Countries and International Cooperation in the Study of
Cosmic Space." In an explanatory memorandum, the Soviet government
linked together outlawing nuclear weapons and a liquidation of foreign military bases as the key elements ensuring the security of all countries and
allowing the advances of science to be turned to international benefits (Ref.
20, p. 300).

The reason why the Soviet leadership insisted on the liquidation of military bases was rooted in different threat perceptions by the Soviet Union and
the United States. The Soviet Union, lacking natural barriers such as seas or
mountains along its borders, had been historically very vulnerable to an
attack from land, whereas the United States had a deeply rooted belief in its
invulnerability because of its surrounding oceans and the absence of aggressive neighbors. After the beginning of the Cold War, the United States sur-
rounded the Soviet Union with a number of military air bases near Soviet
borders, thus threatening the Soviet homeland, while the Soviet Union could
not respond with an equal threat to the United States.

Eisenhower's position was clearly expressed in his letter to Premier
Nikolai Bulganin on 15 February 1958, where he considered the possibility
of use of outer space for military purposes as a "terrible menace": "The time
to deal with this menace is now. It would be tragic if the Soviet leaders were
blind or indifferent toward this menace as they were apparently blind or
indifferent to the atomic and nuclear menace at its inception a decade ago."[25]

Khrushchev, who replaced Bulganin as premier in March 1958, linked in
his response space cooperation with disarmament and with the liquidation of
bases in foreign territories above all:

> We cannot ignore the fact that the atomic and hydrogen weapons can
> be delivered to the target not only by means of intercontinental rockets
> but also by means of intermediate and short-range rockets as well as by
> means of conventional bombers stationed at the numerous American
> military bases located in areas adjacent to the Soviet Union.[26]

Khrushchev's approach to the problem made some sense in terms of creating equal security conditions for the Soviet Union and the United States. However, the existence of the overseas military bases was part of the U.S. commitment to its Western allies and also part of the containment policy. The acceptance of Khrushchev's proposal required a revision of the whole U.S. strategic doctrine that was impossible in the conditions of the Cold War.

INITIAL MOVEMENT TOWARD COOPERATION

The positions expressed in the letters of the two heads of state reflected the approaches that the Soviet Union and the United States followed toward space cooperation and that both sides maintained through the end of Eisenhower's presidency.

> The U.S. approach stressed the parallel need to continue to reach agreement to assure the use of outer space solely for peaceful purposes together with efforts to promote international cooperation in the peaceful uses of outer space. This approach was in direct contrast to the Soviet contention that there could be no international cooperation in the peaceful uses of outer space unless accompanied by a ban on the use of outer space for military purposes, which was in turn to be accompanied by an elimination of all foreign military bases (Ref. 20, p. 304).

Thus, both sides found themselves deadlocked and subsequently abandoned their high-level approaches toward each other. Indicative of the termination of Soviet and American efforts to come to some agreement about demilitarization of outer space was the fact that neither Khrushchev nor Eisenhower raised issues related to international control or cooperation in space during their summit in Washington, D.C., and Camp David in September 1959 (Ref. 13, p. 22).

This does not mean, however, that Moscow bluntly rejected all attempts to get the Soviet Union involved in international cooperation in space. For a short period of time Moscow's position suggested that the benefits of its satellites could indeed be shared with other participants in the IGY. Premier Bulganin in his letter of 10 December 1957 to President Eisenhower considered the launch of the Sputnik as a Soviet contribution to the purposes and activities of the IGY.[27] However, when it came time for action, very little was accomplished. The Soviet Union refused to provide orbital elements for the Soviet satellites to other countries during the course of the satellites' lifetimes, to provide precision radio tracking data for satellites, or to agree to an automatic dispatch of basic data to the world data centers. On the last point, the Soviet Union provided no guarantee whatsoever that the rest of the world would ever see any of the desired data.[28] Subsequently, Arnold Frutkin, a director of the NASA Office of International Programs and a former director of Information of the U.S. National Committee for the IGY, came to the following conclusion: "Soviet compliance with even the modest IGY requirements in space science was largely pro forma. Attempts to improve the situation, though made through the international scientific machinery of the IGY, were totally unavailing" (Ref. 13, p. 47).

AD HOC COMMITTEE ON THE PEACEFUL USES OF OUTER SPACE

In December 1958, trying to separate space issues from the overall context of Cold War relationships, the United States advanced the idea of creating an ad hoc committee under the aegis of the United Nations. This step was strongly supported by Senator Lyndon Johnson. The Soviets opposed the idea, again attempting to make U.N. action in outer space contingent on disarmament.[29] When confronted with the Soviet position, the U.S. administration tried to make the Soviets understand that they should join an international space effort or stand aside and see the United States enjoy the benefits that would go with exclusive leadership of such an effort.[30]

The United States collected enough votes to make the adoption of resolution 1348, which established the Ad Hoc Committee on the Peaceful Uses of Outer Space and set its first order of business: studies and reports looking toward the establishment of a permanent framework within the United Nations to promote peaceful activities and effect international cooperation. Eighteen members were elected to membership on the committee, of which only the Soviet Union, Poland, and Czechoslovakia were from the Eastern bloc. However, the Soviet Union turned down its membership in the committee on the grounds that "it has a one-sided character and is not consistent with an objective consideration of this important problem" (Ref. 13, p. 42). After the Soviet boycott, India and the United Arab Republic, which had also been elected to membership, refused to participate on grounds that nothing could be accomplished unless both the Soviet Union and the United States participated (Ref. 31, p. 187).

PERMANENT COMMITTEE ON THE PEACEFUL USES OF OUTER SPACE

Despite the obstructionist position of the Soviet government, the United States advanced the idea of creating of a permanent U.N. Committee on the Peaceful Uses of Outer Space, which, in contrast to the ad hoc committee, had significantly restricted responsibilities. The committee was created on 12 December 1959. Although the Soviet Union had agreed to the membership of the committee, organizational and procedural problems continued to be matters of negotiation among the committee members without any result. Finally, on 14 November 1961, the Soviet Union refused to participate in the committee. The Soviet position was that, because only the United States and the Soviet Union were capable of deep space probes, decisions of the committee should be reached only on the basis of U.S.–Soviet agreement. According to the Soviet representative Zorin, the United States was "trying to put the Soviet Union in an obviously unequal position," because the Soviet Union "in spite of its accomplishments in space, would be outvoted by the Committee's membership which represented 12 western bloc nations, 5 neutrals, and 7 Communist countries" (Ref. 31, p. 197). The Soviets believed that sharing their space achievements with other countries that did not have space exploration facilities would just fritter away the Soviet advantage, which could otherwise be used as a trump card in bargaining with the United States about a set of problems, including disarmament.

OTHER EARLY COOPERATIVE ATTEMPTS

Apart from the United Nations, the United States made a few attempts to reach some space-related cooperative agreement with the Soviets. Among them were cooperation proposals made in November 1959 by Dryden to Soviet Academicians Sedov, A. A. Blagonravov, and V. I. Krassovsky during their visit to a meeting of the American Rocket Society in Washington, D.C.; an offer to use the U.S. tracking system, made by NASA Administrator Glennan; a letter written by Dryden to Sedov, suggesting a wide variety of cooperative possibilities, including activities related to scientific measurements of telemetered data from the U.S. satellite Explorer VII; U.S. experiments using Soviet spacecraft; and cooperative experiments utilizing the projected U.S. communications satellite Echo. All of these suggestions remained without response (Ref. 13, pp. 44–45).

The growing disparity between American and Soviet space capabilities and accomplishments, together with the real and imagined implications for the relative power positions of the United States and the Soviet Union, made Americans extremely concerned about their security. Hopes and expectations about cooperation with the Soviets in space turned out to be illusory. Space became one of the trump cards in the Cold War, which meant a change in American space policy. The ability to compete successfully with the Soviet Union in space became as important a task as to cooperate with it.

ATTEMPT AT BREAKTHROUGH (KHRUSHCHEV–KENNEDY PERIOD)

KENNEDY'S APPROACH TO A RELATIONSHIP WITH THE SOVIET UNION

The election of John Fitzgerald Kennedy as the president of the United States in 1960 created new opportunities for Soviet–American cooperation in space. Kennedy demonstrated a different approach from his predecessor to a relationship with the Soviet Union. This new approach was based on a few assumptions. First, there was a new balance of world forces, and the Soviet military might was equal to that of the United States. This idea made Kennedy revise the Eisenhower–John Foster Dulles "position of strength" policy toward the Soviet Union. Another result of the revision was Kennedy's belief in the need to develop areas of common interests with the Soviet Union as a means to moderate Cold War tensions. "Wherever we can find an area where Soviet and American interests permit effective cooperation, that area should be isolated and developed,"[32] said Kennedy in one of his pre-election interviews. The second assumption was that competition between two superpowers is not limited to just Soviet–American relations, but has a global character. Kennedy viewed it as a battle "between freedom and tyranny," and "the great battlefield for the defense and expansion of freedom" was "the whole southern part of the globe—Asia, Latin America, Africa, and the Middle East—the lands of rising people" (Ref. 33, 1962, pp. 396–406). A third assumption was that science and technology should play at least an equal role with the one played by weapons in the outcome of this battle.

The second reason for providing new opportunities for Soviet–American cooperation in space was that Kennedy, unlike Eisenhower, had a totally different view of space exploration and its role in international relations. He thought of it not only as a means to restore American prestige and pride damaged by Soviet space feats such as Sputnik and Yuri Gagarin (Ref. 33, 1962, pp. 396–406) and a vehicle to assure future American economic, political, and strategic leadership,[34] but also as a possible area of common interest with the Soviet Union (Ref. 32, p. 347). However, this kind of approach to space exploration challenged Kennedy with three dilemmas.

1) To restore prestige and pride, it was necessary to create an expensive space program aimed at winning a space race against the Soviet Union. To win such a race, Kennedy was ready to put aside even Eisenhower's concern about the incompatibility of the liberty and dynamism of civil society with the centralized mobilization and planning necessary for the accomplishment of an ambitious space program. He believed that such a program "would demand sacrifice, discipline, and organization: the nation could no longer afford work stoppages, inflated costs, wasteful interagency rivalries, or high turnover of key personnel" (Ref. 1, pp. 139, 303). However, cooperation with the Soviet Union would undercut the rationale for the race and consequently for the crash efforts of the nation in space.

2) If the Soviet Union and the United States cooperated in second-rate projects while competing in first-rate space projects, would they still be involved in the strategic race in space with the respective negative

implications for bilateral relations, or could space yet become a sphere of common interest?

3) What was the priority? Should the relationship between the Soviet Union and the United States be improved at least tentatively before the initiation of any large-scale cooperative space project, or should the project come first?

These dilemmas probably presented one of the toughest challenges to the Kennedy administration in terms of space policy. The state of Soviet–American relations as well as Soviet and American space achievements suggested various solutions for the dilemmas during the period of Kennedy's presidency.

EMPHASIZING COOPERATION OVER COMPETITION

In the beginning of his presidency, Kennedy was obviously prioritizing space cooperation over a space race. In his "First Message to the Congress on the State of the Union" he specifically noted that

> This Administration intends to explore promptly all possible areas of cooperation with the Soviet Union and other nations "to invoke the wonders of science instead of its terrors." Specifically, I now invite all nations—including the Soviet Union—to join with us in developing a weather prediction program, in a new communication satellite program and in preparation for probing the distant planets of Mars and Venus, probes which may someday unlock the deepest secrets of the universe.... Today this country is ahead in the science and technology of space, while the Soviet Union is ahead in the capacity to lift large vehicles into orbit. Both nations would help themselves as well as other nations by removing these endeavors from the bitter and wasteful competition of the Cold War (Ref. 33, 1962, pp. 26–27).

There were a few reasons why competition in space was lower than cooperation in Kennedy's agenda.

1) The President was not really convinced that the United Sates would be able to catch up with the Soviet Union and then lead it in space. He was concerned about the outcome of the race emphasizing further Soviet superiority.

2) He was not sure about the scientific value of major manned space flight efforts in the light of the Report of the Committee chaired by Jerome Wiesner, who later was appointed Special Assistant to the President for Science and Technology. The report characterized the Mercury program as "marginal" and likely "to bring discredit to the nation and more particularly to the Kennedy administration" (Ref. 13, p. 62).

3) He did not disregard Eisenhower's warning about the influence on state politics of a military–industrial complex and scientific–technological elite.
4) The president was concerned about the space race intensifying the arms race and as a result aggravating the Cold War.
5) Kennedy was worried about the cost of the space program and its possible impact on other programs especially "in light of narrowness of his election mandate" (Ref. 13, p. 64).

There is one more explanation for Kennedy prioritizing cooperation over competition in space in the beginning of his presidency. There is good reason to believe that even before the presidential elections Kennedy knew that Khrushchev had more sympathy toward him than for the Republican candidate, Richard Nixon (Ref. 11, pp. 88–89). One of the manifestations of this sympathy was that Khrushchev had refused to discuss with Henry Cabot Lodge, who was running for the vice-presidency in the Nixon administration, the fate of the survived crewmembers of the U.S. RB-47 reconnaissance aircraft downed on 1 July 1960 by Soviet fighters in the North Sea. Their meeting took place just before the elections, and Lodge was eager to have the pilots released while Republican President Eisenhower was still in power, so that it would add votes to the Republican candidate Nixon. Khrushchev refused to do so and released the pilots right after Kennedy's inauguration. Khrushchev mentioned this fact during his first meeting with President Kennedy in Vienna, and the president thanked him for this support (Ref. 11, p. 90). According to a former First Deputy Foreign Minister of the USSR, Georgi Kornienko, another manifestation of Khrushchev's intention to develop a special relationship with the new president of the United States was the fact that the Soviet premier made numerous, although unsuccessful, attempts to meet with Kennedy before his inauguration (Ref. 35, pp. 55–56).

The initial reaction of the Soviet leadership to Kennedy's proposal for pooling American–Soviet efforts in space exploration projects did not differ much from the previous attitude of the Soviet leadership toward the issue. Khrushchev welcomed it, although he stressed as usual that the Soviet Union was interested in cooperation only in the context of disarmament.[36] The president in his reply to Khrushchev's response made cooperation in space between the two countries contingent on "harmonious relations" between them (Ref. 33, 1962, pp. 93–94).

EARLY U.S. PROPOSALS FOR SPACE COOPERATION

An important role in the formation of the Kennedy administration's attitude toward cooperation in space with the Soviets was played by the Joint NASA–President's Science Advisory Committee–Department of State Panel for the purpose of conducting "a preliminary study of the possibilities for international cooperation" (Ref. 13, p. 66). The panel worked on this from early February through mid-April 1961. During this time it issued a number of memoranda prepared by consultants, such as Philip Farley, Special Assistant to the Secretary of State for Atomic Energy and Outer Space, or by Dr. Richard Porter of the General Electric Company. The panel also held a number of meetings, each one ending in a report concerning international cooperation in space. All of the panel's suggestions, summarized in a *Report of the Panel on International Cooperation in Space Activities*, dated 20 March 1961, and followed by *Draft Proposals for US–USSR Space Cooperation* issued on 14 April 1961,* could be boiled down to three of the most important findings.

The first finding was concerned with the studious avoidance of any military connotation of cooperative or joint projects. This was the first lesson learned from Eisenhower's unsuccessful attempts to involve the Soviet Union in space cooperation, when the Soviet leadership had perceived the president's initiatives as attempts to realize his Open Skies project using reconnaissance satellites.

The second finding discussed the need to determine the exact magnitude of the potential cooperative project to overcome the existing political–technological obstacles in U.S.–Soviet relations. It should be either so simple that it would avoid the existing obstacles, or on the other hand, should be so vast and dramatic as to overcome the existing obstacles by sheer weight of public opinion. The panel specifically mentioned a "Rendezvous on the Moon" undertaking and a joint U.S.–Soviet project to place an unmanned vehicle on Mars, which by sheer cost should persuade the Soviet Union to cooperate with the United States.[†] This recommendation also seemed to be a lesson learned from the Eisenhower period of U.S.–Soviet space relationships: Eisenhower's offer to the Soviets to cooperate in data collection using satellites under the aegis of IGY did not correspond with the Soviet leadership's ideal of either simplicity or magnitude.

The third finding suggested that the United States should seek direct cooperation with the Soviet Union "since these are only two nations capable of launching space vehicles at the present time." This recommendation was also a clear disengagement from Eisenhower's policy, which wanted to involve the Soviets in cooperation in the context of international agreements and controls.

The first attempt to involve the Soviets in space cooperation, made by Kennedy in his State of the Union Message on 30 January 1961, was fol-

---

*For more information about these documents, see Ref. 13, pp. 66–73.
[†]For more information, see Ref. 13, pp. 66–73.

lowed by two more: one on the occasion of the launching of the Soviet probe to Venus, and another on Gagarin's flight. However, these space feats made Soviets overconfident in their space leadership. Krushchev later recalled,

> Of course, we tried to derive maximum political advantage from the fact that we were first to launch our rockets into space. We wanted to exert pressure on American militarists—and also influence the minds of more reasonable politicians—so that the United States would start treating us better.[37]

He personally asked Korolev to launch a second man—German Titov—into space on 7 August, just a few days before the beginning of the construction of the Berlin Wall, with the obvious intention of reminding Americans about Soviet rocket power (Ref. 11, p. 125). In his diaries published long after his death, Lieutenant General Nikolai Kamanin, who was responsible at that time for the cosmonauts' training, confirmed that space launches were arranged for different political events in the Soviet Union.[38] Academician Keldysh, who was the president of the Soviet Academy of Sciences, said a month after Gagarin's flight that "American space achievements are of such a low value that can't even be compared to Soviet ones."[39] Arnold L. Horelick wrote in his memoirs,

> No event since the death of Stalin had been so widely publicized by the USSR as the April 12, 1961, flight of Major Yuri Gagarin in the spaceship Vostok I and his return to earth after completing one circuit around the globe. The event was headlined in special editions of central Soviet newspapers, rare in a country where events ordinarily become news not when they occur but when they are officially announced. Gala celebration meetings were held in Moscow and other large cities and were carried by both Soviet and foreign television. The Soviet radio, domestic as well as international services, virtually excluded all other subjects for several days in order to transmit bulletins and commentaries devoted to the last Soviet achievement.[40]

Besides the wish to derive maximum political advantage from the Soviet space feats, one other factor stood in the way of cooperation. The promotion of Academician Keldysh to president of the USSR Academy of Sciences in 1961 had meant "the marriage of the academy as the headquarters of pure, basic science to the military-industrial complex" (Ref. 10, p. 147). The academy, which had already been a powerful institution at that time, started gaining even more influence under Keldysh, who "promoted a policy to convert the academy into the formal headquarters of all technological and scientific development in the country" (Ref. 10, p. 148). At some point Keldysh had even been able to challenge party authorities. He dismissed a protest, forwarded by the Space Studies Institute's Communist Party bureau and supported by Moscow city party authorities, against appointing a non-party member (Roald Sagdeev), as director of the Space Studies Institute. Sagdeev got the appointment (Ref. 10, pp. 161–162).

On one hand, Soviet space achievements encouraged Soviet leadership to take full political advantage of them and disinclined the Kremlin to talk about any kind of space cooperation with the United States. On the other hand, public opinion in the United States, as well as the opinion of both Democrat and Republican members of Congress (Ref. 13, pp. 74–75; Ref. 41) became increasingly critical of Kennedy's passivity in the face of Soviet space achievements, which were perceived by the majority of U.S. society as a challenge to U.S. security and world leadership. The Bay of Pigs fiasco in April 1961 made Kennedy especially sensitive to his presidential image in his own country.

U.S. DECISION TO GO TO THE MOON

These developments made Kennedy rethink his space policy. The new approach toward U.S. activities in space should have reached two goals. The first and foremost goal was to restore American prestige and pride. This task could be accomplished only via the successful realization of a project that would overshadow all previous Soviet space achievements. "Now, let's look at this," said Kennedy impatiently. "Is there any place we can catch them? What can we do? Can we go around the moon before them? Can we put a man on the moon before them? What about Nova and Rover? When will Saturn be ready? Can we leapfrog?" Then Kennedy said quietly, "There is nothing more important" (Ref. 15, p. 106). The moon landing project had been already under consideration in NASA. All evidence, including classified intelligence, made James Webb, the NASA administrator at the time, feel that a lunar landing was "the first project we could assure the president that we could do and do ahead of the Russians, or at least had a reasonable chance to do" (Ref. 42, p. 96). In advancing this idea, Webb was defending NASA institutional interests. He understood that NASA had been a child of the Cold War and realized that future opportunities rested with Soviet–American rivalry. Maintaining U.S. prestige in the world required matching and ultimately surpassing the Soviet Union's space successes (Ref. 42, p. 86). The agency reached the conclusion "that it was not only feasible but also, and more important, an excellent focal point around which could be organized most of the activities needed for the development of the broadly based capabilities the United States would have to have to achieve leadership in space" (Ref. 13, p. 76).

This kind of conclusion underlay Kennedy's decision to go to the moon, which was proclaimed by him in a special message "Urgent National Needs" to the Congress on 25 May 1961. In this message Kennedy made it clear that he viewed manned flight to the moon as a means to win a space race with the Soviet Union. "No single space project in this period will be more impressive to mankind, or more important for the long range exploration of space; and none will be more difficult to accomplish," he said in his speech (Ref. 33, 1962, pp. 403–404).

Just a few days after the announcement, in early June, Kennedy and Khrushchev were scheduled to meet in Vienna. The first summit meeting between the two heads of state could have given Kennedy a perfect opportunity to reach a second goal of his new space policy—to use a moon landing project, which by its magnitude could have overcome the existing obstacles in Soviet–American relations, to involve the Soviet Union in space cooperation with the United States, and consequently to make cooperation in space an area of common interest for the two countries. The need for such an area was especially urgent in the view of Soviet–American relations aggravated over the Berlin issue. For this reason Kennedy did not make space cooperation contingent on the improvement of bilateral relations. Although the effect of the meeting was "to increase rather than lower the level of tension" (Ref. 6, p. 244), Kennedy still made a proposal, in casual rather than in formal form, to Khrushchev to go the moon together.

Khrushchev's initial reaction seemed to be positive, but then he stressed again that space cooperation with the United States would be impossible without disarmament. This position brought Soviet–American space relations into a stalemate until the end of September 1961.

"ALL NATION" APPROACH REVISTED

Even after the discouraging results of the Vienna meeting, Kennedy kept his pledge "to continue to attempt to engage the Soviet Union in a common (space) activity."[43] To do this, he changed his tactics and tried to use Eisenhower's "all-nation" approach through the United Nations, aimed at excluding military activities in outer space. He announced it in his address before the General Assembly on 25 September 1961. The idea was basically the same as Eisenhower's—to lead other countries in space through cooperation and therefore to convince the Soviet leadership to join in international efforts. This kind of approach had failed in the past; however, it could have made sense both in the existing context of Soviet–American relations and with innovations introduced in it by President Kennedy.

After the Vienna summit, the U.S. policy toward the Soviet Union was strongly influenced by Secretary of State Dean Acheson, who recommended a tough approach toward the Soviet Union until it softened its policy on the Berlin issue. Acheson discouraged the United States from making any cooperative moves toward the Soviet Union because, in his view, they could have been considered as a sign of American weakness by the Soviet leadership (Ref. 35, p. 65). Clearly, a direct proposal made by the United States to the Soviet Union to join efforts in space at that time would not have fit the general context of Soviet–American relations.

However, the old approach through the United Nations included new ideas. First, the United States agreed with the relationship between cooperation in space and disarmament—something the Soviet Union was insisting on, although the United States viewed disarmament as contingent on successful cooperation. According to U.S. Representative Adlai Stevenson, "if put into operation without delay, it can help lay the basis for a relaxation of tensions and facilitate progress elsewhere toward general and complete disarmament."[44]

Second, the U.S. proposal was aimed not only at "reserving outer space for peaceful purposes," but also at "establishing a global system of communications satellites linking the whole world in telegraph and telephone and radio and television" (Ref. 33, 1962, pp. 622–623). Thus it followed one of the recommendations made by the Panel on International Cooperation in Space Activities to cooperate in "projects that avoid the difficulties connected with a high degree of involvement."

Finally, the Kennedy administration, even though operating through the United Nations, emphasized a special role played by the Soviet Union and the United States in international cooperation in space, thus highlighting the bipolar character of world space activities. The United States initiated a compromise within the Committee on the Peaceful Uses of Outer Space to meet a Soviet requirement to adopt decisions among members of the committee by consensus, i.e., without voting. The Soviet concern was that the Soviet Union, "in spite of its accomplishments in space, would be outvoted by the Committee's membership, which represented 12 western bloc nations, 5 neutrals, and 7 Communist countries" (Ref. 31, p. 197). The U.S. position was expressed by an Australian representative, who said that "in practical terms agreement is desirable, and in many cases necessary, between the Soviet Union and the United States if international work and decisions by the United States are to have any effect."[45] The essence of the compromise was the following: The committee would try to reach agreement by consensus; if voting is required, however, the decision would be made by majority vote (Ref. 31, pp. 199–200). As a result the U.S.-sponsored General Assembly Resolution 1721 (XVI) "International Cooperation in the Peaceful Uses of Outer Space" was unanimously adopted on 20 December 1961 (Ref. 31, pp. 199–200), which was the first meaningful action on space to get through the United Nations after almost three years of effort (Ref. 13, p. 83).

The fact that the Soviet Union agreed with this resolution appeared to be a manifestation of a new Soviet approach toward cooperation in space, in particular with the United States. First, the Soviet Union followed the U.S. initiative and then accepted the separation of international cooperation in space from disarmament issues. Finally, the Soviet Union agreed with the resolution endorsing international cooperation in space in the areas where the United States, but not the Soviet Union, was "already prepared for a dominant role: the use of satellites in the development of global weather and communications systems" (Ref. 13, pp. 85–86).

Besides flexibility and the good will demonstrated by the Kennedy administration, there were also other reasons for the Soviet Union to agree with the U.S. approach toward international cooperation in space. A developing conflict between the Soviet and Chinese leaderships, together with a growing reliance of the Soviet Union on grain supplies from the United States, made the Kremlin reconsider its policy toward the United States from confrontation to negotiation.

The next step made by the Soviet leadership that indicated its willingness to at least consider joining efforts with the United States in space was Khrushchev's congratulatory telegram to Kennedy on 21 February 1962 on the occasion of John Glenn's space flight. In this telegram he admitted that if the Soviet Union and the United States

> pooled their efforts—scientific, technical and material—to master the universe, this would be very beneficial for the advance of science and would be joyfully acclaimed by all peoples who would like to see scientific achievements benefit man and not be used for the "cold war" and arms race.[47]

The willingness of President Kennedy to exploit a change in the Soviet approach toward space cooperation with the United States was clearly indicated in his response to Khrushchev's telegram, which he gave on the same day he received the congratulation: "We of course believe also in strong support of the work of the United Nations in this field (space exploration) and we are cooperating directly with many other countries individually. But obviously special opportunities and responsibilities fall to our two countries" (Ref. 33, 1962, p. 158). This response was followed by a concrete proposal that Kennedy sent to Khrushchev on 7 March 1962. The proposal contained a list of five areas where the Soviet Union and the United States could begin their cooperation immediately: establishing an early operational weather satellite system, obtaining operational tracking services from each other's territories, mapping the Earth's magnetic field in space by using Soviet and American satellites, sharing information on communication satellites, and working together on space medicine (Ref. 33, 1962, pp. 244–245). Moreover, Kennedy emphasized that he was thinking in terms of strictly Soviet–American cooperation, and he did not suggest involvement of any other countries in the Soviet–American space decision-making process (Ref. 13, p. 89).

MOON RACE BEGINS

Kennedy did not repeat his offer to Khrushchev to go to the moon together, although he did not exclude the possibility of the two countries cooperating in the unmanned exploration of the lunar surface and other planets of the solar system, and he expressed a cautious wish for both countries to share tasks and costs, thus minimizing risks for "brave men who engaged in space exploration" (Ref. 33, 1962, pp. 244–245). The work on the moon project had already been underway in the United States, and the American people as well as politicians were thinking of the mission to the moon in terms of a race. Even if he had proposed a joint Soviet–American expedition to the moon to Khrushchev, the chances of this proposal meeting a positive response from the Soviets still would have been very slim. At the February 1962 meeting between top Soviet leadership and Soviet rocket designers, Korolev's N-1 moon rocket project had finally been approved; the Soviet Union became officially involved in the moon race (Ref. 11, p. 162). The main reason for making the decision was to beat Americans to the moon.[48] As in the case of Sputnik, the moon landing was only a by-product of the development of a powerful launch vehicle whose main purpose was to put into orbit a heavy orbital station armed with nuclear bombs, which could be discharged at any place on Earth. For this reason Khrushchev even asked Korolev to increase the maximum payload of N-1 from 20–40 to 75 metric tons (Ref. 12, p. 718).

SOVIET WILLINGNESS TO COOPERATE IN OTHER AREAS

Khrushchev's response to Kennedy's proposal of 7 March indicated Soviet willingness to cooperate in almost all of the areas of space exploration mentioned by the president, including communication satellites, weather observation satellites, mapping of the Earth's magnetic field in outer space by means of satellites, and space medicine. He also tacitly agreed with Kennedy's proposal to cooperate in the unmanned exploration of deep space by offering cooperation in "observation of objects launched in the direction of the moon, Mars, Venus, and other planets of the solar system."[49]

Khrushchev did not respond to Kennedy's proposal to cooperate in the field of tracking stations. To give Americans access to the Soviet tracking stations, although on a reciprocal basis, was in Khrushchev's mind the same as allowing U.S. specialists to spy officially on Soviet military facilities. This part of Kennedy's proposal had not been accepted. At the same time Khrushchev offered cooperation in two areas: "aid in searching for and rescuing space ships, satellites, and capsules that have accidently fallen," and in the "solution of the important legal problems with which life itself has confronted in the space age," including the avoidance of a situation where someone creates "obstacles for space study and research for peaceful purposes." There is good reason to believe that Khrushchev, while making these proposals, was motivated by something other than a wish to find more areas for cooperation in space between the two countries. In the early 1960s American reconnaissance satellites became a growing concern of the Soviet leadership. "Spy-in-sky satellites" were part of Kennedy's "no-cities" counterforce strategy aimed at the destruction of selected targets, as opposed to Eisenhower's "massive retaliation" doctrine.[50]

By September 1961 reconnaissance satellites had provided complete photographic coverage of the Soviet Union. Without this comprehensive mapping of the Soviet Union—the ICBM and intermediate-range ballistic missile (IRBM) bases, the submarine ports, air defense sites, Army and Air Force bases—the "no-cities" version of counterforce strategy could not have been implemented.[50] Khrushchev even considering writing a protest to Kennedy but then dropped the idea, having realized the futility of this kind of action (apparently having recalled that only the downing of a U-2 spy plane stopped reconnaissance flights over Soviet territory [Ref. 11, p. 53]). A few of the satellites' landing capsules had made accidental landings on Soviet territory and had been recovered, although in poor condition owing to the curiosity of local citizens who ruined them to see what they had inside (Ref. 11, pp. 53–56). Khrushchev issued a special decree ordering any "strange object" found to be delivered immediately to local authorities.

Thus, perhaps by mentioning "accidently fallen" capsules and satellites, Khrushchev wanted to remind Kennedy of the possibility of being caught on a spy mission and possibly facing a U-2 kind of scandal. As for the second area of proposed cooperation, Khrushchev was referring not only to the so-far successful cooperation between the Soviet Union and the United States in the development of the first principles of space law, but also Project West Ford.*

One of the most important points made by Khrushchev in his letter to Kennedy was the recognition of the difference between the two issues—space cooperation and disarmament. Although he noted that "the principles

---

*Project West Ford, a U.S. Air Force program conceived at the Massachusetts Institute of Technology's Lincoln Laboratory, involved launching into Earth orbit 350 million copper threads (17.78 mm long and 0.254 mm in diameter), which would serve as reflector antennas for short wavelength communications (8000 MHz). The experiment promised to make global radio coverage invulnerable to jamming. Project West Ford, approved on 4 October 1961 by the White House, met with mixed international scientific reactions, being criticized by many scientists as a possible threat to the study of radioastronomy or as an alteration to the environment of space, but the project was praised by NATO politicians as a significant deterrent defense system. On 10 May 1963, a second attempt to orbit the disputed payload was successful; the dipoles ejected and formed a compact cloud, circling the Earth every 166 min in a near-polar orbit at a height of 3704 km. *Science,* on 16 December 1963, reported that nearly all of project West Ford's dipoles had reentered the atmosphere (as described in Ref. 51, p. 43).

for the designing and producing of military rockets and space rockets are the same," he emphasized that the Soviet Union and the United States "should try to overcome any obstacles which might arise in the path of international cooperation in the peaceful conquest of space." Secretary of State Dean Rusk, in his memorandum to President Kennedy on 15 May 1962, pointed out that "the Soviets continue to cite the need for disarmament as a precondition to intimate and extensive space cooperation but not necessarily to more modest cooperation" (Ref. 13, p. 95).

BLAGONRAVOV–DRYDEN DISCUSSIONS AND RESULTS

The preliminary exchange of ideas at a low level was followed by practical steps on both sides. The Soviet Union and the United States appointed Academician Blagonravov and NASA Deputy Administrator Dryden to discuss the concrete questions of cooperation in space exploration. It was the first time since the beginning of the space age that the two countries were about to start discussing cooperation.

The first Blagonravov–Dryden period in the history of Soviet–American cooperation in space lasted from 1962 through 1963. It consisted of five meetings and covered a period of Khrushchev–Kennedy interactions regarding space issues. The first meeting took place in New York City on 27–30 March 1962; the second, in Geneva from 29 May to 7 June 1962; the third, in Rome in March 1963; the fourth, in Geneva in May 1963; and the last one in September 1963.* Despite the promising beginning, when the Soviet side basically demonstrated its willingness to talk about cooperation in practical terms, the meetings brought few practical results. Academician Blagonravov first refused to discuss cooperation in space medicine and then in any kind of spacecraft tracking operations (Ref. 52, p. 95). The two sides, however, managed to sign a three-part bilateral agreement on 8 June 1962. The first part provided for coordinated launchings of experimental meteorological satellites by the two countries, for the exchange of data thus obtained, and for the exchange of conventional meteorological data, prior to and on a secondary basis, during the exchange of satellite data. The second part provided for the launch by each country of an Earth satellite equipped with absolute magnetometers and a subsequent exchange of data to arrive at a map of the Earth's magnetic field. The third part provided for joint communications experiments by means of the U.S. passive satellite ECHO II.[53] The agreement was followed by a "First Memorandum of Understanding to Implement the Bilateral Agreement of June 8, 1962." According to Frutkin, then director of International Programs for NASA, the main result of the

---

*For more detailed information about the meetings, see Ref. 13, pp. 92–102 and Ref. 51, pp. 42–48.

negotiations regarding the agreements and other areas listed in Khrushchev's and Kennedy's letters

> was not progress toward getting cooperative enterprises underway but (1) a further shrinkage, presumably in order to adjust to realities, of the already very narrow areas in which the two countries were committed to cooperate and (2) reaffirmation in more concrete terms of intentions to cooperate within the newly established limits (Ref. 13, p. 99).

This interaction did not involve any exchange of classified or sensitive data, equipment, or funds (Ref. 52, pp. 100–101). Both sides also kept ruling out any cooperation on manned lunar flights for political reasons.[54]

The first period of the Blagonravov–Dryden talks did not generate any tangible results because of the Soviet reluctance to having them. At the same time, Khrushchev agreed with Kennedy's proposal to pursue cooperation in some areas of space exploration and even approved negotiations on the USSR Academy of Sciences–NASA level. The reason for such ambiguity was Khrushchev's intention to reach a twofold goal. On one hand, he wanted to stress the open and peaceful character of the Soviet space program. The making of such an image became especially important after 20 May 1962, when the decision to install missiles in Cuba had been made. On the other hand, Khrushchev was concerned about the possible disclosure of the Soviet missile bluff as a result of contacts between Soviet and American specialists (Ref. 11, p. 459). When he decided that enough of such contacts had been made to prove the Soviet leadership's willingness to negotiate, he created a situation in which a continuation of talks was impossible and traditionally blamed the United States for this. On 24 May 1963, just a few days after the fourth Blagonravov–Dryden meeting, the Soviet U.N. representative sent a letter to the secretary general entitled "Dangerous United States Activities in Space." This letter referred to Project West Ford, which was used by the Soviets as a target for attacking U.S. foreign policy, and the U.S. space program in particular. The letter accused the United States

> [of] deliberately and consciously sabotaging agreement on basic legal principles governing the activities of States in outer space.... The facts make it unmistakably clear that in space, just as on earth, the United States is guided in its actions not by the interests of peace and improved relations between States but by its policy of preparing for war.... Needless to say, the Soviet Union and other peace-loving States, concerned with safeguarding their security, must draw the appropriate conclusions from the dangerous activities of the United States in outer space (Ref. 13, p. 104).

Although all of the previous attempts of his administration to involve the Soviets in cooperation in space had failed, Kennedy was still pursuing his idea to find and develop together with the Soviet Union areas of common interest. He made it clear in an address at American University on 10 June 1963:

> Some say that it is useless to speak of world peace or world law or world disarmament—and that it will be useless until the leaders of the Soviet Union adopt a more enlightened attitude.... I am not referring to the absolute, infinite concept of universal peace and good will of which some fantasies and fanatics dream. I do not deny the value of hopes and dreams, but we merely invite discouragement and incredulity by making that our only and immediate goal.
>
> Let us focus instead on a more practical, more attainable peace— based not on a sudden revolution in human nature but on a gradual evolution in human institutions—on a series of concrete actions and effective agreements which are in the interest of all concerned... (Ref. 33, 1963, pp. 460–462).

U.S. SKEPTICISM OVER COOPERATION IN SPACE

Kennedy did not repeat his proposal to the Soviets to cooperate in space exploration. There was growing skepticism in Congress after it had been initially well-disposed toward cooperation with the Soviet Union. Senators Kerr and Margaret Chase Smith expressed their disappointment about the state of Soviet–American cooperation in space.[55] Smith even questioned the need to continue attempts to involve the Soviets in space cooperation, especially because the United States was committed to achieving superiority over the Soviet Union in space.[56] As long as public opinion polls in Ohio could be reflective of a general public mood in the United States, a slight majority (47%) of American people were against cooperation with the Soviet Union in space, while others, with the exception of those who had no opinion about the issue, cautiously favored it (about 40%) so long as it did not hinder reaching a main American goal in space—to get a man to the moon and bring him back safely by the end of the 1960s.[57]

There were also other reasons for Kennedy to drop the idea of cooperation for a while. When he set a moon landing goal, he was not quite sure of the success of the enterprise (and this was one of the reasons why he had sought cooperation with the Soviets), but as the U.S. space program was successfully developing, "Kennedy increasingly identified personally with the achievements and with the astronauts involved" (Ref. 13, p. 109). One of the manifestations of his growing confidence in American strength in space was that he started making cooperation with Soviets in moon landing contingent on the improved relations between the two countries. He said in mid-1963,

> The kind of cooperative effort which would be required for the Soviet Union and the United States together to go to the moon would require a breaking down of a good many barriers of suspicion and distrust and hostility which exists between the Communist world and ourselves.... Obviously, if the Soviet Union were an open society, as we are, that kind of cooperation could exist and I would welcome it. I would welcome it, but I don't see it as yet, unfortunately (Ref. 33, 1963, pp. 567–568).

Moreover, as Kennedy

> became more thoroughly educated as well as involved in space affairs, he began to see the development of space capabilities not just as a means of manifesting the nation's strength, but as a means of further building and extending that strength. The true importance of the lunar goal would lie not in simply beating the Russians in a particular space endeavor but in providing a focal point around which to organize and carry forward the range of activities necessary to achieve capabilities with which to beat the Russians in the attainment of all-round preeminence in space, and not for a fleeting moment but enduringly (Ref. 13, p. 110).

It is not to say that Kennedy totally turned down the idea of cooperation with the Soviets, but he was interested in it only if it could result in the real improvement of relations between the two countries (Ref. 13, p. 111). His new approach to pooling efforts with the Soviets in space was tested in the latter half of 1963, when, according to Frutkin, "by all odds the strangest chapter" in the history of Soviet–American cooperation in space began (Ref. 51, p. 49).

SOVIET UNION OUT OF THE MOON RACE?

On 17 July 1963, British astronomer Sir Bernard Lovell indicated in reports of his meetings with Soviet scientists that the Soviets were considering dropping out of the moon race. In a letter to Frutkin, Lovell presented the following reasons for the Soviets prioritizing unmanned over manned exploration of the moon, as it was explained to him by Academician Keldysh:

1) Soviet scientists could see no immediate solution to the problem of protecting the cosmonauts from the lethal effects of intense solar outbursts.
2) No economically practical solution could be seen for launching sufficient material on the moon for a useful manned exercise with reasonable guarantee of safe return to Earth.

3) The Academy was convinced that the scientific problems involved in the lunar exploration could be solved more cheaply and quickly by their unmanned, instrumented lunar program (Ref. 13, p. 115).

If the news about the Soviet Union dropping out of the race was a maneuver designed to create confusion and to weaken domestic support in the United States for the moon landing, the Soviet leadership partially reached its goal. The rumor that the Soviet Union was no longer a participant in the moon race resulted in substantial cuts in the NASA budget. The Kennedy administration did not manage to restore part or all of the $600 million (Ref. 51, p. 368). However, there is a reason to believe that it was more than just one of the Kremlin's maneuvers. According to Sergei Khrushchev, his father, in a private conversation with him at the end of the summer of 1963, recalled a few times the talk he had with Kennedy in Vienna about the Soviet Union and the United States pooling together their efforts in space. Khrushchev turned down Kennedy's proposal to go to the moon together because he was afraid that Americans would know that the real number of Soviet ICBMs was significantly lower than the Kremlin wanted it to appear. Moreover, their launch readiness was considerably worse than that of American ICBMs. In Khrushchev's view, the information about Soviet missile inferiority could have pushed some U.S. politicians to resolve all of the problems with the Soviet Union by military means.

The situation was different in 1963. The Soviet missile program had been successfully realized. Khrushchev believed that if the United States became aware of the significant amount of Soviet ICBMs, it would discourage American politicians from talking to the Soviet Union from a position of strength.

There was also another reason for Khrushchev to consider cooperation with the United States in a moon landing. Korolev's N-1 program turned out to be too expensive. Further participation in the moon race would have burdened a troubled Soviet economy even more. Khrushchev started to believe that if the Soviet Union had been unable to maintain leadership in space, it would have made sense to pool the efforts of the two countries in space (Ref. 11, pp. 459–460).

Whatever Khrushchev had in mind, he did not realize it in practice. Rumors about the Soviets possibly considering cooperation with the Americans had been further fueled by a statement made by Academician Blagonravov during his fifth meeting with Dryden in New York in September 1963. He admitted that "Lovell's statement might be true as of today" (Ref. 13, p. 119), but this statement had not been followed by any concrete cooperative proposal from the Soviets.

However, the Kennedy administration still had reason to believe that what the Soviets told Lovell was more than plain bluffing. At the end of July 1963 representatives of the United States, the Soviet Union, and Great Britain initiated a treaty prohibiting nuclear weapons tests in the atmosphere, space, and under water. Taking these facts together with Lovell's reports, Frutkin came to the conclusion that the Soviet leadership might have been sending a message to the U.S. government, indicating its willingness to accept a cooperative proposal from the Kennedy administration. If an agreement would have been arranged this way, nobody would believe that the Soviet Union initiated a cooperative effort because of its inability to win the moon race (Ref. 13, p. 120). Thoughts like these, together with growing public concern about the lack of a national coherent space policy demonstrated by NASA and the U.S. government's inability to make a final decision about whether to cooperate or to compete with the Soviets in space (Ref. 13, pp. 120–121), pushed the Kennedy administration to clarify its position on its space relationship with the Soviet Union. This clarification was expressed in a memorandum from McGeorge Bundy to the president:

> 1) If we compete, we should do everything we can to unify all agencies of the United States Government in a combined space program which comes as near to our existing pledges as possible.
> 2) If we cooperate, the pressure comes off, and we can easily argue that it was our crash effort in '61 and '62 which made the Soviets ready to cooperate.
> I am for cooperation if it is possible, and I think that we need to make a really major effort inside and outside the government to find out whether in fact it can be done (Ref. 13, p. 121).

This kind of opinion, expressed by one of Kennedy's closest advisers, was coupled with support that a U.S. cooperative initiative received from the Congress (see, e.g., Ref. 58). This support was brought about by two factors, both of which were consequences of the Cuban missile crisis. On one hand, "both Russians and Americans sought to avoid future confrontations on the scale of what had happened in 1962; as a result, there emerged more congruities of interest than either side thought existed" (Ref. 6, p. 253), and on the other, Kennedy could more actively pursue areas of common interest with the Soviets "without provoking outcries of 'appeasement' and 'softness toward communism,' not just from Republicans but from members of the president's own party as well" (Ref. 6, p. 255). This kind of an internal political situation, as well as that in September 1963 when, in the U.N. Committee on the Peaceful Uses of Outer Space, the Soviets showed a willingness to soften their previous insistence on the illegality of reconnaissance satellites,[59] encouraged Kennedy to address the

Soviets with one more proposal to pool efforts in space. The proposal was about a joint flight to the moon and it had to be made through the United Nations. There were three reasons for choosing a moon landing as a cooperative project and the largest international forum as a podium for the announcement of the decision.

1) A new initiative in the view of previous unsuccessful attempts to engage Soviets in space cooperation could have presented the United States as a "bagger" soliciting for cooperation from the Soviets, with possible negative implications for American national prestige. To avoid this Kennedy had to present his cooperative initiative as an organic part of his global peace-seeking efforts aimed at the creation of "a constructive and conciliatory climate which could lead to a serious discussion between the United States and the Soviet Union on the basic political issues of the cold war."[59]

2) Kennedy wanted to emphasize that the American lunar program had been motivated not only by egoistical considerations of national prestige, but had also been seeking scientific goals, which when reached would be a contribution to the progress of all mankind. In doing this he was responding to criticism later expressed in the words of Nobel Prize winner and British physicist Sir John Cocroft, who said to Americans: "We smile as we see your flights on television. Your efforts represent a distortion of science in the name of competition with the Soviet Union."[60]

3) By using the United Nations as a forum for proposing that the Soviet Union and the United States go to the moon together, Kennedy wanted the world to see who was responsible for the continuation of the moon race. If the Soviets had declined his proposal, it would have been easier for him to justify the continuation of the moon race both inside the country and in the eyes of international community. Because he had not made a joint moon mission contingent on any changes in Soviet–American relations as a precondition for such an endeavor, he obviously preferred his proposal to be accepted by the Soviet leadership.

Kennedy expressed his view on Soviet–American cooperation in a moon landing in his address to the U.N. General Assembly on 20 September 1963. He said,

> [I]n a field where the United States and the Soviet Union have a special capacity—in the field of space—there is a room for new cooperation, for further joint efforts in the regulation and exploration of space. I include among these possibilities a joint expedition to the moon....

Why, therefore, should man's first flight to the moon be a matter of national competition? Why should the United States and the Soviet Union, in preparing for such expeditions, become involved in immense duplications of research, construction, and expenditure? Surely we should explore whether the scientists and astronauts of our two countries —indeed of all the world—cannot work together in the conquest of space, sending some day in this decade to the moon not the representatives of a single nation, but the representatives of all our countries (Ref. 33, 1963, p. 695).

The proposal never met any response from the Soviet side, except for an address made by Yuri Gagarin before the United Nations in October 1963. In his speech he talked about some areas of possible Soviet–American cooperation in space, including the exchange of scientific knowledge, the tracking and rescue of downed spacecraft, a global weather study, and an international communications system. Gagarin indicated that more clarification with relation to manned space flights would be required before the two countries could cooperate in this area of space exploration, and he did not mention Kennedy's proposal to go together to the moon.[61]

As to the official reaction of the Soviet leadership to Kennedy's proposal, it could have been boiled down to two arguments advanced by Khrushchev against a joint manned moon mission. The first was the need to prepare carefully for the flight to avoid unnecessary risk for a crew,[62] although a week later Khrushchev clarified that this did not mean the Soviet Union would abandon attempts to go to the moon (Ref. 31, p. 160). The second argument was the need to reach disarmament as a precondition for cooperation in space.[63] Successful development of the Soyuz spacecraft, designed specifically for a circumlunar flight, and Korolev's confidence in the proximity of a successful realization of the moon mission (in November 1963 the chief designer made a suggestion to General Kamanin to begin training cosmonauts for the upcoming expedition to the moon and received approval for his idea [Ref. 38, p. 388]) again tempted Khrushchev to use space exploration as a trump card for bargaining with the Kennedy administration. The Soviet leadership's position clearly demonstrated that hopes inspired by Lovell's reports, Blagonravov's hint that the information contained in the reports could possibly be true, successful initiation of the Test Ban Treaty, changed Soviet position on reconnaissance satellites, and Kennedy's address of 20 September were nothing more than illusions (Ref. 13, p. 126).

END OF THE KENNEDY ERA

Meanwhile, Kennedy's proposal, as well as the Soviet rebuff of it, evoked a complex public and political reaction in the United States. There was a considerable endorsement of the recent U.S. cooperative initiative. Those who shared such feelings believed that this initiative, even if unwelcomed by the Soviets, could still have a positive impact on the American space program in terms of terminating a space race. A more gradual approach toward space exploration would allow both to avoid mistakes that could be caused by rushing and, according to Senator William Fulbright, to release "funds for important domestic programs, such as education, employment, urban renewal, and conservation of resources" (Ref. 13, p. 128).

However, there was considerable discontent caused by Kennedy's proposal. Representative Albert Thomas was "very anxious to know whether or not this national goal [flight to the moon] is being abandoned or changed." He supported the program "believing we can be the first nation to put a man on the moon and knowing that we must achieve this goal if we are to help establish the fact that space will be used for peaceful purposes." He stressed the importance of the success of the program for American national security and the security of the rest of the world. Representative Thomas Kelly accused Kennedy of undermining rationale not only for the moon landing but for all manned space flights by offering cooperation with the Soviets. In Senator Clinton P. Anderson's view, "it would be foolish indeed if this country were to attempt to reorient its vast space program each time the Soviet Union made some pronouncement about its goals." Representative Wyman believed that "the mere fact that the President has suggested such a possibility [to cooperate with the Soviets] infects the entire Apollo program with fiscal uncertainty,"* which turned out to be the truth when the House and the Senate passed the following amendment to the NASA appropriation bill for the fiscal 1964:

> No part of any appropriation made available to the National Aeronautics and Space Administration by this act shall be used for expenses of participating in a manned lunar landing to be carried out jointly by the United States and any other country without the consent of Congress.[64]

Trying to respond to criticism from those who favored cooperation and those who were opposed to it, Kennedy defined his attitude toward the cooperation/competition dilemma in the following way:

> If cooperation is possible, we mean to cooperate, and we shall do so from a position made strong and solid by our national effort in space. If

---

*The interpretation of the senators' and representatives' positions is based on their statements reprinted in full in Ref. 13, pp. 127–129.

cooperation is not possible—and as realists we must plan for this contingency too—then the same strong national effort will serve all free
men's interest in space, and protect us also against possible hazards to
our national security.[65]

There is good reason to believe that Kennedy had not been totally discouraged by the continuing resistance of the Soviet leadership to engage in
space cooperation with the United States. Despite that in his last public
speech about space made on 31 October 1963, he emphasized the competitive, as opposed to the cooperative, character of U.S. space activities by saying that the successful realization of the U.S. program had killed "any fear in
[the] free world that a Communist lead in space will become a permanent
assertion of supremacy and the basis of military superiority" (Ref. 33, 1963,
pp. 831–832). In National Security Memorandum No. 271, he asked James
Webb "to assume personally the initiative and central responsibility within
the Government for the development of a program of substantive cooperation with the Soviet Union in the field of outer space, including the development of specific technical proposals."[66] These proposals should have been
developed as a direct outcome of Kennedy's 20 September proposal for
broader cooperation between the United States and the Soviet Union in outer
space.[66] However, Kennedy never saw these proposals, which were submitted in the form of report on 31 January 1964. Ten days after the memorandum had been issued, he was assassinated.

CONCLUSION

The issue of Soviet–American space cooperation grew along with the
space age. The cooperative initiative during the Eisenhower and Kennedy
administrations belonged to the United States. Each administration had its
own particular approach to cooperation with the Soviets and different rationales for seeking it.

Eisenhower did not attach great importance to space exploration at all. He
did not think of international cooperation in space in terms of a bilateral
Soviet–American space relationship and made the United Nations the main
instrument for implementing his cooperative projects. Eisenhower sought
international space cooperation for five reasons.

1) He wanted to create a legal framework for the operation of spy satellites.
2) He was concerned about the ballistic missile race being accelerated by
   the space race.
3) He wanted to moderate the space race with cooperation to avoid centralized mobilization and planning.

4) He did not want an excessive strengthening of the military–industrial complex and the scientific–technological elite, which could have resulted from a space race.
5) He was seeking a less costly type of leadership in space, which was leadership through cooperation. To facilitate cooperation in the field of space exploration, Eisenhower deliberately separated the American satellite project from the ICBM program.

The Soviets did not welcome U.S. cooperative approaches basically for three reasons.

1) The organic link between the Soviet space and ICBM programs covered all Soviet activities beyond the atmosphere with excessive secrecy.
2) The Soviet Academy of Sciences and Soviet rocketry engineering elite used the space race as a justification for a major Soviet effort in space.
3) The Kremlin tried to derive maximum political advantage from Soviet firsts in space.

As a result of the Soviet's intransigent position, no cooperative project between the two countries was realized during this time.

Kennedy's election as president introduced revolutionary changes in the U.S. attitude toward cooperation with the Soviet Union. There were a number of factors behind the new approach. First, unlike Eisenhower, Kennedy believed that space exploration was very important from a political standpoint. Second, he was actively trying to moderate tensions in the Soviet–American relationship. Third, he believed that one of the ways to moderate these tensions would be to develop areas of common interest between the two countries, and in his view space could become one of these potential areas. Fourth, he was not sure at the beginning of his presidency of an American victory in the space race, of the validity of the manned space program, and whether the United States could afford such an expensive space enterprise. Finally, like Eisenhower, he was concerned about strengthening the military–industrial complex and did not want the arms race to be intensified by a space race.

To increase the efficiency of his cooperative efforts, Kennedy sought Soviet–American cooperation not as a part of global international cooperation in space, but as a bilateral Soviet–American endeavor with concrete and clearly defined projects, even when addressing the Soviet leadership with cooperative proposals through the United Nations. One of these projects was a mission to the moon, which was also being realized in the Soviet Union. In Kennedy's view the moon landing program was expensive and challenging enough to persuade the Soviets to realize it through joint Soviet–American efforts. He was so persistent in pursuing bilateral space cooperation, even to the point of being ready to sacrifice the prestige of a U.S.-only moon landing, for the joint Soviet–American mission to the moon.

For a while the Soviet leadership, seeking accommodation with the United States, appeared to agree to cooperate with the United States in joint space projects, excluding the prestigious moon landing. However, as time passed, it became clear that the same factors that hindered cooperation with the United States during the Eisenhower presidency stood in the way of bilateral cooperation during the Kennedy administration.

REFERENCES

[1]McDougal, W. A., *The Heavens and the Earth: A Political History of the Space Age*, Basic Books, New York, 1985.

[2]Sentner, D., "Ike Scores Again with Satellite Revelation," *Los Angeles Examiner*, 1 Aug. 1955. NASA Historical Reference Collection, NASA History Office, NASA Headquarters, Washington, DC.

[3]"Satellite Only in 'Planning Stage'; Aircraft Industry Not Consulted," *Aviation Week*, 8 Aug. 1955, p. 23. NASA Historical Reference Collection, NASA History Office, NASA Headquarters, Washington, DC.

[4]"Space Lawers," *New York Times*, 27 Sept. 1956. NASA Historical Reference Collection, NASA History Office, NASA Headquarters, Washington, DC.

[5]Hall, R. C., "The Origins of U.S. Space Policy: Eisenhower, Open Skies, and Freedom of Space," *Colloquy*, Dec. 1993, p. 23.

[6]Gaddis, J. L, *Russia, the Soviet Union and the United States: An Interpretive History*, McGraw–Hill, New York, 1990.

[7]Hillaby, J., "Flight into Space by 1970 Expected," *The New York Times*, 2 Aug. 1955. NASA Historical Reference Collection, NASA History Office, NASA Headquarters, Washington, DC.

[8]"We'll Launch 1st Moon, and Bigger, Says Russ," *Los Angeles Examiner*, 3 Aug. 1955. NASA Historical Reference Collection, NASA History Office, NASA Headquarters, Washington, DC.

[9]Schauner, W., *The Politics of Space*, Holmes and Meier, New York, 1976, p. 32.

[10]Sagdeev, R., *The Making of a Soviet Scientist*, Wiley, New York, 1994.

[11]Khrushchev, S., *Nikita Khrushchev: Krisisy e Rakety* [Nikita Khrushchev: Crises and Missiles], Vol. II, Novosti, Moscow, 1994.

[12]Golovanov, Y., *Korolev: Fakty e Miphy* [Korolev: Facts and Myths], Science, Moscow, 1994.

[13]Harvey, D. L., and Ciccoritti, L. C., *U.S.-Soviet Cooperation in Space*, Univ. of Miami, Coral Gables, FL, 1974.

[14]"U.S. Experts Deny a Satellite Race," *The New York Times*, 8 Oct. 1955. NASA Historical Reference Collection, NASA History Office, NASA Headquarters, Washington, DC.

[15]Logsdon, J. M., *The Decision to Go to the Moon*, MIT Press, Cambridge, MA, 1970.

[16]*National Aeronautics and Space Act of 1958*, Public Law 85-568, approved July 29, 1958, 85th Cong., 2nd sess., H. Rept. 12575. NASA Historical Reference Collection, NASA History Office, NASA Headquarters, Washington, DC.

[17]McCurdy, H. E., *Inside NASA: High Technology and Organizational Change in the U.S. Space Program*, Johns Hopkins Univ. Press, Baltimore, 1990, p. 84.

[18]Divine, R. A., *The Sputnik Challenge: Eisenhower's Response to the Soviet Satellite*, Oxford Univ. Press, New York, 1993.

[19]U.S. Congress, House, Committee on Science and Astronautics, *A Report on Chronology of Missile and Astronautics Events*, 87th Cong., 1st sess., 1961, H. Rept. 67, p. 40.

[20]U.S. Congress, Senate, Committee on Aeronautical and Space Sciences, *International Cooperation in Outer Space: A Symposium*, No. 92-57, 92d Cong., 1st sess., 1971.

[21]Baker, R., "Earth Satellite Already Stirring Political Battle," The New York Times, 31 July 1955. NASA Historical Reference Collection, NASA History Office, NASA Headquarters, Washington, DC.

[22]U.S. Congress, Senate, Special Committee on Space and Astronautics, *Final Report*, No. 100, 85th Cong., 1st sess., 1958, pp. 10–11.

[23]U.S. Congress, House, Select Committee on Astronautics and Space Exploration, House Report on International Cooperation in the Exploration of Space, No. 2709, 85th Cong., 2nd sess., 1959, pp. 7–8, 15–16.

[24]United Nations, General Assembly, 11th Session, 12 Jan. 1957, *United States Memorandum Submitted to the First Committee of the General Assembly*, A/C. 1/783, Agenda Item 22, pp. 5–6.

[25]Public Papers of the Presidents of the United States: Dwight D. Eisenhower, U.S. Government Printing Office, Washington, DC, 1958, pp. 154–155.

[26]Khrushchev, N., "Letter to President Dwight D. Eisenhower, 22 April 1958," U.S. Department of State Bulletin, 19 May 1958, p. 814.

[27]Bulganin, N., "Letter to President Dwight D. Eisenhower, 10 December 1957," U.S. Department of State Bulletin, Jan. 27, 1958, p. 128.

[28]Newell, H. E., Jr., "IGY Conference in Moscow," *Science*, Jan. 9, 1959, p. 80.

[29]United Nations, General Assembly, 13th section, 1st Committee, Nov. 18, 1958, A/C. 1/L. 219/Rev. 1.

[30]Lodge, H. C., "United Nations Establishes Committee on Peaceful Uses of Outer Space," *U.S. Department of State Bulletin*, 5 Jan. 1959, p. 26.

[31]U.S. Congress, Senate, Committee on Aeronautical and Space Sciences,

Staff Report on International Cooperation and Organization for Outer Space, S. Doc. 56, 89th Cong., 1st sess., 1965.

[32]"An Interview with John F. Kennedy," *Bulletin of the Atomic Scientists*, Nov. 1960, p. 347.

[33]*Public Papers of the Presidents of the United States: John F. Kennedy*, 1961, U.S. Government Printing Office, Washington, DC, 1962, 1963.

[34]Kennedy, J. F., "If the Soviets Control Space, They Can Control Earth," *Missiles and Rockets*, 10 Oct. 1960, pp. 12–13.

[35]Kornienko, G., *Kholodnaya Voina: Svidetelstvo Eyo Uchastnika* [Cold War: A Testimony of Its Participant], International Relations, Moscow, 1994.

[36]*U.S. Department of State Bulletin*, 13 March 1961, p. 369.

[37]Khrushchev, N., *Khrushchev Remembers: The Last Testament*, Little, Brown and Co., Boston, 1974, p. 53.

[38]Kamanin, N., *Skritiy Kosmos* (1960–1963) [Hidden Cosmos], Infortext, Moscow, 1995, p. 15.

[39]Protocol #4 of the General Meeting of the USSR Academy of Sciences, held on the occasion of the first manned space flight, 19 May 1961. Documentary Collection on Science and Technology, Russian Academy of Sciences, Moscow.

[40]Goldsen, J., (ed.), *Outer Space in World Politics*, Frederic A. Praeger, New York, 1963, p. 57.

[41]U.S. Congress, House, Committee on Science and Astronautics, *Discussion of Soviet Man in Space Shot*, 87th Cong., 1st sess., 1961, pp. 7, 9, 13–14, 26.

[42]Lambright, W. H., *Powering Apollo: James. E. Webb of NASA*, Johns Hopkins Univ. Press, Baltimore, 1995.

[43]Webb, J., "International Relations in Space," American Political Science Association, St. Louis, MO, NASA News Release, 6 Sept. 1961.

[44]United Nations, General Assembly, First Committee, 14th sess., 4 Dec. 1961, *Address of US Representative Adlai Stevenson*, A/C.1/PV. 1210, pp. 6, 16, 17.

[45]United Nations Doc. A/C.1/PV.1211, 5 Dec. 1961, pp. 26–27.

[46]United Nations, General Assembly, Resolution 1721 (XVI): *International Cooperation in the Peaceful Uses of Outer Space*, 20 Dec. 1961.

[47]*U.S. Department of State Bulletin*, 12 March 1962, p. 411.

[48]Mishin, V., "Pochemu My Ne Sletali Na Lunu" [Why We Did Not Fly to the Moon], *Cosmonautics, Astronomy*, Dec. 1990, p. 23.

[49]U.S. Congress, Senate, Committee on Aeronautics and Space Science, *Documents on International Aspects of the Exploration and Use of Outer Space*, 1954–1962, S. Doc. 18, 88th Cong., 1st sess., 1963, pp. 248–251 (see Krushchev's letter).

[50]Ball, D., "United States Strategic Policy Since 1945: Doctrine, Military-

Technical Innovation and Force Structure," *Strategic Power: USA/USSR*, edited by C. G. Jacobsen, Macmillan, London, 1990.

[51]Ezell, E. C., and Ezell, L. N., *The Partnership: A History of the Apollo-Test Project*, NASA, Washington, DC, 1978.

[52]Frutkin, A., *International Cooperation in Space*, Prentice–Hall, Englewood Cliffs, NJ, 1965.

[53]The White House Conference on International Cooperation, Nov. 28–Dec. 1, 1965, National Citizens' Commission, *Report of the Committee on Space*, Washington, DC, 1965, p. 7.

[54]*Washington Post*, 21 March 1963. NASA Historical Reference Collection, NASA History Office, NASA Headquarters, Washington, DC.

[55]U.S. Congress, Senate, Committee on Aeronautical and Space Sciences, *NASA Authorization for FY 1963, Hearings Before the Committee on Aeronautical and Space Sciences, Senate on H.R. 11737*, 87th Cong., 2nd sess., 1962, p. 47.

[56]U.S. Congress, House, Committee on Science and Astronautics, *Report on Astronautical and Aeronautical Events of 1962*, 88th Cong., 1st sess., 12 June 1963, p. 46.

[57]*Congressional Record*, 87th Cong., 1st sess., 18 April 1962, Vol. 108, p. A3035.

[58]*Congressional Record*, 18 Sept. 1963, pp. 16536–16537.

[59]Gaddis, J. L., "Evolution of Reconnaissance Satellite Regime," *U.S.-Soviet Security Cooperation: Achievements, Failures, Lessons*, edited by A. L. George, P. J. Farley, and A. Dallin, Oxford Univ. Press, New York, 1988, p. 358.

[60]"An Interview with Sir John Cocroft," *Bulletin of the Atomic Scientists*, Oct. 1966, p. 13.

[61]*New York Times*, 17 Oct. 1963. NASA Historical Reference Collection, NASA History Office, NASA Headquarters, Washington, DC.

[62]*New York Times*, 26 Oct. 1963. NASA Historical Reference Collection, NASA History Office, NASA Headquarters, Washington, DC.

[63]*Pravda*, 2 Nov. 1963.

[64]U.S. Congress, Senate, Report No. 641, 88th Cong., 1st sess., 13 Nov. 1963.

[65]Kennedy, J. F., "Letter to Representative Albert Thomas," *Houston Post*, 27 Sept. 1963. NASA Historical Reference Collection, NASA History Office, NASA Headquarters, Washington, DC.

[66]"National Security Memorandum No. 271," 12 Nov. 1963. NASA Historical Reference Collection, NASA History Office, NASA Headquarters, Washington, DC.

[67]Leskov, S., "Mir Mozhet e Dolzhen Rabotat Dalshe" [Mir Can and Must Continue Its Operation], *Izvestia,* 20 Oct. 1998.

Chapter 3

Stalemate: Khrushchev–Johnson and Brezhnev–Johnson Periods

Johnson's View of Space Exploration and Cooperation

Lyndon B. Johnson, like Kennedy, was a space-oriented president.* As a Democratic leader in the U.S. Senate at the time of Sputnik in October 1957, Johnson continually dealt thereafter with space affairs. Senator Johnson spearheaded Senate actions leading to the passage of the NASA Act of 1958 and to the creation of the Senate Committee on Aeronautical and Space Sciences. On 19 November 1958, he presented U.S. proposals for the international control of outer space before the Political Committee of the United Nations. As vice president in the Kennedy administration, he served as chairman of the National Aeronautics and Space Council (Ref. 1, p. 57). Moreover, Johnson had a strong personal motivation for deep involvement in space issues. He

> was vitally concerned with national security and believed that the Eisenhower response to Sputnik gravely underestimated the political loss the United States had suffered. He believed that the Congress, which was controlled by the Democratic party had a responsibility to develop alternatives to the policy of the Republican administration and that the Soviet space firsts provided an opportunity for such opposition. And he also wanted to become president. Johnson could sense public reaction to space, and concluded that this issue was a means of becoming better known and respected outside the Senate.[2]

Like Kennedy, Johnson had been an advocate of different countries pooling together their efforts in space (Ref. 1, pp. 57–90). He believed that "the end is peace. The means to that end is international cooperation," because "men who worked together to reach the stars are not likely to descend into the depths of war and desolation."[3] In one of his first conceptual statements on international cooperation in space, Johnson said,

> Thus far the space age has been characterized as a period of competition. It is important, however, that we keep in mind the fact that the

---

*Johnson's background in space affairs is given in Ref. 1.

51

competitiveness is between political systems, not between national scientific communities. In the world of science the logical instinct is toward cooperation without regard to political boundaries. This impulse we must preserve. The real challenge of the space age is for the politician to tear down the walls between men which have been erected by his predecessors and contemporaries in the political field, rather than to raise its barriers higher into the free and peaceful vastness of space. If the potentials of the age of space are fully realized, this period will someday be known—and blessed by all people on earth—as the Golden Age of Political Science.[4]

However, despite similarities, there were differences between Kennedy's and Johnson's approaches toward international cooperation in space— Soviet–American cooperation in particular. Unlike Kennedy, Johnson had never been persistent in inviting the Soviets to pool their efforts with the Americans in space exploration. As president he did not make any direct personal approach to the Kremlin on space cooperation. He made seven indirect approaches, two during his first term and five more during his second term. The first one was made on his behalf at the United Nations by U.S. Representative Adlai Stevenson on 2 December 1963. Stevenson said that Johnson had instructed him to reaffirm Kennedy's proposal to go to the moon together with the Soviet Union.[5] The second approach was made in Johnson's article, "The Politics of the Space Age," published in *The Saturday Evening Post* on 29 February 1964, where he praised "the recent United States and Soviet cooperative space effort in communications" in what may be "an auspicious beginning for better East–West understanding through the advancement of knowledge. We would like to see this understanding broadened by joint space endeavors in meteorology, astronomy, lunar and interplanetary exploration."[4]

The third approach was made after Johnson had been reelected, during his news conference on 25 August 1965. Jonhnson expressed his confidence that the United States "will continue to hold out to all nations, including the Soviet Union, the hand of cooperation in the exciting years of space exploration, which lie ahead for all of us" (Ref. 6, 1965, Vol. II, p. 918). He made his fourth approach in remarks during a visit to Cape Kennedy with Chancellor Ludwig Erchard of Germany, when he confirmed the U.S. determination to seek cooperation in space with the Soviet Union (Ref. 6, 1966, Vol. II, p. 1075). The fifth approach was made in Johnson's interview published in *America Illustrated* on 27 September 1966 for distribution in the Soviet Union. In this interview Johnson expressed his opinion that the Soviet Union and the United States "must work toward progress in the field of disarmament and in greater cooperative efforts between our countries in space exploration."[7] The sixth approach was in fact a hint Johnson made on possible Soviet–American cooperation in communication satellites in his Special Message to the Congress on Communications Policy on 14 April 1967: "The Soviet Union is a leader in satellite technology. I am advised that there is no insurmountable technical obstacle to an eventual linking of the Soviet MOLNIYA system with the INTELSAT system" (Ref. 6, 1967, Vol. II, p. 769). Finally, in his remarks at the ceremony marking the entry into force of the Outer Space Treaty on 10 October 1967, Johnson expressed his hope that the coming decade would increasingly bring a partnership involving the Soviet Union and America (Ref. 6, 1967, Vol. II, p. 920).

Johnson had not only shown a lack of persistence in trying to involve the Soviets in space cooperation with the Americans; he actually deemphasized any special significance of the Soviet–American bilateral space relationship by stressing the need for equal participation of all space-exploring countries in international cooperation in space. In the beginning of his presidency, Johnson criticized financial restrictions imposed by Congress on a moon landing project if it were carried out through cooperative efforts rather than by the United States alone. He was concerned because he thought it might "raise some doubts as to our willingness to work cooperatively with other nations in the most important space effort of this decade" (Ref. 6, 1967, Vol. I, pp. 72–73). The *Report of the Committee on Space* prepared within a framework of the White House Conference on International Cooperation stressed that "the United States has no desire to establish a bipolarity in space matters with the Soviet Union. Rather, it wishes to insure that these initial cooperative projects are, from the outset, open to other countries and will serve the general interest" (Ref. 8, p. 8). Even in some of his afore-mentioned indirect approaches to the Soviet leadership on space cooperation, Johnson made it clear that for him Soviet–American cooperation in space was just part of global international space efforts. In his special message to the Congress on a communications policy, Johnson urged "the Soviet Union and the nations of Eastern Europe to join with the United States and our 57 partners as members of INTELSAT" (Ref. 1, p. 83), and in his remarks on the Outer Space Treaty he expressed his hope that "the next decade should increasingly become a partnership—not only between the Soviet Union and America, but among all nations under the sun and stars..." (Ref. 6, 1967, Vol. II, p. 920).

DIFFERENCES IN ATTITUDE TOWARD INTERNATIONAL COOPERATION IN SPACE

A number of factors account for the difference between Kennedy's and Johnson's attitude toward international cooperation in space. First, Johnson had been identified with the drive for a strong U.S. national program "to be first in space" to the extent that this identification exceeded even that of

Kennedy (Ref. 6, 1967, Vol. II, p. 133). Summarizing all that he did for the U.S. space program shortly before the end of his presidency, he said, upon presenting the NASA Distinguished Service Medal to the Apollo 8 astronauts on 9 January 1969, "there were those men in our Government who 10 years ago fought to guarantee America's role in space. And I am glad that I was one of them" (Ref. 6, 1968–1969, Vol. II, pp. 1247–1248). Thus, there is a good reason to believe that Johnson was thinking about space not so much in terms of cooperation, but in terms of competition.

Second, in the mid-1960s, space programs of other countries had been rapidly rising. With the establishment of the European Space Research Organization and the European Launcher Development Organization, European countries became potential challengers of U.S. leadership in space, assuming that the United States would finally win the space race with the Soviet Union. The fact that Johnson realized it could be confirmed by the words he spoke during his tour of U.S. space facilities in September 1964: "As long as I am permitted to head this country I will never accept a place second to any other nation in this [space] field."[9] The way to maintain leadership would be not only to boost growth and development of the U.S. space program but also to play a leading role in the organization and guidance of international cooperation in space. Emerging space powers would probably be a more lucrative target for cooperation because they would be more willing than the Soviet Union to accept American leadership in cooperation. As authors of the *Report of the Committee on Space* put it,

> The rapid growth of space activity abroad is a challenge to American leadership in the field of cooperation. Foreign scientists and engineers will find increasing opportunities in their own national and regional programs for space research of the sort that our cooperative projects have thus far supplied. We must therefore be alert to possibilities for more considerable and advanced cooperative efforts, going beyond the appeal of national programs, if we are to extend the technical and political advantages of cooperation, for others as well as ourselves (Ref. 8, p. 24).

The third reason for the greater emphasis that the U.S. administration placed on cooperation with Europeans was the changed international environment that had affected U.S.–European and U.S.–Soviet relations. As Soviet–American tensions began to decline after 1962, friction developed within NATO and the Warsaw Pact, because both "depended for their cohesion in large part on shared perceptions of external threat."[10] One of the manifestations of these internal frictions in NATO was President Charles DeGaulles's 1966 decision to withdraw France (one of the most rapidly rising European space powers) from military participation in NATO (Ref. 10,

p. 266). Space cooperation with Europe, particularly with France, was supposed to strengthen the shaken NATO cohesion.

As for the U.S.–Soviet relations, they were negatively affected by two factors:

1) The new Soviet leadership, which had come to power in October 1964, felt it necessary to achieve strategic equivalency with the United States to compensate for the lack of balance caused by Kennedy's strategic buildup. As Robert McNamara, secretary of defense in the Kennedy administration, later admitted, it might have slowed down the progress of détente (Ref. 10, p. 261).
2) The increasing American military involvement in Vietnam put Soviets and Americans on different sides of the front and strained Soviet–American relations.[11]

The fourth reason why Johnson deemphasized a bilateral cooperation with the Soviets was because of the criticism his space policy faced from the Republicans. *The New York Herald Tribune*, a very influential newspaper identified closely with the Eisenhower–Rockefeller wing of the party, accused Johnson in June 1964 of losing his enthusiasm for the Soviet–American space race. This kind of accusation was based on a number of facts, such as the continuing NASA–USSR Academy of Sciences negotiations and Hugh Dryden's statement that the United States would like "to cooperate to as great [an] extent as possible with the Soviets in exploring space."[12] The Democratic space cooperation policy with which Johnson was associated had faced specific criticism from the Republican presidential candidate Barry Goldwater in October 1994:

> The great enterprise of exploring the Moon and the planets is not a job for America alone. It is rather an area for international cooperation.... But in such cooperation, I would give first priority to joint work with our Allies—the advanced free democracies—rather than with the Communist nations. It is with our Allies that maximum advantage can be obtained from the free exchange of ideas.
>
> It is a mistake to attempt international cooperation in every instance as we have been doing, by first inquiring whether the men in [the] Kremlin deign to cooperate with us. They have said often and clearly that such cooperation is subject to their long-range goal of world mastery and the destruction of free governments.[13]

The fifth reason was the change of leadership, first in the United States in November 1963, and then in the Soviet Union in October 1964, which did not contribute to Soviet–American cooperation in space. Khrushchev and Kennedy were leaders "capable of perceiving shared interests through the distractions created by ideological differences, unwieldy bureaucracies, dissimilar backgrounds, and the allurements of pride and prestige. Such had not always been the case in the history of Russian–American relations, nor would it invariably be so in the future" (Ref. 10, p. 257). According to Sergei Khrushchev, his father trusted Kennedy and was ready to negotiate a joint Soviet–American mission to the moon with him. Kennedy's death, in Khrushchev's view, made such a mission impossible, something that Khrushchev had really regretted. He had trusted a concrete person, not the U.S government, and this person had ceased to exist (Ref. 15, p. 465).

Finally, it is very likely that Johnson was following recommendations of the study made by NASA on U.S.–USSR cooperation in space research programs. The final version of the study was called *A Report on Possible Projects for Substantive Cooperation with the Soviet Union in the Field of Outer Space*, signed by NASA Administrator James Webb. One of the recommendations follows.

> As a tactical device, calculated to put pressure upon the Soviet Union, demonstrate our serious intentions, and gain good will from certain nations, consideration should be given to means by which "other countries" than the Soviet Union might be further identified with our lunar programs.[16]

NASA's REPORT OF POSSIBLE PROJECTS

CHANGE IN DIRECTION

The report was prepared following the aforementioned assignment given by Kennedy to Webb in National Security Memorandum No. 271. Whatever Kennedy had in mind by making NASA a leading American institution responsible for the further development of Soviet–American cooperation in space, this decision put an end to any boldness in U.S. cooperative approaches toward the Soviet Union. Unlike Kennedy, NASA was not really interested in making space an area of common interest. This is not to say that NASA was against cooperation with the Soviet Union, but the agency's corporate interests would have been better served not by cooperation in space, but by the race that at that time had been associated with the moon-manned program. This is why the report suggested that cooperation in a moon landing should necessarily be preceded by "significant new relationships with the Soviet Union" (Ref. 16, p. 172). Plus, the report advanced conditions for cooperation in manned moon missions that were less than acceptable for the Soviet leadership. Among them there was an obligation for the Soviet Union "to present a considerable amount of information not previously made available" regarding the Soviet manned space program and to verify this information "through independent sources." As the authors of the report acknowledged themselves, "this step would not place an undue burden upon us because of the publicity already given to our own intentions, but it would for the first time require the Soviet Union to describe its conceptual approach to the lunar landing problem" (Ref. 16, pp. 5–6). Two of the boldest variants of possible Soviet–American cooperation in manned moon mission—USSR booster/U.S. spacecraft integration and the Turner proposal*—had been rejected—the first because it had not been consistent "with the U.S. objective of achieving a leading space capability to delegate the development of an adequate booster to the Soviet Union" (Ref. 15, p. 16), and the second because its realization would have meant "an unacceptable interdependence, prejudicing seriously our ability to proceed with our own program in the event that the Soviets do not live up to their agreement over the extended period of years required to implement it" (Ref. 15, p. 18).

The authors of the staff report had also considered the interchange of astronauts because "the U.S. would have far more to gain than to lose from such reciprocity in view of the relative secrecy of the Soviet program to date." However, they themselves realized that to make this kind of offer to the Soviets, implying that the latter should give "extensive access to training facilities and programs, flight hardware and systems, launching sites and so forth" to American astronauts, would have been "politically premature" (Ref. 15, pp. 18–19). This proposal was not included in the final version of the report, signed by Webb. At the same time the Staff Report recommended cooperation in second-rate projects, such as unmanned flights to support a manned lunar landing, including a study of micrometeoroid density in space between the Earth and the moon, the radiation and energetic particles environment between the Earth and the moon, the character of the lunar surface, and the selection of lunar landing sites.[17,18] The report did not recommend any new high-level U.S. initiative until the Soviet Union had a further opportunity to discharge its current obligations under the existing NASA–USSR Academy agreement (made between Dryden and Blagonravov), or "in the alternative, until the Soviets respond affirmatively to the proposal already

---

*A Republic Aviation engineer, Thomas Turner, had proposed in *Life* (11 October 1963) the following cooperative effort: The United States would forego the development of a large booster and concentrate simply on placing its lunar excursion module (LEM) in Earth orbit. The Soviet Union would at the same time place a very large and powerful spacecraft in Earth orbit. The two would rendezvous, then utilize the Soviet spacecraft's propulsion to transfer to a lunar orbit, at which time the LEM would separate and descend to the lunar surface with both a Soviet and an American aboard. It would then return to lunar orbit, the occupants would transfer to the Soviet spacecraft, abandon the LEM, and return to Earth. The only requirements the project had to face would have been common docking hardware and a communications agreement. (As described in Ref. 15, pp. 17–18.)

made by President Johnson in the U.N."[17] Apparently Johnson followed this suggestion because he did not mention Soviet–American cooperation in space at all during the celebration of the United Nations's anniversary in San Francisco.

INFLUENCE OF THE REPORT DURING THE JOHNSON YEARS

The report played an important role in the formation of the U.S. approach toward cooperation with the Soviet Union in outer space during the Johnson administration. This came about for two reasons. The first was rooted in Johnson's policy aimed at giving a new strength to "the Government's buffeted and beleaguered space agency."[19] One of the ways to give this strength was to increase NASA's decision-making role in space issues. The second reason was probably the fact that the "ties between Mr. Kennedy and space administrator James Webb were neither long-standing nor particularly warm."[19] The Johnson–Webb relationship was a different case. Webb, who had business and political roots in the southwest, had been named as NASA chief on the recommendation of Johnson; their association dated back to Truman administration days when Webb had served as Budget Director and Undersecretary of State. In short, Webb and Johnson knew how to talk to each other.[19]

The overall effect

> of Johnson policy was to place squarely on the NASA team responsibility for conducting the U.S. effort for direct cooperation with the USSR; to give the team a near free hand within the framework of the broad guidelines set forth in the Webb report, to initiate and react to cooperative proposals; and to provide the team with assurance of Presidential support and follow-through should its continuing efforts lead to substantial advances or promise substantial advances (Ref. 20, p. 142).

In practice it meant that, from then onward, the United States would have proceeded with cooperation with the Soviet Union through the existing NASA–USSR Academy of Sciences channel.

WORKING THROUGH THE NASA–SOVIET ACADEMY OF SCIENCES LINK

Cooperation through this channel went through a number of ups and downs. The most significant projects the Soviet Union and the United States were supposed to cooperate on according to the "Bilateral Space Agreement of June 8, 1962" and the "First Memorandum of Understanding of March 20 and May 24, 1963" were experiments involving the American communication satellite ECHO II. The Soviets did not agree to share with their American colleagues radar observations of the inflation phases of the satellite, but agreed only to receive and not to send signals in the experiments and tests. However, after the launch of the satellite in January 1964, the Soviets not only observed the critical inflation phase of the satellite optically and forwarded the data to the Americans, but also provided recordings and other data of their reception of the transmissions via ECHO from the Jodrell Bank Observatory. Moreover, the Soviets expressed readiness to start operation of the Washington–Moscow weather data link, providing the Americans with "data that was not previously available in the United States," and even proposed to extend cooperation to the field of space biology and medicine.[21]

The last proposal was probably related to the foundation of the Institute of Biomedical Problems in 1963, the leading Soviet/Russian space medicine and biology research institution. Cooperation in this field concerned the joint writing of a book on space biomedical matters. However, a lack of understanding between Soviet and American specialists on how the book should be written stopped the realization of the project. While Americans proposed that U.S. and Soviet authors write on the same topic, "each emphasizing the work going on in his own country" (Ref. 20, p. 143), Soviets wanted to "assign the writing of each chapter to the most outstanding specialist in the appropriate field from among the USSR and U.S. specialists" (Ref. 20, p. 144). However, Americans believed that it "would bring uneven results which quite possibly would reflect inadequately the work done on one side or the other, since each author would be essentially unfamiliar with work outside his own country" (Ref. 20, p. 145). Ultimately, neither biomedical, nor weather data link, nor geomagnetic surveys projects were realized. As Dryden summarized it: "Experience has taught us that the Soviets prefer general discussions and agreements which relieve them of the need to be specific and allow them to gain credit for international cooperation without partaking of its substance" (Ref. 20, p. 141).

Despite such a pessimistic conclusion, Americans made a few more approaches to the Soviets, including offers to cooperate in avoiding planetary contamination, to use the Soviet MOLNIYA I satellite for joint experiments and tests, and to send unmanned probes to Mars and other planets of the solar system. However, the Soviets never responded to any of these initiatives. Moreover, Blagonravov made it clear that the Soviets were backing away from the bilateral relationship itself (Ref. 20, p. 147).

POLITICAL CHANGES IN THE SOVIET UNION

The negative changes in Soviet–American cooperation in space were a result of political changes in the Soviet Union. In October 1964 Nikita Khrushchev was ousted. The new Soviet leadership led by Leonid Brezhnev held different views from its predecessors on Soviet domestic and foreign policy. According to George Arbatov, who was one of the political consultants to Khrushchev and later to Brezhnev,

> it soon became apparent that the country was being pushed to the right by Brezhnev's rivals, as well as some of his close associates.... Many wanted to turn him into the leader of the right-wing agenda, which meant the rehabilitation of Stalin and Stalinism and a return to the old dogmas of domestic and foreign policy (Ref. 22, p. 126).

During the first substantial discussion on foreign policy to take place in the Central Committee Presidium,* a group of progressive-minded consultants and high officials made proposals and advanced initiatives aimed at improving relations with the United States and the Western European countries. However, these proposals and initiatives had been attacked fiercely for their insufficient "class position," "class consciousness," and excessive "leniency toward imperialism" by representatives of the right wing of the party. As a result of the discussion, the proposed new approach toward relations with the United States and Western Europe was sunk (Ref. 22, p. 115). Another approach, which became dominant in Soviet foreign policy, at least through the end of the 1960s, could be illustrated by what happened at the traditional October Revolution holiday reception at the Kremlin just a month after the new leadership had come to power in the Soviet Union. "Rodion Malinovsky (the minister of defense) had drunk too much and proposed a cocky anti-American toast which offended the U.S. ambassador, Foy Kohler" (Ref. 22, p. 114). According to Yuri Andropov, who was then a "mere" secretary of the Central Committee,† it was a manifestation of the new leadership's policy (Ref. 22, p. 114). Given "the xenophobic policies and trends that marked the new Soviet leadership, there was obviously little room for even limited space cooperation with the U.S." (Ref. 20, p. 151).

A turn to the right in Soviet politics also affected the Soviet Academy of Sciences. One of the key players in space cooperation between the two countries, the president of the Academy, Mstislav Keldysh, joined the crusade against the "unfaithful," or dissidents—people who disagreed with political oppression inside the country and its Iron Curtain foreign policy (Ref. 20, p. 136). Clearly, this could only hinder space cooperation between the two states.

---

*The supreme ruling body of the Communist Party of the Soviet Union was their Congress, which usually got together every four or five years. However, between the congresses the supreme ruling body of the party was its Central Committee, with its two upper structures the Presidium and Politburo.

†In 1982 he became supreme Soviet leader, having ascended to the position of General Secretary of the Central Committee of the USSR's Communist Party after Brezhnev's death.

The new Soviet leadership also emphasized the need to strengthen unity with Soviet "natural allies" and "class brothers" (Ref. 22, p. 115), which was, in particular, a reflection of growing Soviet–Chinese competition for the influence in Third World countries. One of these countries was Vietnam. Brezhnev and his team, even more than Khrushchev, denounced the U.S. armed involvement in Vietnam, especially after the Americans started bombing the northern part of Vietnam in February 1965. It happened precisely when the new head of the Soviet government, Alexei Kosygin, was there, which naturally led to increased mistrust and hostility (Ref. 22, p. 117).

INFLUENCE OF THE SOVIET MILITARY–INDUSTRIAL COMPLEX

Another factor, which did not contribute to the development of Soviet–American cooperation in space during the first five years of Brezhnev's rule, was the growing influence of the Soviet military–industrial complex (MIC). Brezhnev wanted "to provide more for the population without skimping on military expenditures and without making fundamental changes in the Soviet system."[23] One of the examples of the MIC's growing influence was the SS-18–SS-19 ICBM case. The SS-18 missile was designed by the "successors of the huge empire of Sergei Korolev and Mikhail Yangel,"* and SS-19 by Vladimir Chelomey.† Two opposite groups within the Soviet MIC associated with different factions within the Soviet political elite were supporting different designs. Brezhnev, according to Sagdeev, was "a softy" when it came to dealing with the MIC and "instead of making the critically important principal decision, he chose to compromise." As a result both projects were adopted in parallel.[24]

The increased influence of the Soviet MIC on Soviet politics had a twofold negative consequence for Soviet–American cooperation in space. On one hand, the MIC needed an external enemy to justify huge military expenditures. This search for an enemy resulted in the overall cooling of the Soviet attitude toward the United States, promoting a freezing impact on bilateral space cooperation. On the other hand, the broadened and intensified MIC activities, particularly in the field of rocketry, required more secrecy to cover them. Secrecy became a serious obstacle in the way of cooperation between the two countries in the area of dual-use technology.

---

*Mikhail Yangel (1911–1971) was a Soviet rocketry chief designer, famous mostly for his development of military missiles.

†Vladimir Chelomey (1914–1984) was a Soviet rocketry chief designer. His most famous rocket is the Proton launch vehicle used for putting modules of Soviet/Russian space stations into orbit. It is currently one of the most widely used commercial launch vehicles and will also be used for the assembly of the ISS.

SOVIET SUPERIORITY IN ALL AREAS

Another reason for secrecy was "the Soviet means the best" policy of the new Soviet leadership. The essence of this policy was to convince the Soviet people as well as the rest of the world of the social, economic, cultural, and technological superiority of Soviet society. Space achievements were supposed to, and in fact did, play a key propagandistic role in this policy. However, Soviet space officials as far back as early 1962 realized that the Soviet Union was losing the space race to the United States (Ref. 25, pp. 90–91). This was later confirmed by Khrushchev himself at the wedding of two Soviet cosmonauts—Andrian Nikolaev and Valentina Tereshkova—in November 1963 (Ref. 25, p. 377). Cooperation with the Americans could have revealed to the world that Soviet space achievements had not been as outstanding as the Soviet leadership claimed and that the United States had gotten ahead in the space race.

VIEW TOWARD SPACE EXPLORATION

Finally, the attitude of the new Soviet leadership toward cooperation in space with the United States reflected to some extent the overall attitude of the new Soviet rulers to space exploration in general. Desirous of consolidating his power and promoting his policies, Khrushchev set up the space program much as Stalin had set up aviation, as a personal trophy. Hence he depended on the technicians, Korolev most of all, to provide him with glory and leverage, even as he frustrated them and twisted their efforts to serve his personal rule.[26]

The situation changed when Khrushchev was removed from power. The new leadership team of Brezhnev, Kosygin, and Nikolai Podgorni was at best not a positive gain for the Soviet space program (see Ref. 14, p. 514). Unlike Khrushchev, they had not been closely associated with Soviet space triumphs in the late 1950s and early 1960s, although Brezhnev as secretary of the Central Committee of the USSR's Communist Party had been one of the high officials in charge of the Soviet space industry.[27,28] Neither were they given to the dramatic gestures such as space coups that so delighted their colorful predecessor. Moreover, the new leaders evidently favored a more rational division of technical talents and research funds throughout the whole economy, particularly to applied research in the industrial sector for immediate benefits. Responsible as they were for a rapid buildup of Soviet strategic and conventional forces, the new Soviet leaders may have felt that the space program was absorbing resources required for national defense.[29]

TERMINATION OF MUCH SPACE COOPERATION

All of these political developments in the Soviet Union had virtually terminated any kind of Soviet–American interactions in the field of space exploration. A few more U.S. attempts aimed at reanimation of bilateral cooperation failed to bring any substantive results. Blagonravov agreed to meet with Dryden during a session of the International Committee on Space Research (COSPAR) at Mar Del Plata, Argentina, in May 1965, but their negotiations had an informal and inconcrete character (Ref. 20, p. 153). Blagonravov explained the lack of progress in the weather data link by the fact that "the quality of satellite pictures available to the Soviets was not as equal to that of American satellite pictures," and that many of the Soviet space scientists had not managed to attend a COSPAR meeting for the simple reason of "not being able to round up" for the event (Ref. 20, p. 154). In September 1965 Johnson extended a personal invitation to a high-ranking Soviet scientist to watch the launching of America's next Gemini satellite. The invitation was rejected by the Kremlin, although Keldysh restated that Soviet scientists still "positively evaluate" cooperation with the United States. The rejection had been ascribed "not only to the cool Moscow–Washington relations accompanying the Vietnam war, but also to the Soviet Union's assumption that if it accepted, it would have to reciprocate by inviting American scientists to witness one of its super-secret launchings."[30] On 6 January 1966, NASA administrator James Webb, who had become the U.S. representative in the NASA–USSR Academy of Sciences relationship after Dryden's death, sent a letter to Blagonravov asking for a description of experiments conducted by Soviet Venus probes, so that the United States would complement rather than duplicate Soviet accomplishment. Blagonravov indicated in his response that he did not have the authority to release information on the experiments. In March and May of the same year, Webb wrote Blagonravov a letter in which he gave initiative to the Soviets in proposing areas for discussion with a view to extending cooperation between the two countries. Blagonravov's response was that the Soviet Union was not yet ready for further cooperation.

Despite the intransigent approach of the Soviet leadership and the Soviet Academy of Sciences toward space cooperation with the United States, three developments were perceived by Americans as encouraging signs indicating that the Soviets might reconsider their position with relation to bilateral cooperation.

1) The Soviets suddenly began to supply Americans with meteorological data through the long inoperative Washington–Moscow weather data link line.

2) A final agreement to an Outer Space Treaty, on which negotiations had been conducted in the United Nations for four years, was signed by representatives of the United States, the Soviet Union, and the United Kingdom on 27 January 1967.

3) The Americans believed that the deaths of U.S. astronauts Grissom, Chafee, and White in the Apollo fire in January 1967, coupled with the death of Soviet cosmonaut Vladimir Komarov in April 1967 (who died on impact during his Soyuz 1 crash landing), would induce Soviets to pool their efforts with the Americans to prevent similar tragedies from happening again.

All of these hopes turned out to be illusory; Moscow did not change its position. The Soviets did not respond to the offer made in September 1966 by the U.S. representative to the United Nations to provide tracking coverage from the United States territory for Soviet launches. Keldysh did not give any constructive reaction to the U.S. National Academy of Sciences proposals to exchange data resulting from lunar surface experiments conducted by the two countries and to exchange data on planetary exploration obtained by Soviet Venus and American Mariner probes. President Johnson's call on the Soviets to resume bilateral cooperation, made on the occasion of the signing of the Outer Space Treaty, was not responded to either.* This was the last effort undertaken by the Johnson administration to involve the Soviet Union in bilateral cooperation with the United States.

CONCLUSION

Political changes in the United States and the Soviet Union, as well as in the international political environment in 1963–1964, introduced some innovations in the Soviet–American space relationship. The new U.S. president, Lyndon B. Johnson, had another approach to Soviet–American cooperation in space. He was still pursuing the idea of pooling Soviet and American efforts in space, but unlike Kennedy, deemphasized a bilateral space relationship and offered cooperation only in second-rate projects, excluding a moon landing as the most prestigiously rewarding. His approach came about for a number of reasons.

1) Johnson was more personally associated than Kennedy with achieving the goal of American leadership in space.
2) He was concerned about possible future challenges to U.S. leadership in space from emerging space programs in countries other than the Soviet Union and wanted to ensure American leadership through cooperation.

---

*For a more complete account of Soviet–American space interactions in 1966–1967, as well as sources used for reconstructing these interactions, see Ref. 20, pp. 156–159.

3) He thought of intensive cooperation in space with Europe as a vehicle for strengthening U.S.–European relations.

4) He wanted to avoid criticism from Republicans of overemphasizing bilateral space cooperation with the Soviet Union to the neglect of cooperation with other countries.

5) He wanted to diversify American space relations as a tactical device, calculated to put pressure on the Soviet Union.

The Soviet attitude toward space cooperation with the United States did not undergo any significant changes under Brezhnev's leadership. The preceding factors negatively affecting Soviet–American cooperation in the past were added to by three more—increased conservatism in Soviet domestic and foreign politics, the strengthened role of the Soviet military–industrial complex, and the war in Vietnam, which further strained Soviet–American relations. The combination of all of these factors rendered impossible any substantial interaction between the two countries in the field of space exploration.

REFERENCES

[1]Congress, Senate, Committee on Aeronautical and Space Sciences, *Statements by the Presidents of the United States on International Cooperation in Space: A Chronology, October 1957–August 1971*, No. 92-40, 92d Cong., 1st sess., 1971.

[2]Logsdon, J. M., *The Decision to Go to the Moon*, MIT Press, Cambridge, MA, 1970, p. 21.

[3]U.S. Department of State, *Press Release*, 17 Nov. 1958, No. 695.

[4]Johnson, L. B., "The Politics of the Space Age," *Saturday Evening Post*, 29 Feb. 1964, p. 23.

[5]*U.S. Department of State Bulletin*, 30 Dec. 1963, p. 1011. NASA Historical Reference Collection, NASA History Office, NASA Headquarters, Washington, DC.

[6]*Public Papers of the Presidents of the United States: Lyndon B. Johnson*, U.S. Government Printing Office, Washington, DC.

[7]*Weekly Compilation of Presidential Documents*, 3 Oct. 1966, p. 1361.

[8]The White House Conference on International Cooperation, Nov. 28–Dec. 1, 1965, National Citizens' Commission, *Report of the Committee on Space*, Washington, DC, 1965.

[9]Thompson, R. E., "Johnson Vows to Keep U.S. First in Space," *Los Angeles Times*, 16 Sept. 1964. NASA Historical Reference Collection, NASA History Office, NASA Headquarters, Washington, DC.

[10]Gaddis, J. L., *Russia, the Soviet Union and the United States: An Interpretive History*, McGraw–Hill, New York, 1990, p. 265.

[11]Kornienko, G., *Kholodnaya Voina: Svidetelstvo Eyo Uchastnika* [Cold War: A Testimony of Its Participant], International Relations, Moscow, 1994, p. 123.

[12]Loory, S. H., "U.S. Talking Down the Space Race," *New York Herald Tribune*, 9 June 1964.

[13]"Views of the Presidential Candidates on the Future of the U.S. in Space," *Missiles and Rockets*, 26 Oct. 1964, p. 17.

[14]Khrushchev, S., *Nikita Khrushchev: Krisisy e Rakety* [Nikita Khrushchev: Crises and Missiles], Vol. II, Novosti, Moscow, 1994, p. 465.

[15]*U.S.-USSR Cooperation in Space Research Programs*, Jan. 1964. NASA Historical Reference Collection, NASA History Office, NASA Headquarters, Washington, DC.

[16]Webb, J. E., *U.S.-USSR Cooperation in Space Research Programs*, 31 Jan. 1964, *Exploring the Unknown: Selected Documents of the U.S. Civilian Space Program*, Vol. 2, edited by J. M. Logsdon, D. A. Day, and R. Launis, External Relationships, NASA History Office, NASA Headquarters, Washington, DC, 1996.

[17]Webb, J. E., Letter to President Lyndon Johnson with a summary of "Report on Possible Projects for Substantive Cooperation with the Soviet Union in the Field of Outer Space," 31 Jan. 1964. NASA Historical Reference Collection, NASA History Office, NASA Headquarters, Washington, DC.

[18]Dryden, H. L., "Letter to President Lyndon Johnson," 21 Jan. 1964. NASA Historical Reference Collection, NASA History Office, NASA Headquarters, Washington, DC.

[19]Spivak, J., "New President Likely to Give Strong Support to Beleaguered NASA," *Wall Street Journal*, 2 Dec. 1963. NASA Historical Reference Collection, NASA History Office, NASA Headquarters, Washington, DC.

[20]Harvey, D. L., and Ciccoritti, L. C., *U.S.-Soviet Cooperation in Space*, Univ. of Miami, Coral Gables, FL, 1974.

[21]Dryden, H. L., Testimony in U.S. Congress, Senate, *NASA Authorization for Fiscal Year 1966: Hearings Before the Committee on Aeronautical and Space Sciences*, part 1, *Scientific and Technical Programs and Program Management*, 89th. Cong., 1st sess., March 1965, S. 927, pp. 60–61.

[22]Arbatov, G. A., *The System: An Insider's Life in Soviet Politics*, Time Books, New York, 1992.

[23]Tucker, R. C., *Political Culture and Leadership in Soviet Russia: From Lenin to Gorbachev*, Norton, New York, 1987, p. 128.

[24]Sagdeev, R., *The Making of a Soviet Scientist*, Wiley, New York, 1994, pp. 205–206.

[25]Kamanin, N., *Skritiy Kosmos (1960–1963)* [Hidden Cosmos], Infortext, Moscow, 1995.

[26]McDougal, W. A., *The Heavens and the Earth: A Political History of the Space Age*, Basic Books, New York, 1985, p. 248.

[27]Alexandrov-Agentov, A. M., *Ot Kollontai do Gorbachyova* [From Kollontai to Gorbachev], International Relations, Moscow, 1994, p. 117.

[28]Medvedev, R., *Lichnost e Epokha: Politicheski Portret L. I. Brezhneva* [The Personality and the Era: Brezhnev's Political Portrait], Hovosti, Moscow, 1991, pp. 86–87.

[29]Schauer, W., *The Politics of Space*, Holmes and Meier, New York, 1976, p. 170.

[30]Furguson, E. B., "Reds to Shun Gemini Shot," *Baltimore Sun*, 9 Sept. 1965. NASA Historical Reference Collection, NASA History Office, NASA Headquarters, Washington, DC.

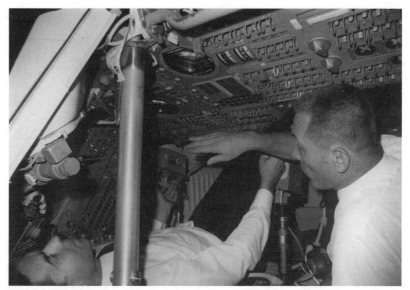

After performing a successful docking in the flight simulator, on 22 October 1969, Soviet cosmonaut Georgi Beregovoy listens to explanations from Jack Swigert (later a hero of the Apollo 13 mission). (Photo courtesy of Marina Grigorovich-Barskaya.)

On 1 November 1969, Astronaut Bill Anders and cosmonauts Georgi Beregovoy and Konstantin Feoktistov place flowers on the graves of U.S. astronauts Virgil (Gus) Grissom and Roger Chaffe. (Photo courtesy of Marina Grigorovich-Barskaya.)

Natalie Wood, Goldie Hawn, and Kirk Douglas welcome the Russian cosmonauts to the Douglas home on 24 October 1969. This event recognized the heritage of Russian–American celebrities. Georgi Beregovoy is standing with his back to the camera. (Photo courtesy of Marina Grigorovich-Barskaya.)

From left: U.S. astronauts Buzz Aldrin and Neil Armstrong welcome Soviet cosmonauts Vitaly Sevastyanov and Nikolai Andrianov to Washington, D.C., 18 October 1970. (Photo courtesy of Marina Grigorovich-Barskaya.)

U.S. astronaut Russell (Rusty) Schweickart (left) explains to cosmonauts Vitaly Sevastyanov and Adrian Nikolayev the deployment of the space telescope on Skylab after the canopy is removed. 21 October 1970. (Photo courtesy of Marina Grigorovich-Barskaya.)

Adrian Nikolayev (right) conversing with Wernher von Braun through an interpreter (left) at the AIAA banquet in 1970. Von Braun was one of the most active advocates of the ASTP. (Photo courtesy of Marina Grigorovich-Barskaya.)

Adrian Nikolayev, a fighter pilot who had never before flown a multi-engine aircraft, is ready to take off in a B-747. Vitaly Sevastyanov is videotaping the "historic" moment in Seattle. October 1970. (Photo courtesy of Marina Grigorovich-Barskaya.)

Vitaly Sevastyanov is about to begin his neutral buoyancy training in a U.S. EVA spacesuit during the 1970 visit. (Photo courtesy of Marina Grigorovich-Barskaya.)

U.S. astronaut Thomas Stafford gestures at the Apollo 9 command module. Next to him from right to left: Andrian Nikolayev, Vitaly Sevastyanov, and Donald (Deke) Slayton. Stafford later becam an Apollo commander and Slayton was the docking module pilot during the ASTP flight in 1975. (Photo courtesy of Marina Grigorovich-Barskaya.)

President Richard Nixon and Premier Alexei Kosygin sign the five-year agreement on cooperation in the fields of science and technology in the Kremlin, 24 May 1972. The agreement included the rendezvous and docking in Earth orbit of an American and a Soviet spacecraft and a coordinated effort to explore and share information on space. (Photo courtesy of NASA.)

The prime crew members of the ASTP mission visit President Gerald Ford in the White House on 7 September 1974. From left: Vladimir Shatalov, chief of cosmonaut training for the Soviet Union; Valeriy Kubasov, engineer on the Soviet crew; Alexei Leonov, commander of the Soviet crew; Anatoly Dobrynin, Soviet ambassador to the United States; President Ford; George Low, NASA deputy administrator; Thomas Stafford, commander of the American crew; Donald (Deke) Slayton, docking module pilot on the American crew; and Vance Brand, command module pilot on the American crew. (Photo courtesy of NASA.)

The Soyuz spacecraft was launched on 15 July 1975 from Baykonur Cosmodrome in Kazakhstan, 7 1/2 hours before the Apollo–Saturn liftoff from NASA Kennedy Space Center in Florida. The two spacecraft were joined in Earth orbit 17–19 July 1975. (Photo courtesy of NASA.)

The Apollo–Saturn 1B space vehicle on the pad in early July 1975 during ASTP prelaunch preparations. (Photo courtesy of NASA.)

On 17 July 1975, at the Mission Control Center in Kaliningrad outside Moscow, Charles Lewis, head of the U.S. consultants team (second from right), and the Soviet ASTP flight director, cosmonaut Alexei Eliseev, congratulate each other on the successful completion of the Apollo–Soyuz flight. (Photo courtesy of Russian Academy of Sciences.)

Chapter 4

APOLLO–SOYUZ TEST PROJECT AND AFTERWARD

CHANGED ENVIRONMENT: DÉTENTE

The period in the history of Soviet/Russian–American cooperation in space that lasted from 1969 through 1975 and eventually led to the Apollo–Soyuz Test Project (ASTP) is the most puzzling and intriguing. Neither Brezhnev nor Richard Nixon, who were the heads of state when the idea of ASTP was born, had ever been the space advocates that their predecessors—Khrushchev, Kennedy, and Johnson—were. Moreover, at least at the beginning of their respective terms neither Brezhnev nor Nixon wanted to make space a Soviet–American area of common interest or to promote actively bilateral space cooperation. Although Nixon considered building a network of mutually advantageous relationships, such as cooperation in the field of science and technology, space was considered only part of this kind of cooperation. However, it was during the Brezhnev–Nixon and later the Brezhnev–Gerald Ford periods that Soviet–American cooperation in human space flight suddenly moved from the field of expectations into the field of reality and reached its peak in the form of ASTP. Why did this happen? The answer to this question would constitute a valuable contribution to the understanding of mechanisms driving Soviet/Russian–American cooperation in space.

The set of factors that influenced Soviet–American cooperation in space in the late 1960s and early 1970s can be divided into two categories. The first includes a variety of U.S. and USSR domestic and international factors affecting the general bilateral relationship between the Soviet Union and the United States. The second category consists of factors related to changes in the space industries and space policies of the two countries.

The analysis of the previous period of Soviet–American space relations has shown that one of the major obstacles in the way of cooperation was the very cold political climate in the relationship between the Soviet Union and the United States. Thus the first precondition necessary to end the stalemate in bilateral space cooperation should have been an improvement in the general context of Soviet–American relations.

Many experts consider 1969 to be a threshold in the evolution of the Cold War; beyond the threshold lay détente. This phenomenon was due above all to internal political developments in the Soviet Union and the United States. By the end of the 1960s the Soviet Union reached parity with the United States in intercontinental military forces, certainly in numbers, if not yet in capability. By proposing strategic arms limitation talks to the Soviet Union, American leadership had admitted de facto military equality between the Soviet Union and the United States. It meant that at least one of the goals of the Soviet leadership—one that played an important role in the justification of the Soviet system, i.e., creation of a superpower equal to that of the United States at least in terms of military might—had been achieved. It allowed the Kremlin to concentrate more on the domestic economic problems of the Soviet Union, the solutions of which had been hampered by the arms race. The economic reforms launched in the Soviet Union in the mid-1960s were gradually degenerating by the end of the decade as a result of the strengthening of the administrative-command system.[1] Brezhnev's regime needed foreign investment to maintain economic growth, or at least to prevent economic decline in the country. Détente contributed to the development of trade relations between the Soviet Union and Western countries—the United States in particular.

Détente was used by Brezhnev as a tool in the Kremlin internal political struggle. In the early 1970s he removed his rival, former First Secretary of the Communist Party of Ukraine Pyotr Shelest, from the Politburo. One of the formal reasons for this was Shelest's opposition to détente (Ref. 1, p. 518).

The international political environment also contributed to détente. In Europe in the late 1960s there was a general movement toward détente. This movement was actively supported by the Soviet leadership, who aspired to repair a peace-loving image of the Soviet Union damaged by the Soviet invasion of Czechoslovakia. Also, the Kremlin was monitoring with increasing concern the unfavorable development for the Soviet Union of the international political situation near its eastern borders—aggravating Sino–Soviet conflict and the growing Sino–American rapprochement. Accommodation with the United States could have prevented, if not further aggravation of Sino–Soviet relations, then at least a formation of an anti-Soviet alliance consisting of military, economic, and human resources of the two major countries. On the same day, 20 October 1969, when the Soviet Union started negotiations with China concerning border disputes, Soviet Ambassador to the United States Anatoly Dobrynin met with President Nixon and informed him of the Soviet willingness and readiness to begin preliminary strategic arms limitation talks with the U.S. administration.

Facing growing economic problems, the Nixon administration was also considering the reduction of the U.S. military budget. The end of the arms race with the Soviet Union and disengagement from direct military involvement in the war in Vietnam could have helped to achieve this goal. Nixon realized that the United States and the Soviet Union were entering a new era of strategic parity. In his view the United States could not restore the military superiority over the Soviet Union it had enjoyed over the last two decades. However, the United States still retained considerable strength, enough to convince the Soviet Union to begin negotiations with the United States. Nixon knew the Soviet Union better than any of his predecessors (in particular because of his visits to that country in 1959 as vice president, and in 1965–1967 as a private citizen). Moreover, he realized that his credentials as a conservative, a Republican, and an anticommunist politician would neutralize opposition from the right, thus giving him more freedom than his Democratic predecessors had in dealing with the Soviet Union (Ref. 2, pp. 9–29).

Nixon, like Brezhnev, employed détente for the advance of his own political goals. In his inaugural address made in January 1969, he characterized the moment when he was assuming his duties. He said, "for the first time, because the people of the world want peace and the leaders of the world are afraid of war, the times are on the side of peace."[3] Moreover, he expressed his confidence that

> [t]he greatest honor history can bestow is the title of peacemaker. This honor now beckons America—the chance to help lead the world at last out of the valley of turmoil, and onto that high ground of peace that man has dreamed of since the dawn of civilization…. After a period of confrontation, we are entering an era of negotiation.[4]

This kind of address was quite an unexpected move on the part of Nixon, who had been far from friendly toward the communist world. However, it was a calculated step because by abandoning the protracted ideological rigidity a politician could enhance his reputation for statesmanship. Nixon did not miscalculate because détente certainly contributed to his reelection in November 1972 (Ref. 2, p. 14). Moreover, if the process of reconsidering traditional ideological values alienated old supporters, "then the influx of newfound and pleasantly surprised allies can more than compensate for them."[4] The newfound ally became Henry Kissinger, who, like Nixon, "grasped the significance of the moment" and was determined to use it to bring about an end to the Cold War. The incorporation of Henry Kissinger into Nixon's team was not only a contribution to détente—Kissinger also became one of the major supporters of ASTP.

SPACE RAPPROCHEMENT

The developments in space industries, space achievements and failures, and the changing space policies of the two countries also contributed to Soviet–American cooperation in space. After Apollo 11 landed successfully on the lunar surface, the most prestigious component of the space race—the race to the moon—was over. The Soviets lost it. Cooperation with the Americans would help them make up for their loss of face by portraying themselves as equal space partners with the United States (Ref. 5, p. 142). The Soviet cosmonaut Konstantin Feoktistov* declared an "official" end to the space race during his October 1969 visit to the United States together with another cosmonaut, General Georgi Beregovoy.[†] Feoktistov said he did not believe the Soviet Union and the United States were in a space race any longer. "This was the first phase of space flight. I think the programs of space research are now in their second phase, and I think we can say that in that phase Soviet and American scientists are intensely helping each other."[6] Taking into consideration that heroes of space were selected on a very strict basis of utmost political loyalty and coordinated all of their actions and statements with the official party line (Ref. 7, p. 227), it would be right to consider Feoktistov's words as a reflection of the changed attitude of the Soviet leadership toward its space relationship with the United States.

REASONS FOR SOVIET ACCEPTANCE OF COOPERATIVE OVERTURES

Although one of the reasons for the Soviet space officials to respond positively to U.S. space cooperative overtures was to make up for the loss of face, they did not want to create the impression of seeking cooperation from weakness. A space endurance record set by Andrian Nikolayev and Vitaly Sevastyanov,[‡] who orbited the Earth in their Soyuz-9 spacecraft for 18 days in June 1970, and Soviet success in bringing back their own lunar soil sample via the unmanned Luna 16 in September 1970 enabled the Soviets to repair partially a damaged image of equality with the Americans after the lost moon race. This "undoubtedly...made it easier for the Russians to consider cooperation with the U.S."[8] In the same month when Sevastyanov and Nikolayev set their space record, Academician Keldysh promised to give "maximum attention" to the U.S. proposals about space cooperation with the Soviets.[9] According to Western diplomats, Keldysh's remarks appeared to

---

*Konstantin Feoktistov was a flight engineer of the world's first multiman Voskhod-1 mission. His other crewmembers were Vladimir Komarov (mission commander) and Boris Yegorov (a physician). The flight took place in October 1964. It was the only flight made by Feoktistov, who was also one of the chief designers of the Vostok and Soyuz spacecraft.

[†]Georgi Beregovoy was a Soyuz-3 cosmonaut. He was launched into space in October 1968. The goal of his only mission was to test the redesigned Soyuz vehicle after the major Soviet space disaster—the death of Soyuz-1 cosmonaut Vladimir Komarov upon crash landing in April 1967.

[‡]Andrian Nikolayev's first space flight in Vostok-3 lasted from 11–15 August 1962. He made his second and last space flight from 11–19 June 1970 in Soyuz-9 as a mission commander. His flight engineer was Vitaly Sevastyanov, who made his first space flight as a flight engineer together with Andrian Nikolaev in Soyuz-9 in June 1970. He made his second and last space flight in the Soyuz-18/Salyut-4 orbital space complex together with Petr Klimuk, who was a mission commander. Their flight lasted from 24 May through 26 July 1975.

carry more authority than any previous Soviet comments on possible space cooperation with the United States.[10] Although the Soviets had formerly turned down offers to participate in analyzing lunar surface materials brought back by Apollo 11, in January 1971, just four months after the Luna 16 success, they agreed to exchange moon samples with the Americans.[11]

Another important factor that could have contributed to the Soviet–American space rapprochement was Korolev's death in January 1966. The chief designer had been an ardent advocate of continuing the space race with the United States. Korolev's successor, Vasili Mishin, did not have either Korolev's obsession with space firsts or his charisma and ability to convince the Soviet top policymakers to support his plans and projects. One of the projects that Korolev did not have time to complete before his death was the docking of two spaceships in outer space.[12] Later this idea was converted into the project of creating permanently inhabited space bases—space stations. However, realization of the idea was impossible without a solution for docking and crew transfer problems. For a while the Soviets tried to cope with the challenges without any apparent breakthrough. Cosmonauts were supposed to transfer from one cabin to another by crawling through airlocks and walking through space to the other spaceship holding onto special rails. This kind of experiment was performed first in January 1969 by the crews of Soyuz-4 and Soyuz-5 spaceships. Soyuz-4 was flown solo by Vladimir Shatalov and Soyuz-5 by Alexei Yeliseev, Evgeny Khrunov, and Boris Volynov. After the two spaceships docked, Yeliseev and Khrunov donned pressure suits and made a spacewalk from Soyuz-5 to Soyuz-4, returning to Earth with Shatalov. However, this technique was already obsolete.

Although the docked ships were considered by Soviet mass media as the world's first space station, the act of docking and the way it was realized could hardly obscure one fact—American spacecraft had performed orbital rendezvous and docking operations three years earlier and a crew transfer had been made from one module to another through an internal tunnel linking the two spacecraft. This seemingly less sophisticated procedure involving crew transfer (it did not require astronauts to put on spacesuits) in fact required more advanced spacecraft with more sophisticated docking mechanisms. For everybody in the Soviet space industry, the American superiority in this kind of space operation was obvious.

In October 1969 three Soviet spacecraft were placed in orbit. Even though various experiments were performed during their flight, none of the ships linked, which was regarded by some Western observers as a failure in their flight program. Thus, cooperation with the United States in the field of docking and rendezvous operations could have enriched Soviet space technology with the U.S. knowledge in this area of space hardware design (see Ref. 5, pp. 98–99 and Refs. 13 and 14). It was indirectly confirmed in Moscow in

October 1970 by Arnold Frutkin, who said after the end of Soviet–American talks on standardized spacecraft that Soviet "plans for a future transfer system are very similar to the present Apollo system."[15]

Some changes also took place in the leadership of the Soviet Academy of Sciences that played a crucial role in Soviet–American space cooperation. Apparently the views of "the military–industrial czar" Mstislav Keldysh, president of the Academy, had undergone certain changes. There were some indirect signs showing his growing dissatisfaction with the military-linked bureaucracy at the Academy—the very bureaucracy he himself had strengthened—and with the growing impact of the military–industrial complex on almost all sides of life in the Soviet society. Cooperation with the Americans was something that could have helped him move the Academy out of its stagnation and probably strengthen the nonmilitary part of the space program (Ref. 7, p. 215).

ACHIEVEMENT AND CHANGES IN U.S. SPACE POLICY

The American achievements in space and the subsequent changes in American space policy also contributed to a renaissance in Soviet–American space relations. In January 1969 President Richard Nixon took up his duties. He was no stranger to either space exploration or the U.S. space program. In the Eisenhower administration, Vice President Nixon was one of the first to point to the challenge of Sputnik. It would be a mistake, he said, "to brush off this event as a scientific stunt" rather than accepting it as "a grim and timely reminder" that the Soviet Union "has developed a scientific and industrial capacity of great magnitude." As President of the Senate, the vice president signed the National Aeronautics and Space Act before it was signed by President Eisenhower on 29 July 1958. In the vice president's campaign for the presidency in 1960, Nixon foresaw a lunar landing in 1970–1971 (Ref. 16, p. 93).

NIXON'S MISGIVINGS. Although not an opponent of the vigorous U.S. space program and space competition with the Soviets, Nixon, "while not above wining and dining astronauts as American heroes to further his political purposes, never exhibited the personal enthusiasm or expansive commitment for the space program that Kennedy and Johnson had shown." This can be explained partly by the fact that he did not need to use U.S. space achievements to prove himself in dealing with the Soviets as Kennedy and Johnson apparently thought they did (Ref. 17, pp. 172–173). In the beginning of the space age, he expressed his doubts about the existence of a rationale for the space race by saying "militarily, the Soviet Union is not one bit stronger today than it was before the satellite was launched..." (Ref. 16, p. 93).

Nixon's moderate enthusiasm for the space program became skepticism after he assumed the presidency. He had inherited too many problems from the previous administration, among them, the war in Indochina, the "Great Society" social reform effort, and the civil space program. By the end of the 1960s all three became a noticeable burden on the U.S. budget. None of them justified its cost in the eyes of American taxpayers. As the smallest of these inheritances and considerably devaluated as a vehicle for the advance of U.S. national interests after the victory in the moon race, the space program was the easiest to target for cuts by the new economy-minded administration because it had the least broad public constituency (Ref. 17, p. 170). Whereas before "Americans had felt excitement about the prospects of space, now there was only apathy and boredom. Where once there had been images of technological marvels in space, now there were images of technological failures on earth."[18] For this reason space was a very minor issue in Nixon's campaign in which the emotional issues of "law and order" and Vietnam were paramount.[19] Nixon also realized that the end of the space race could have a negative impact on U.S. space efforts. In his view one of the main troubles of the program was that the Soviets had not been flying dramatic missions for a long time. He believed that the unfortunate truth was that new Soviet spectacular achievements were what the public needed to get interested in increased U.S. space activities.[20] As it was expressed in *The Wall Street Journal,*

> some of the arguments advanced for the space program deserve careful examination. One is simply the old contention that if the U.S. does not keep moving fast, it will lose "space supremacy" to Russia. While that may indeed be a valid national goal, there are indications that Russia itself realized several years ago that the damn-the-torpedoes approach was neither scientifically nor financially very sound.[21]

PRIORITIZING DEFENSE APPLICATIONS OF SPACE TECHNOLOGY. Nixon always prioritized the defense applications of space technology, such as ICBMs, reconnaissance satellites, and antiballistic missiles (ABMs) (Ref. 17, p. 200). Saving on the U.S. civil space program would have allowed him to avoid drastic cuts in the U.S. military space budget. By the time that Nixon assumed his presidency in January 1969, the NASA budget had been reduced to approximately $4 billion from its $5.2 billion peak.[22] Toward the end of Nixon's presidency, the NASA budget fell to $3.04 billion.[23] At the same time the military space budget was not curtailed as dramatically as the civil one and even experienced some increase by the time Nixon stepped down from office.[23]

NIXON'S VIEWS ON SPACE PARTNERSHIP

While campaigning for the presidency in 1968, Nixon was consistent in his advocacy of a U.S. space program second to none. He said,

America must be first in space.... Space is something more than an area to be explored. It is a new dimension, an adventure to be shared, a promise of benefits to mankind. Lose the race for space and we lose something more than a race; we lose our impetus to greatness which is the driving force of great nations... (Ref. 16, p. 93).

The end of the moon race, Nixon's plans to move from confrontation to negotiation, and his wish to save on the space program increased his interest in international cooperation in space. On 22 July 1969, when Apollo 11 was already on its way back to Earth, he expressed his hope that "when the next great venture into space takes place that it will be one in which Americans will be joined by representatives of other countries so that we can go to the new world together."[24] However, in the beginning of his first term, Nixon had no concrete ideas about space partnership in general or about doing business in space along with the Soviets in particular. His presidential statements about international cooperation in space were vague and reflected his peacemaker's ambitions more than any well-considered plans about how to join efforts with other countries in space. Space, in his view, was supposed to have a healing moral effect on international relations. In his Inaugural Address on 20 January 1969, Nixon talked about space exploration in terms of emphasizing that "man's destiny on earth is not divisible...that however far we reach into cosmos, our destiny lies not in the stars but on Earth itself, in our own hands, in our own hearts..." (Ref. 25, p. 4). In his statement following the launching of Apollo 9 on 3 March 1969, Nixon expressed his hope that space exploration would "bring humanity together by dramatically showing what men can do when they bring to any task the best of man's mind and heart" (Ref. 25, p. 177). The Apollo 11 crew message "from the Sea of Tranquility...inspires us to redouble our efforts to bring peace and tranquility to earth" (Ref. 25, p. 530). Nixon made the same kind of statements during the visits he made in July and August 1969 to the Philippines, Indonesia, Thailand, Vietnam, India, Pakistan, Romania, England, and Germany (see Ref. 16, pp. 101–106). Nixon stressed not only the unity of mankind as a single part of the universe; he also mentioned the existing ideological differences that might hinder such unification. Talking about the TV broadcast of the Apollo 11 landing, he expressed his regret that "there was approximately one-half the world that did not see it: the whole of Communist China and the world of the Soviet Union."[24] However, he made it clear that venturing into space involved a danger for human life that could help overcome these differences. In his statement announcing the inclusion on the Apollo 11 flight of symbols commemorating earlier Soviet and American space heroes, Nixon said:

There is no national boundary for courage. The names of Gagarin and Komarov, of Grissom, White and Chaffee, share the honor we pray will come to Armstrong, Aldrin, and Collins.

> In recognizing the dedication and sacrifice of brave men of different
> nations, we underscore an example we hope to set: that if men can reach
> the moon, men can reach agreement (Ref. 25, p. 521).

Speaking about the troubled journey of Apollo 13, Nixon stressed that
"not just Americans but people all over the world, not just people in the free
world but people in the Communist world...were on that trip with these
men" (Ref. 26, Vol. 6, pp. 548–549). The Soviets not only expressed their
solidarity with the crew in trouble but also offered concrete help.*

For the first time since the beginning of his presidency, Nixon mentioned
what kind of practical results he expected from international cooperation in
space on 7 March 1970, in his "Statement on the Future of the United States
Space Program." He said, "Unmanned scientific payloads from other nations
already make use of our space launch capability on a cost-shared basis; we
look forward to the day when these arrangements can be extended to larger
applications satellites and astronaut crews." He also mentioned the countries
that were supposed to become the U.S. primary partners in space. These
were the countries of Western Europe, Canada, Japan, and Australia. The
Soviet Union was not on the list (Ref. 26, pp. 328–331). However, unlike his
predecessors Nixon made the main emphasis not on the need for the United
States to maintain "leadership through cooperation," but rather to be "a partner instead of a patron."[20] By changing accents in his view on international
space cooperation, Nixon created important preconditions for the future
involvement of the Soviet Union, which considered itself a space power
equal to the United States, in joint cooperative projects with the United
States.

Despite the lack of formal interest to cooperation in space with the
Soviets, shown by Nixon in the beginning of his presidency, he was not foreign to the idea of pooling efforts with the Soviets in space. According to
former Ambassador to the Soviet Union Foy Kohler, it was Richard Nixon
and not John F. Kennedy who first proposed a joint space project with the
Soviet Union. Kohler said that Vice President Nixon stated in a TV–radio
broadcast to the Soviet people in 1959: "Let us cooperate in our exploration
of outer space. As a worker told me in Novosibirsk, let us go to the moon

---

*In their thank-you letter to Soviet cosmonaut Konstantin Feoktistov,[27] sent three weeks after their
splashdown, James A. Lovell, Apollo 13 Commander; John L. Swigert, Command Module Pilot; and Fred
W. Haise, Lunar Module Pilot, wrote the following:

"Thank you for your concern of Apollo 13. There were times during the flight when the outcome of
our safe return was in doubt. Although we did not have a chance to explore the moon, we did learn a lot
about returning home under emergency conditions.

On behalf of the Apollo 13 crew, please thank those people who were responsible for insuring non-
radio interference during recovery operations. The Astronauts here appreciate the Soviet Union's offer of
assistance and hope that in the future there will be more cooperation in space activities."

together."[28] In March 1968, while already campaigning for the presidency, Nixon said:

> Outer space should be seen as the focus for ever increasing United States–Soviet collaboration rather that as the site of an endless series of increasingly expensive prestige races. Because our society is open, so much is known about our space program that inviting Soviet participation in the non-military projects would be unlikely to endanger national security. By insisting upon reciprocal privileges we would acquire much additional knowledge about their space efforts, thus achieving a net gain for the United States security.[29]

This kind of statement shows that Nixon, when he was still a presidential candidate, had a rather pragmatic approach to cooperation with the Soviets in space. He did not prioritize the significance of Soviet–American space partnership as a vehicle for improvement of the bilateral relationship. If he had, he probably would have reduced the number of supporters of the U.S. space program even further because, according to the results of a national Trendex poll conducted at the end of 1968, only 18% of the respondents viewed reducing world tensions as a benefit of the space program.[30] On one hand, Nixon was obviously thinking of getting access to Soviet space technology. He showed a deep interest in it as far back as 1959 during his first visit to the Soviet Union as vice president in the Eisenhower administration.[31] On the other hand, Nixon was apparently responding to the demand of the American public for Soviet space data. Americans were no longer buying "the arguments of prestige and international competition with the Russians as once they did—no matter how valid such arguments may be in the long run" (Ref. 32, p. 74), and thus wanted to see—"is there really a space race, or is the competition an illusory one which America has already won by default."[33]

U.S. SPACE COMMUNITY RENEWED INTEREST IN COOPERATION

However, Nixon's wish to know more about the Soviet space program could not per se explain the renewed interest of the U.S. space community toward space cooperation with the Soviets. Moreover, this interest had not been instigated by presidential initiative. It was only on 25 February 1971 that Nixon officially incorporated cooperation with the Soviet Union into the U.S. space policy in his report to Congress on "U.S. Foreign Policy for the 1970s: Building for Peace":

> I have also directed NASA to make every effort to expand our space cooperation with the Soviet Union. There has been progress. Together

with Soviet scientists and engineers we have worked out a procedure for the development of compatible docking systems.

In January we reached a preliminary agreement with the Soviet Union which could serve to bring much broader cooperation between us in the space field. I have instructed NASA and the Department of State to pursue this possibility with utmost seriousness... (Ref. 26, Vol. 7, p. 374).

The previous period in the history of Soviet–American space cooperation showed that presidential statements about the need to cooperate with the Soviets usually preceded any kind of space interactions between the two countries. The report to Congress indicated that January 1971 was a time when Soviet–American cooperation in space was formally resumed under the Nixon presidency. However, by the time Nixon made his report, the Soviet and American space officials had already been involved in almost two years of discussion concerning possible joint U.S.–Soviet projects in space. This unprecedented phenomenon in the history of Soviet–American space relations was a product of a number of developments in the U.S. space policy and program.

One of them was the fact that in a totalitarian state, such as the Soviet Union, all decisions were made by the top leadership and were not subject to consideration by the public, while in a democratic country, such as the United States, any high-spending state project is subject to constant evaluation and revision by a legislature that reflects the needs and interests of its constituency. This kind of project may eventually be significantly reduced or even canceled as soon as the rationale behind it becomes weaker or eventually vanishes. Although 42% of the respondents in December 1968 believed that the United States was spending enough on the space program, with 35% thinking that the United States was spending too much on space (Ref. 30, pp. 197–198), in 1969 it was already 41 vs 40%, respectively. In 1973 those who did not want the U.S. space program to be cut in funds were outnumbered two to one (29 vs 59%) by those who wanted the U.S. space budget to be reduced.[34] As a result, starting from the early 1970s, after the race to the moon was over, NASA was losing one mission after another to White House budget cuts. Five lunar missions and a second Skylab were canceled.[35] The future of a vigorous U.S. space program was seriously jeopardized.

Even though "both the U.S. and the USSR were experiencing the pressure of domestic needs and would have welcomed some sort of cooperative space effort which could have helped reduce the funding of space programs in each country,"[36] the Soviet space program did not face the same kind of political and financial difficulties as the U.S. program. In the view of the Soviet leadership, the Soviet Union lost the race to the moon, but not the space race, and space achievements continued to play an important role in the justification of the Soviet system. According to the study of the Soviet Space Program prepared by the Senate Committee on Aeronautical and Space Sciences in early 1972, the Soviet space program was a "strong and growing enterprise" with apparently few of the budgetary restraints that hampered the American program.[37] Whereas the interest of the Soviet space officials in cooperation with Americans was caused by their wish to get access to the U.S. docking experience and technology, and also to some extent by their attempts to create an image of parity by pooling efforts of the two countries in space, the interest of the U.S. space community toward cooperation with the Soviets was generated by factors that were of vital importance to the future of the U.S. space program. It explains why the pressure for a joint project was stronger from the United States, and why the Americans played a more active role in trying to involve the Soviets in bilateral space cooperation. The overall growing interest of the Americans in cooperation with the Soviets consisted of a number of components.

First, the cancellation of a few missions to the moon left a number of Apollo surplus command modules. From NASA's standpoint, a joint space flight could have used at least one of them. Second, with Apollo ending in 1972 and the next program not due until 1978, it would fill a gap during which Americans would otherwise have no manned space missions. Most of $300 million—the estimated cost for the United States to fund the first joint docking mission—was supposed to "be spent to keep the manned flight facilities intact at Houston, Huntsville, Alabama, and Cape Kennedy."[38] The other possible worthwhile alternatives to joint flight were considerably more expensive than the Apollo–Soyuz project and would have faced difficulties fitting the shrunken NASA budget. They could have included: a second Skylab project in 1976 that would have come to $800 million and a two-week manned moon mapping flight in lunar orbit during 1976 or 1977 that would have cost $400 million.[39]

Third, taking into consideration the general improvement of the political climate between the two countries, the agency had good reason to expect that the U.S. political community would view a joint Soviet–American space project as politically desirable, thus facilitating its funding.[35] Fourth, after Apollo 13's failed mission, the chances for the joint project involving the development of a rescue system based on Soviet and American space hardware, to meet positive public, political, and financial reaction, were even higher.[40] Fifth, public interest in the U.S. space efforts fell off after the Apollo lunar orbiting and landing missions of the late 1960s and early 1970s. A forthcoming joint space flight could spark new interest for both its political and technological significance.[41] Sixth, the joint space manned mission would have provided NASA time to define more fully its requirements for

Shuttle-era subsystems (Ref. 42, p. 164). Finally, the goal of the first Soviet–American space flight was to build "mutual confidence and trust" to ensure future joint operations in space.[43]

Another important development in U.S. space policy that contributed to the increased interest of the U.S. space community toward cooperation with the Soviets was Nixon's contradictory plans on one hand to save on space exploration and on the other to go beyond the Apollo program with some other ambitious space project. As Nixon, when he was still a Republican candidate, expressed it himself in May 1968:

> I believe that space is one of the areas that will have to be in the President's recommendations for budget-cutting. I think that what we have to do here is to concentrate on those areas that offer the greatest possibility for breakthrough; but with the immense financial crisis which currently confronts the United States, we will have to make some cuts. I would support those cuts. But once we get this country back again on a sound basis and we can move forward with space and reclamation programs, then space should have a very high priority because no great power can afford to be second in exploring the unknown (Ref. 32, pp. 73–74).

The top U.S. officials, like Vice President Spiro Agnew, considered the increased international participation in a large-scale space project, based on sharing the cost as well as benefits, as a precondition for its successful realization.[44] Whereas the development of the Space Shuttle, which required new technology with potential military application, became a goal for U.S. cooperation with its friends and allies such as Western Europe and Canada,[45] the design and building of a space station—the project advocated by NASA, which could be based on already existing Soviet and American technologies and hardware—could have been realized within the framework of cooperation between the two countries.

POSSIBILITY OF A JOINT SPACE STATION

The space station project had its opponents. The report submitted to Nixon by a special task force before his inauguration opposed the space station idea on the ground that it was "not obviously an effective way of continuing to demonstrate for prestige purposes our manned space capability" despite the need for "prestige" in the international arena.[46] Nixon apparently did not disregard one piece of the advice contained in the report, which was to work harder on getting some international efforts off the ground, particularly in conjunction with the Soviet Union. Although the authors of the report stressed unmanned planetary exploration as the most promising field for joint Soviet–American activity in space, the rationale they presented for cooperation in this area—"one in which Soviet competence matches our own, and with obvious savings in both countries"—could have well underlined Soviet–American cooperation in manned orbital flights.[46]

The Americans were well aware of the Soviet efforts aimed at the development of the space station. Joe Califano wrote to President-elect Nixon in particular that "considerations of the pace of the Soviet lunar landing program, manned spacecraft systems and launch vehicle availability suggest to us that the Soviets have the capability for an early space station consistent with extended earth-orbital operations."[47] This made the Soviets natural potential partners of the Americans in the development and building of a manned space laboratory.

Plans for cooperating with the Soviet Union in this sort of project had already been under consideration in the United States for some time. The Johnson administration space spokesman Dr. Edward C. Welsh said in May 1968 that the U.S. policy was to cooperate in space with the Soviet Union when such action would be of mutual advantage and suggested that "a joint manned lunar surface laboratory" might be a good subject for such a cooperative undertaking.[48]

PRELIMINARY DISCUSSIONS ON COOPERATION

However, before Soviet–American talks concerning the pooling of efforts in space reached the point of practical discussion, they went through a number of preliminary stages. These stages were reflected in correspondence between NASA Administrator Thomas Paine and President of the USSR Academy of Sciences Mstislav Keldysh; the exchange of letters started in the spring of 1969. In one of them, Paine asked Keldysh to consider possible Soviet participation in the U.S. Viking mission (which was eventually launched in 1975) as well as coordination between American and Soviet planetary programs. It was apparently a calculated step aimed at maximum facilitation for the beginning of space cooperation between the two countries. Deep-space exploration using planetary probes was the only kind of activity in the Soviet Union dedicated to nonclassified open space science (Ref. 7, p. 212). Another letter assured Keldysh that NASA would welcome proposals from Soviet scientists concerning the analysis of lunar samples. However, Paine apparently had something more in mind than just joint space exploration using planetary probes. He forwarded to Keldysh copies of the Space Task Group Report, suggesting that it might be an appropriate time for a meeting to discuss the possibilities of a complementary or cooperative space program. Because Paine wanted cosmonauts/astronauts to participate in these discussions, it is obvious that he was considering these talks as a prelude for the consideration of joint manned space projects.[49]

Keldysh's first substantive response to the series of Paine's overtures came on 12 December 1969. He did not rebuff American proposals in his usual manner. He agreed that Soviet–American cooperation in space "bears a limited character at the present time and that there is a need for its further development." He accepted Paine's suggestion that they should meet on this question but deferred further discussion of the time and place for three or four months. He declined Paine's specific proposals for the discussion of possible Soviet experiments aboard NASA planetary probes, advocating instead an "arm's length" relationship in which NASA and the USSR Academy would have coordinated planetary goals and exchange results of unmanned planetary investigation.[50] He apparently did not make any reference to possible joint space manned mission discussions, but de facto did not disregard such possibility. Two Soviet cosmonauts, Konstantin Feoktistov and Georgi Beregovoy, came to the United States in October 1969, and two other cosmonauts, Andrian Nikolaev and Vitaly Sevastyanov visited the United States in October 1970, while the Soviet Union was visited by Neil Armstrong and Frank Borman. Although neither the American astronauts nor Soviet cosmonauts shared any military or commercially sensitive data with each other during these visits, they exchanged information about their respective space programs, which played an important role in the building of a mutual understanding between the space communities of the two countries.[51] Between the two visits of Soviet cosmonauts to the United States, the Second National Convocation on the Challenge of Building Peace was held in New York City at the end of April 1970, and Mikhail D. Millionshchikov, a vice president of the Soviet Academy of Sciences, commented that the time was favorable for the Soviet Union and the United States to negotiate an agreement on cooperation in space exploration and said that he thought such talks could be "very fruitful and useful." At the convocation, other Soviet sources indicated that discussions on space cooperation could begin in the near future and could cover a number of activities. A possible cooperation in manned space flight, although not on the top of the agenda, was present in the list in the form of the development of an international rescue system for astronauts or cosmonauts who might get into trouble in space. Other areas of possible cooperation included an exchange of scientific information and some active cooperation on other space projects. Although Millionshchikov's remarks could not have been considered official in the true sense, and might have been only his personal opinion, it was generally felt in the United States that this cooperation was unlikely since he was a leading Soviet scientist.[36]

The fact that Millionshchikov's words reflected the new approach of the Soviet leadership and Soviet Academy of Sciences toward cooperation with the United States in space was proved by two subsequent events. The first was the decision made by the Soviet government at the beginning of September 1970 to turn over to the U.S. government an experimental capsule from the Apollo spacecraft found by Soviet fishermen three weeks earlier in the Bay of Biscay, off France. The second event, and the most important development, came in the October 1970 meeting in Moscow, when Soviet and American specialists involved in the development of space hardware and equipment got together to discuss the possibility of joint space projects. This meeting, which took place 26–27 October, marked the beginning of a five-year period in the history of Soviet–American space cooperation, which culminated in July 1975 during the ASTP joint space flight.

The seriousness of intentions on both sides was confirmed by the composition of the U.S. and Soviet delegations that participated in the talks. The first U.S. delegation consisted of Robert Gilruth, director of the Manned Spacecraft Center; Arnold Frutkin, assistant NASA administrator for International Affairs; Caldwell Johnson, a specialist in mechanical and electrical questions related to the development of a compatible docking system; Glynn Lunney, a flight control specialist; and George Hardy, chief of program engineering and integration for Skylab. The Soviet delegation consisted of Academician Boris Petrov; a former cosmonaut and then one of the leading designers of Soviet spacecraft, Konstantin Feoktistov; Vladimir Syromyatnikov, a mechanical design expert for docking systems; V. V. Suslennikov, a Soyuz radio guidance equipment specialist; and Iliya Lavrov, a life support systems specialist (Ref. 42, pp. 97–123; Ref. 52). Among the participants in the talks, no Soviet or American top space officials were present; however, space hardware and systems designers were present. It showed that the leaders of the Soviet and U.S. space programs on one hand were not rushing to establish a formal bilateral relationship and were merely probing each other's intentions, yet on the other hand that they were already considering cooperation in practical terms.

CONVERGENCE OF PROFESSIONAL INTERESTS

The significance of this period in the history of Soviet–American cooperation in space, which lasted from 1970 through 1975, consists of the fact that the Apollo–Soyuz project became a product of convergence of purely professional interests by the Soviet and American space officials with very little interference from the U.S. and Soviet politicians. On 10 July 1970, President Nixon publicly confirmed his interest in pursuing discussions on space cooperation but stressed that they should be conducted at the technical agency level (Ref. 42, p. 97). This kind of attitude was a manifestation of an overall Nixon–Kissinger leadership style. "Both men were predisposed to be suspicious and distrustful of governmental bureaucracy. Although it may seem odd, both men believed the bureaucracy of even their own administration was not to be trusted" (Ref. 2, p. 78). With relation to space, this kind of policy was aimed at maximum facilitation of the pursuit of joint interests by the respective space agencies. As expressed in 1972 by McGeorge Bundy, special assistant to the President for National Security Affairs during the Kennedy and Johnson administrations until March 1966 when he became president of the Ford Foundation, "many of our past relations with the Soviet Union [in the field of science and technology] have been defined by the cultural agreements negotiated between diplomats, and it has not been very frequent that the two governments have allowed direct and businesslike negotiation between operating agencies" (Ref. 53, p. 6). Apparently the Soviet leadership adhered to the same tactics. According to Anatoly Dobrynin, a former USSR ambassador to the United States, the staff of the Soviet embassy in Washington, D.C., was not involved at all in the ongoing negotiations between the U.S. and Soviet space specialists.[54] The October 1970 talks in Moscow succeeded when most aspects of Washington–Moscow relations "were roiled by tensions ranging from the Middle East to the continued imprisonment in Russia of two American generals whose plane had strayed across the Turkish border."[55]

DIPLOMATIC TENSION

The only occasion during the initial phase of Soviet–American negotiations, which resulted in the Apollo–Soyuz decision, when U.S. politicians almost exercised some influence on the space relationship between the two countries, happened in January 1971. Before departing for Moscow for another round of talks, NASA Acting Administrator George Low was briefed by Under Secretary of State Alexis Johnson on the heightening diplomatic tension between the Soviet Union and the United States because of the death sentences meted out by the Soviet court to Jewish airplane highjackers. The trial was viewed in the United States as having anti-Semitic overtones even though two of the Soviet Jews charged with the crime appealed their death sentences. The Jewish Defense League undertook a campaign of bombing Soviet installations and intimidating Soviet personnel in Washington and New York. Soviet Ambassador Dobrynin delivered a note to the State Department accusing the U.S. government of "connivance" in these hostile acts and warned that the Soviet government could not guarantee the safety of American officials and businessmen in Moscow. Johnson cautioned Low to check with the embassy in Moscow before making a favorable release to the press if the diplomatic situation were to worsen (Ref. 42, pp. 125–126). However, the outcome of the meeting between Soviet and U.S. space officials met expectations completely, and there is no evidence of any kind of impact on the complicated diplomatic situation in the course or result of these negotiations. It appeared that "the desire to cooperate in space exploration outweighed any extraneous political events" (Ref. 42, p. 127).

SCIENCE VS POLITICS

It would not be accurate to characterize the attitude of Soviet and American political leadership toward developing space cooperation between the two countries as indifferent or lacking in interest. The ongoing negotiations corresponded with the overall climate of détente between the Soviet Union and the United States. The U.S. politicians viewed the significance of the formal space agreement made in 1972 in the promise it held "for the reduction of tensions between the two signatories. It is difficult to shake hands and fists at the same time" (Ref. 53, p. 1). However, in the beginning of the space talks that coincided with the onset of détente, and apparently remembering the long history of an uncooperative space relationship between the Soviet Union and the United States, both the Kremlin and the White House tried to avoid formal commitments of cooperation in the field of space technology. The reason was probably because the leaders of the two countries were not sure whether another attempt to cooperate would succeed. Thus, if it failed in the beginning, it would have been the fault of the respective space agencies, but not of the political leaders of the two countries.

In January 1971, at the same time that Low was briefed by Under Secretary of State Johnson, Frutkin and Low had a meeting with the President's Foreign Policy Adviser Henry Kissinger. In response to Low's request for the administration position on a proposed test mission using Apollo and Soyuz spacecraft, Kissinger replied that as far as the White House was concerned Low had total freedom in negotiating in any area that was within NASA's responsibility. Kissinger had only one request for Low; he would not like NASA officials to contribute to the misleading notion that if they could reach technical agreements they could also solve political problems if given the opportunity. In the past, some of the astronauts tried to suggest that, because it was easy to negotiate with the Soviets on space topics, it should be equally easy in other areas. Such political naivete and poorly based optimism on the part of highly publicized individuals only hampered the work of the U.S. and Soviet diplomats. Kissinger summarized the attitude of the U.S. leadership toward cooperation in space with the Soviets in the following words: "As long as you stick to space, do anything you want to do. You are free to commit—in fact, I want you to tell your counterparts in Moscow that the President has sent you on this mission" (Ref. 42, pp. 126–127). Following Kissinger's advice Low conveyed President Nixon's desire to expand international cooperation in space with the Soviet Union at the January 1971 meeting in Moscow.[56]

U.S. PARTY DIFFERENCES

Whereas the Republican executive power preferred to wait and see how the space relationship between the two countries would develop before determining its formal position, the Democratic legislature, in Kennedy's tradition, wanted the United States to play a more active role in engaging the Soviet Union in bilateral space cooperation. In October 1970, a group of 8 senators and 39 congressmen had called on President Nixon to seek talks with the Soviet Union concerning Russian participation in the Skylab experiments and other future U.S. space projects. All 47 were members of the ad hoc Members of Congress for Peace Through Law organization that concentrated on attempting to cut the defense budget. The significance of this action goes beyond the fact that the lawmakers saw broader benefits from Soviet–American cooperation in the space station program than just sharing the costs of its development. They demonstrated the opposite approach from their predecessors to the initiation of Soviet–American cooperation in space. While the Eisenhower, Kennedy, and Johnson administrations considered cornering the Soviet Union with the development of broadscale cooperation between the United States and other countries, thus leaving the Soviets no choice but to either join these cooperative efforts or become outsiders, the lawmakers under the Nixon administration viewed talks with the Soviets as the steps leading to the enlistment of all nations in a cooperative space effort. The talks with the Soviet Union, suggested by them, should have considered "U.S.–Soviet cooperation in the Skylab and other orbiting methods by which all interested nations may cooperate in the exploration of outer space and in the application of the technology developed to problems confronting mankind on Earth."[57] Thus, by the beginning of the 1970s, changes in the general context of the Soviet–American relationship, and the developments in space industries and the policies of the two countries, created preconditions for what General Thomas Stafford called in July 1975 "the beginning of a new era in space exploration."

PRELUDE TO THE APOLLO–SOYUZ PROJECT

The period of Soviet–American cooperation in space that eventually led to the Apollo–Soyuz joint space flight could be divided into two subperiods:

1) First was the decision-making part, which lasted from October 1970 until the signing of the Nixon–Kosygin agreement on 24 May 1972. This agreement formalized the decision to fly the Apollo–Soyuz mission.
2) Second was the decision-implementation part, which lasted from May 1972 through July 1975, when the joint flight took place.

Even though the 1970–1975 period was mostly concerned with engineering problems and their solutions, it also involved some technical–political issues. The most important of these issues during the 1970–1972 subperiod could be expressed in the form of two questions: what kind of mission to fly and what kind of hardware to use, or if put in more concrete terms, Apollo–Soyuz, Soyuz–Skylab, Apollo–Salyut, or some other mission in the distant future involving Soviet and American hardware that was at that time at the design stage? (Fig. 1 shows a 1972 docking proposal for the possible Apollo–Salyut mission.) The answers given to these questions reflected the approaches of the Soviets and Americans toward cooperation with each other, the developments in the two space industries, and space policies in the two countries.

Although at the beginning of meaningful cooperation, which officially started at the October 1970 meeting in Moscow, neither Americans nor Soviets had any clear idea as to what kind of projects this bilateral cooperation should involve; both were certain about two things: 1) the cooperation should have a long-term character and 2) it should be focused on the most challenging and broadscale space projects—manned space flights. Two goals were of "immediate interest to both nations." On one hand, Soviet and American space officials wanted to know each other's space programs better and to put "a foot in the door toward the establishment of some kind of joint operations in space." On the other hand, both agreed on the need to develop a compatible manned spacecraft, which was the only topic on the

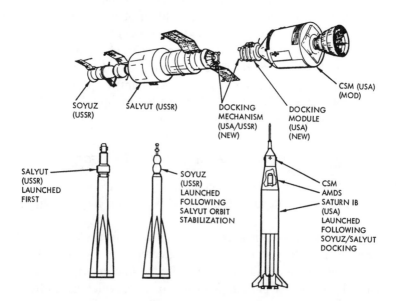

Fig. 1 Apollo–Salyut hardware elements as proposed by Rockwell International.

agenda.[58] The phrase "some kind of joint operations in space" could have implied basically two kinds of projects.

STANDARDIZED UNIVERSAL HARDWARE

The first project involved the creation of standardized universal hardware for future space laboratories that could be serviced and supplied by Soviet and American spacecraft.

A few months before official negotiations began, NASA Administrator Paine had expressed the hope that the Soviet Union might participate in some way in the U.S. Shuttle/space station program, including participation by scientists in space.[59] Some U.S. space officials saw "the time coming when cosmonauts and astronauts study geology, meteorology, ecology, and astronomy together, in preparing themselves for space station missions where Soviet and U.S. spacecraft might dock together for a period of joint operations."[58] Others believed that "if present plans are fulfilled by both East and West, there may be 100 or more men in orbit at the same time...by the end of this decade."[40]

This kind of attitude toward bilateral cooperation met with a positive reaction from the Soviet side. A week before the beginning of the October negotiations in Moscow, Soviet cosmonauts Nikolayev and Sevastyanov had visited the United States and said, "they were hopeful the United States and the Soviet Union can reach agreement on a proposal that would enable the two nations to visit each other's spaceships in orbit," but warned that it would be many years before the technical problems of such magnitude could be solved.[60] According to Frutkin, the October talks in Moscow were "open and forthright," but the Soviets and Americans were "talking about a visit of a Soviet spacecraft to a U.S. space station and vice versa. We're not talking about present systems. Both sides recognized how impractical that would be."[15]

INTERNATIONAL RESCUE SYSTEM CONCEPT

The term *impractical* was applied not only to the creation of manned space outposts based on existing U.S. and Soviet hardware but also to the development of rescue systems involving previously designed and built spacecraft. Rescue in space was the second possible Soviet–American project falling within the category of "some kind of joint operations in space," discussed by the U.S. and Soviet space officials during the October talks. The Soviet–American cooperation in this field of space activity started even before the beginning of negotiations in Moscow. On 8 October at the 21st Congress of the International Astronautical Federation, the Soviet Union agreed to discuss an international rescue system with the United States and 10 other nations.[61]

Although the use of the Apollo and Soyuz spacecraft seemed to be the most logical and least expensive solution to the problem, a few difficulties stood in the way of realizing this sort of project. First, it would have been very difficult for the Soviets to accommodate all three Americans aboard their spacecraft unless they attempted an unmanned rendezvous with Apollo (the American spacecraft could potentially accommodate up to five crewmembers). Second, the Apollo spacecraft had more maneuvering capabilities than Soyuz. The difference was rooted in the designs of the two spacecraft—Soyuz was essentially an Earth orbital spacecraft, while Apollo was designed for lunar missions that involved more sophisticated space flight maneuvers. Taking into consideration that the two spacecraft usually flew in different orbital paths, the opportunities during which Soyuz could provide assistance to Apollo were limited, while the American spacecraft, because of its greater maneuverability, could have provided this kind of assistance to the Soviet spacecraft. However, even in this scenario the Soviets were supposed to provide the American agency with their flight schedules and launch parameters well in advance so the U.S. side could divert the necessary Apollo spacecraft and launch facilities in time for the Soviet missions. Still, such preconditions for a successful rescue operation would have required a high degree of openness on the side of the Soviet space program and thus would have been totally unacceptable to the Soviets at that time. Finally, since the Apollo spacecraft spent most of its time in space out of the Earth orbit, a rescue capability limited only to an Earth orbit would have been of little assistance in an Apollo 13 kind of emergency, which happened when the spacecraft was on its way to the moon.

All of these factors taken into consideration made it clear that rescue possibilities would be limited mostly to an Apollo spacecraft saving the crew of a Soyuz craft in distress (Ref. 42, p. 98). Aside from the fact that for Americans it would not have meant a mutually beneficial cooperation in rescue operations, the Soviets would never have accepted it because it would have implied that Soyuz was inferior to the Apollo spacecraft. Moreover, manned space stations that were supposed to be put into orbit by the end of the 1970s required more sophisticated rescue equipment than what might have been developed for U.S. and Soviet spacecraft. Representatives of the NASA Advanced Manned Mission program, Herbert Schaefer and Jack Wild, stressed during their presentations at the IAF Congress that the future rescue vehicles should be versatile. They should be able to grab a space station that is tumbling out of control and bring it to a halt. They should be equipped to rescue passengers with little special training in space operations, because such passengers (astronomers and other specialists) were expected to ride inside the space station. They should also be equipped to extract passengers safely when the nuclear power plant of a space station ruptured, enveloping the station in radioactive debris.[62] Neither Soyuz nor Apollo could have met any of these requirements.

SOYUZ–SKYLAB FLIGHT CONSIDERED

A possible joint project using hardware already in the design stage was a Soyuz–Skylab flight. Even before the October talks in Moscow, some American space officials suggested that it "may be possible to have a Soviet manned spacecraft link up with an American vehicle in Earth orbit as early as 1973. That is when the American Skylab, an experimental space station, is scheduled for flight."[63] At the beginning of September 1970, Paine wrote a letter to Keldysh in which he proposed a Soyuz rendezvous with Skylab. However, this proposition, at least with relation to the Skylab A mission, was not met with enthusiasm by either the Office of Manned Space Flight (OMSF) or by the Skylab Program Office. The realization of this project would have required considerable joint development in many areas, including interconnection of the ground systems for tracking, of mission control, and of launch control and development of spacecraft-to-spacecraft communication link. All of these proposed measures could have delayed the Skylab program or adversely affected its budget (Ref. 42, pp. 99–100).

CONTINUED TALKS, LITTLE ACTION

In spite of the Soyuz–Skylab proposal, neither the OMSF nor the Skylab Program Office had to introduce changes in their planning. "Soviet and American scientists have ended their first formal talks on cooperation in space technology, with their only apparent accomplishment an agreement to continue talking."[64] The parties agreed to design compatible rendezvous and docking systems not for the existing, but for future manned spacecraft. However, this agreement was more than just a protocol of intentions because the two sides also agreed to a procedure by which they could, "through a combination of independent actions and coordination, arrive at compatible systems" (Ref. 53, p. 31). Furthermore, under a proposal by Konstantin Feoktistov, who became "one of the prime movers behind the Soviet desire to develop complementary systems as soon as practical" (Ref. 42, p. 119), three working groups were formed. Their task was to find the most effective solutions for whatever problems there might be. The first working group had to ensure the compatibility of overall methods and means for rendezvous and docking. Working group 2 was responsible for compatibility of radio guidance, optical, and other guidance systems and communications. Finally, working group 3 had to ensure compatibility of the docking assembly and tunnel. One part of the agreement clearly demonstrated the mutual interest of the U.S. and Soviet sides in a continuous working relationship with each other. Both sides agreed to hold a further meeting of working groups in March and April 1971, and in the meantime they were

> 1) To conduct, by correspondence, during the month of November 1970, a mutual exchange of technical materials on radio guidance and rendezvous systems, on the composition and characteristics of space-craft atmospheres, and on systems of voice communication.
> 2) Each side will prepare its own draft of technical requirements for systems for which it considers it advisable to assure compatibility and, in January-February 1971, will send these documents to the other side for preliminary familiarization.[52]

The next meeting between representatives of U.S. and Soviet space communities took place in January 1971. The composition of the delegations (as shown in Table 1) and the agreements reached at the end of the meeting demonstrated the willingness of the top Soviet and American space officials to establish a formal relationship with each other.[52]

The talks resulted in agreements on the exchange of lunar surface samples and on procedures to produce recommendations for joint consideration on the objectives and results of space research, the improvement of existing weather data exchanges, research with meteorological rockets, techniques for studying the natural environment, and the expended exchange of space biology and medicine. To implement this agreement, five working groups (WGs) were established: WG on Meteorological Satellites; WG on Meteorological Rocket Soundings; WG on the Natural Environment; WG on the Exploration of Near-Earth Space, the Moon and the Planets; and WG on Space Biology and Medicine.[65]

Overall, the outcome of the October 1970 and January 1971 meetings in Moscow had a twofold impact on the Soviet–American space relationship. It established a framework for future productive cooperation, and it encouraged both sides to start considering possible joint projects in more concrete terms. This is when political–technical considerations within the Soviet and U.S. space communities as well as the attitude of Soviet and American leaderships toward bilateral space cooperation, started playing an increasingly noticeable role.

ADVANCING THE IDEA OF JOINT MANNED SPACE FLIGHT

The initiative for a joint manned space flight in the near future using existing hardware belonged to the U.S. specialists. Soon after the October 1970 meeting, the U.S. designers advanced three kinds of projects. All of them implied variations for dockings between Soyuz and Apollo spacecraft and

Table 1 Composition of delegations at the January 1971 meeting

Soviet delegation	U.S. delegation
M. V. Keldysh—president, Academy of Sciences of the USSR	George M. Low—acting administrator, NASA
A. P. Vinogradov—vice president, Academy of Sciences of the USSR	William Anders—executive administrator, National Aeronautics and Space Council
B. N. Petrov—chairman, "Intercosmos" Council, Academy of Sciences of the USSR	John E. Naugle—associate secretary for Space Science and Applications, NASA
G. I. Petrov—director, Institute for Space Research, Academy of Sciences of the USSR	Arnold W. Frutkin—assistant administrator for International Affairs, NASA
I. P. Rumyantsev—member, "Intercosmos" Council, Academy of Sciences of the USSR	Robert F. Packard—director, Office of Space–Atmospheric and Marine Science Affairs, Department of State
I. V. Meshcheryakov—member, "Intercosmos" Council, Academy of Sciences of the USSR	Arthur W. Johnson—deputy director, National Environmental Satellite Service, National Oceanic and Atmospheric Administration
A. I. Tsarev— member, "Intercosmos" Council, Academy of Sciences of the USSR	William Krimer—interpreter, Department of State
M. Ya. Marov—scientific staff member, Institute of Applied Mathematics	
Ye. K. Fedorov—chief, Main Administration of the Hydrometeorological Service, Council of Ministers	
L. A. Aleksandrov—deputy chief, Directorate of the Main Administration, Hydrometeorological Service, Council of Ministers (for Technology)	
N. N. Gurovsky—chief, Directorate of the Ministry of Health	
O. G. Gazenko—director, Institute of Medical–Biological Problems, Ministry of Health	
Yu. A. Mozzhorin—professor, Moscow Physics–Technical Institute	
V. P. Minashin—chief, Main Administration for Space Communications, Ministry of Communications	
I. Ya. Petrov—deputy chief, Main Administration for Space Communications, Ministry of Communications	
Fedoseev—deputy chief, USA Section, Ministry of Foreign Affairs	

the Skylab space station. The desire to cooperate with the Soviets was so strong among the U.S. space engineers and designers that they even considered a crew rotation, with Soyuz crew occupying Skylab after the NASA crew had departed. Other projects included Apollo docking with Soyuz and acting as a propulsion stage to place the Soviet craft in another orbital position, and Soyuz docking with Apollo to prove its ability as the active rendezvous partner (Ref. 42, p. 122).

The idea of developing compatible rendezvous and docking systems for use with Apollo and Soyuz rather than with future spacecraft was proposed by Low and Frutkin to Keldysh and Feoktistov while the January 1971 negotiations were still in progress. Both Keldysh and Feoktistov were intrigued. Although they said that they had no authority to commit their government to such a project, they wanted to pursue this subject further and learn more details about it (Ref. 42, p. 129).

INTERNAL INTERESTS

The way the two sides started working on the problem provided further evidence that the current period of Soviet–American cooperation in space was driven by professional interests within the U.S. and Soviet space communities and that the space officials of the two countries wanted to remain apart from nonspace political concerns. Keldysh suggested to Low that they not make any public statements about rendezvous- and docking-related conversations until the subject developed into a formal topic for negotiation; then both sides would make a public announcement. Apparently Low agreed to this arrangement because all of the relevant information contained in the NASA press release published after the January 1971 meeting mentioned only that Low and Keldysh "took note of the significance of past agreements between them and in particular the understanding of Oct. 28, 1970, with regard to the question of providing for compatibility of rendezvous and docking systems of manned spacecraft and space stations of both countries."[66] Low publicly released information about his conversation with Keldysh and Feoktistov for the first time in his statement during the Hearing before the Subcommittee on International Cooperation in Science and Space in June 1972 (Ref. 53, p. 100). On 20 June 1971, just a few days before the beginning of the second meeting of working groups, the U.S. delegate said that both sides were "not looking for ways to force drastic changes on each other," but were "looking for ways to make things as easy as possible without forcing too many changes." He particularly stressed that they were not "talking about spacecraft that will be flying in the next few years. We'll be talking with the future in mind, with the reusable space shuttle that will be in service 10 years from now."[67]

PATTERN OF PRIVACY

This kind of attitude toward public release of critical information became a pattern. The U.S. and Soviet space officials agreed that they would wait two months before taking any formal public action. During that period of time, they were supposed to report to their governments and set up a schedule for the following session.[68] Eventually the information about the Soviets and Americans working on possible joint manned space missions was leaked to the press. However, even then, the space representatives of the two countries continued stressing the preliminary character of their discussions.

The most logical reason for such tactics as maintained by the U.S. and Soviet space officials until the signing of the Nixon–Kosygin space agreement in May 1972 seemed to be the desire of the space specialists of the two countries to develop the fledgling cooperation into a sound one, thus facilitating the process of its political formalization. Both NASA and the Academy of Sciences agreed "that a common understanding of basic principles for organizing, developing, scheduling, and conducting such a test mission is required as a necessary prerequisite to the possible approval by their governments of such a test mission" (Ref. 53, p. 30). This approach proved to be meaningful because one of the official reasons for the signing of the Nixon–Kosygin space agreement became "the positive cooperation which the parties have already experienced in this area." Moreover, the U.S. and Soviet leaderships committed themselves to the fulfillment of the agreements reached at an interagency level in January 1971 and April 1972 (Ref. 53, p. 27).

SALYUT-1

One of the events that probably had a major influence on the course of the negotiations concerning what kind of joint mission to fly and what type of hardware to use was the Soviets' launch of the world's first space station, Salyut-1. This station was put into orbit on 19 April 1971. Preparation for the flight was so intense that talks scheduled for the spring of 1971 between U.S. and Soviet working groups on problems such as common docking mechanisms for Soviet and American spacecraft had been indefinitely postponed after a request from the Soviet side.[69] However, another likely reason for the postponement—other than that the Soviet specialists had been busy preparing for the future Salyut mission—was the need to make a decision about the kind of Soviet spacecraft to use for the future joint project. It is not a coincidence that the postponed meeting took place in Houston on 20 June 1971, just two weeks after the Soyuz-11 crew, Georgi Dodrovolskiy, Vladislav Volkov, and Viktor Patsayev, had manned the Salyut-1. The Soviet specialists were "basking in the reflected glory of their space station" (Ref. 42, p. 140).

The successful continuation of the Salyut-1–Soyuz-11 flight influenced the U.S. approach toward a possible joint mission profile. They tended to believe that

> because the Soviet Union has demonstrated the first space station experiment, the United States could be the guest of a Salyut mission. Perhaps, it is possible to mount an Apollo Earth orbital mission within the next two years, before the advent of the Skylab. While it may not be possible to achieve the necessary compatibility for docking an Apollo with a Salyut system in that period of time, it should be possible to mount a joint tandem flight whereby the Apollo could fly in formation with a Salyut and a Soyuz for a period of time. In 1973 the United States could return the invitation for a Soviet Soyuz visit to the Skylab during its second or third manned mission. Such a search for joint participation would not diminish the prestige of either, yet would add immeasurably to the stature of both. Each could continue its own competitive way, neither would be slighting or downgrading the other. Both space programs would benefit. The United States would gain supporters it never had.... For those who harbor the thought that the United States can wait until it is mounting its Skylab mission to extend the invitation for such a joint program, let it not be overlooked that the Soviet Union may not be interested. They have the first space station experiment and within the confines of such meetings as are now in Houston, it is proper to suggest or propose that they have the precedence of extending the first invitation.[70]

As a result of such considerations, space talks moved from general discussions to specific and existing spacecraft and experimental space stations of the two countries.[71]

This Salyut-1–Soyuz-11 mission and the way it ended had a major impact on the Soviet decision about what kind of space hardware to use for a joint space flight. The flight was a combination of success and failure. The crew worked in space for 23 days and set a spaceflight endurance world record. Also, the cosmonauts performed a significant number of experiments relating to the study of the prolonged influence of weightlessness on the human body. However, the crew died during reentry because of accidental depressurization of their landing capsule. On one hand, the Soviets had a new type of spacecraft—the space station. By docking it with Apollo they could have attracted more world attention to the new Soviet achievements in the field of science and technology, whereas the docking between Soyuz and Skylab prior to the Apollo and Salyut docking psychologically would have thrown the "orbiters of the first space station into the inferior position of making a dramatic visit, watched worldwide, to an American space station before the Americans visited their station."[72]

On the other hand, doubts were raised about the safety of Soyuz-type spacecraft. Any kind of malfunction of Soyuz during a joint flight with Apollo would have been a major embarrassment for the Soviet space and political leadership. This is why the final choice made by the Soviets at that time was to use the Salyut space station for a joint space project.

This decision met with a positive reaction on the U.S. side. "That mission [Apollo–Salyut]," said one of the U.S. engineers involved in negotiations, "would be the logical step because both the Apollo and the Salyut have flown successfully. We have not yet flown Skylab, and we always feel more comfortable with spacecraft which have a good record behind them."[73] Overall, the U.S. delegation made the following observations at the U.S./USSR Joint Rendezvous and Docking Meeting at NASA Marshall Space Center on 2 July 1971:

a. The Soviets were direct and positive in their approach, and apparently had a great deal of technical flexibility in whatever agreements were reached. They were most cooperative and amenable to suggestions for technical changes.

b. The Soviet delegation exhibited a very strong desire to reach positive conclusion and seek solutions to make things work.

c. The Soviet side was very positive in proposing an early experiment using Salyut and Apollo hardware. They indicated that they were merely responding to our proposal for Soyuz/Apollo docking, and said that this had now been bypassed by the existence of Salyut. They also wanted to consider docking between Skylab and Soyuz, but agreed that to do this with Skylab I would be premature. However, they assumed that we would have additional Skylabs after the first one and, therefore, insisted on discussing the possibility for future Skylab/Soyuz dockings.[74]

FACTORS ENCOURAGING AGREEMENT FOR A JOINT MISSION

The general result of the U.S.–Soviet summer 1971 space talks was the approval of approximately 90% of the plans for a compatible rendezvous and docking system for the spacecrafts of the two nations (Ref. 75).* This kind of agreement raised hopes that the U.S.–Soviet linkup in space could take place as early as the mid-1970s.[77] The follow-up decision to transcend the relative technical material was considered as "the implication that both sides did indeed feel that joint space ventures were in the offing, possibly as early as 1974."[78] The third in a series of meetings between Soviet and U.S. specialists on the design and testing of compatible docking systems, which took place in November and December 1971 in Moscow, was already focused on concrete

---

*For more information about the results of the talks, see also Ref. 76.

issues, such as "the definition of technical requirements for compatible systems in future spacecraft, as well as in planning possible joint test missions, the first of which could link an Apollo-type spacecraft and a Salyut-type space station."[79] Based on the results of these talks, NASA had recommended to the White House that a formal agreement on an Apollo–Salyut mission be included on the agenda for the May summit meeting between President Nixon and Premier Kosygin (Ref. 42, p. 182).

Even though the joint manned project seemed to be a matter for the near future, there was one thing that was standing in the way of a public announcement of the plan to conduct joint U.S.–Soviet manned space flight missions beginning in 1974—the lack of the formal top administration pronouncements of the United States and the Soviet Union. This problem was expected to be resolved by the time of Nixon's visit to Moscow in May 1972.[72] Meanwhile NASA and the Soviet space authorities kept avoiding any formalization of the agreements reached at the interagency level. After the Soviet and American engineers and scientists held a number of meetings where they discussed possible joint missions, the U.S. embassy issued a report that fueled talks about the upcoming joint manned mission. In December 1971, NASA rejected this kind of assumption saying that it was the result of a misunderstanding concerning the report. The agency said the misunderstanding stemmed from just a "routine progress report by the American Embassy in Moscow on discussions between representatives of NASA and the Soviet Academy of Sciences now in Moscow."[68]

Besides the successful development of a working relationship between the U.S. and Soviet space specialists, there were three other factors that helped the approaching formalization of the Soviet–American cooperative efforts in space at a high political level. The first factor consisted in several main strands of policy and events coming together by mid-1972 to create the necessary critical mass to launch Soviet–American détente. They were: European détente, with a Berlin settlement sought by the West and the treaties with West Germany sought by the Soviet Union together with its Eastern European satellites; a Sino–American rapprochement, pursued by the United States (as well as by China), which the Soviet Union could not prevent but wished to counterbalance; a SALT agreement coming to its formalization; a growing mutual understanding of the need to expand trade; the need to cooperate to settle the conflict in Vietnam; and internal political support for détente on both sides (Ref. 2, p. 122).

Another factor was a growing understanding by the Kremlin and the White House of the role that space exploration could play in rapprochement between the two countries. Nixon expressed his wish to work together with the Soviet Union "in creating not just a climate of peace, but an actual structure of peace and cooperation." Nixon believed that a space exploration

agreement would "constitute an important building block for such a structure."[80] Upon his return from Moscow, the president stressed that cooperation resulting from the agreement and strengthened institutional ties would
"create on both sides a steadily growing vested interest in the maintenance
of good relations between our two countries."[81] The Soviet leaders also considered formalization of U.S.–Soviet cooperation in space to be "of major
importance to the development of Soviet–American relations."[80]

The third factor was a growing public recognition of the healing impact that
cooperation in space could have on the bilateral relationship. The result of this
recognition was an increasing public support for cooperation. On the U.S. side,
this kind of support was expressed in an article published in *The New York
Times*. The article stressed, in particular, that "far-sighted observers have long
pointed out that the chief value of space exploration might yet be as domain
where international cooperation and shared peril among astronauts of different
nationalities could help ease political and ideological animosities here on
earth."[82] Overall, according to a survey conducted a little more than a month
before the May 1972 summit in Moscow, 66% of Americans supported the idea
of joint U.S.–Soviet exploration of outer space, with only 23% opposing it, and
11% not having any opinion.[83]

As to articles published in the Soviet press, they indicated not only Soviet
public support to pooling efforts with the Americans in space, but also the
official attitude of the Soviet political leadership toward space cooperation
with the United States. The newspaper *Pravda*, an official press organ of the
Communist Party, which always reflected the position of the Soviet leadership, published a letter from 19 Soviet cosmonauts that said: "We think that
in the interest of peace of friendship between the peoples of our planet, business cooperation between space explorers of different countries, including
the Soviet Union and the United States, should develop and grow stronger."[84]
Another article in *Pravda* had the following statement: "Earth is a planet of
mankind. Cooperation in space paves the road to peace, mutual understanding and good of all the people."[85]

Having taken into consideration all of the political and technical developments in U.S.–Soviet space cooperation in 1970–1971, as well as the general context of the bilateral relationship, Henry Kissinger asked NASA to make
a firm recommendation by 15 April 1972 concerning the feasibility of conducting a joint manned mission (Ref. 42, p. 183). This recommendation was
given to him after another meeting between NASA and the USSR Academy
of Sciences representatives in early April 1972 (Ref. 42, pp. 190–191).
Moreover, realizing the need to gain congressional approval for the approximately $250 million necessary "to finance the mission and provide for the
development of compatible systems," NASA officials made a final estimate
of the "significant domestic impact" the joint manned space flight would

have on the United States. Besides the aforementioned utilization of approximately $100 million of remaining Apollo hardware (Apollo command and service modules and Saturn 1B) and filling a gap in the manned American space program between the Skylab and Space Shuttle projects, the realization of the joint space flight "will have an impact on domestic jobs, stabilizing many that otherwise would be jeopardized and creating as many as 4,400 additional ones. These new jobs will be in addition to the 50,000 new jobs which the space shuttle program is expected to create" (Ref. 53, p. 29).

PROBLEMS TO BE SOLVED

SUBSTITUTING SOYUZ FOR SALYUT

On 9 February 1972, a *Space Business Daily* bulletin expressed the hope that "unless there is something unforeseen developing between now and president Nixon's visit to Moscow, the formal announcement of the [Apollo–Salyut] plan is expected to be made by the date of the trip by the President."[86] However, the unforeseen happened during a 4–6 April meeting in Moscow between the U.S. and Soviet space officials. A member of the Soviet delegation, Academician Vladimir Kotelnikov, told the NASA representatives that in reevaluating the proposed test mission the Soviets had come to the conclusion that it would not be technically and economically feasible to fly the mission using Salyut. Docking of the Apollo spacecraft with the Salyut station would require a docking adapter and transfer tunnel not yet developed at that time. Furthermore, this modification would have required the relocation of the Salyut attitude control thrusters and other critical equipment and would have necessitated removal of the orbit maneuvering engine. Developing the necessary hardware and introducing necessary changes would have been costly in both time and money. Therefore, the Soviets proposed to conduct the test flight using Soyuz, which could have been relatively easily modified for such a mission. Besides, the Soviets assured their American colleagues that there would be no changes in any of the agreements made thus far (see Ref. 42, p. 185; Refs. 87, 88).

While, the time–money factor was undoubtedly an important one, there were at least two more that could explain the sudden switch from the Salyut space station to the Soyuz spacecraft made by the Soviets. The first factor was rooted in the contradictory character of the Soviet space station program. Despite the fact that all of the Soviet space stations were called Salyut, a few of them were actually Almaz—a military orbital laboratory designed by someone other than Korolev's Design Bureau. (At that time Korolev's Bureau was already headed by his successor Vasily Mishin.) Almaz was designed and built by Vladimir Chelomey's team, while Salyut by Mishin's. The reason why all of the stations had the same name was because the international public and the Soviet people already supported the principle that anything related to manned space flight should represent a peaceful mission. At the time when the Soviets and Americans were discussing a possible joint manned flight, Chelomey's plans to put the first military cosmonauts into orbit onboard his Almaz station had already been under consideration. At some point this entered the phase of practical realization, which could well have happened in early 1972, by the time of the U.S.–Soviet April meeting in Moscow. The first Almaz called Salyut-2 was launched into orbit in early April 1973, but it was soon discovered that the station was losing air pressure, and an attempt to dock the crew with it was never made. The station was deorbited in late May of the same year. There were two subsequent launches: Salyut-3 in 1974 and Salyut-5 in 1976. These stations were more fortunate. Salyut-3 operated in orbit for six months and Salyut-5 for more than a year; both were manned. Thus, it is enough to look at the time periods of the launching and operation of the military stations to realize that by the mid-1970s—the time when the joint Soviet–American flight was supposed to take place—a possible candidate for docking with Apollo could have been a military Salyut, which would have been totally unacceptable for the Soviets and Americans even during détente (see Ref. 7, pp. 207–208; Ref. 89, pp. 136, 145, 158). Although Soyuz spacecraft had not made another flight after the Soyuz-11 mission tragedy, it was already determined by that time that the sudden depressurization was caused by a malfunction of the pressure equalizing valve—a defect that was easily remedied.

Another factor was rooted in the conflicting interests of the leading Soviet space hardware designing bureaus. Although Vasily Mishin and Vladimir Chelomey were heads of the competing design bureaus, Mishin was supporting Chelomey's Almaz space stations (for this reason in 1973 a team of employees from Mishin's bureau sent a letter to the Soviet Union's Communist Party's Central Committee with a suggestion to release Mishin from his duty as a chief designer). Mishin supported his competitor's project because he wanted his own design bureau to concentrate on a mission to the moon, which was canceled only in 1974, after Mishin was replaced by Valentin Glushko (Ref. 90, pp. 268, 294). Mishin did not think of the moon project in terms of a race with the Americans, and even after the successful accomplishment of the Apollo 11 mission, he continued working on the realization of the Soviet manned moon landing project.* For this reason Mishin tried to present the Salyut program as inferior to Almaz and any modifications required to adapt Salyut for docking with Apollo (such as the

---

*Even after his retirement Mishin continued to believe that if the N-1 project had not been canceled, the Soviet Union would have landed a man on the lunar surface sometime in the early 1970s and would have had a permanently manned moon base by the end of 1970s.

installation of a second docking port) as too complex and costly. Apparently Mishin realized that positive international exposure of Salyut resulting from the joint space flight would boost political support for the program from the Soviet leadership, making it more difficult to terminate Salyut. In April 1972, both Mishin and Chelomey, in their letter to Minister of General Machine Building Sergey Afanasiev concerning the disadvantages of the Salyut program, mentioned that the Soviet space industry would have to build at least two or three extra Salyut space stations for the purpose of practicing docking procedures with the Apollo spacecraft (Ref. 90, pp. 295–296).

Another factor that could have made Soviets reconsider using Salyut for the joint space mission was their growing confidence in the possibility of making up for the lost moon race by establishing a permanent presence in orbit in the space stations. As the article published in *The New York Times* in March 1972 described:

> But despite all the movements toward cooperation, the Russians still seem to feel keenly that there is a race to be won in space. Much of the visible expansion at Star City is said to center around the new facility for training men for missions on earth-orbiting space stations. And now that the American Apollo moon-landing project is ending and its Skylab space station program will run less than a year, the Russians feel they have a chance once again to dominate manned spaceflights in this decade—as they did in the early 1960s.[91]

Although docking between the Salyut space station and the Apollo space-craft would have attracted world attention to new Soviet space hardware, it could have also blurred the uniqueness of the Salyut-type orbital laboratory. Because both approaches made almost equal sense in terms of the continuation of the space race, it is possible to the assume that the advocates of this approach prevailed over advocates of the first.

NEW RATIONALE FOR THE JOINT MISSION

Even though U.S. space officials quickly agreed to switch from Salyut to Soyuz and found no political implications in the switch, apparently there were some. It would still be possible to exchange crews and to make a joint flight a public show. Moreover, this kind of mission, involving only Soyuz and Apollo spacecraft would have given the Americans an extra advantage— not attracting attention to the fact that the Soviets already had an operational space station, while NASA did not (Ref. 42, p. 185). However, it was necessary to find a new rationale for the joint mission. Because "the unilateral Soviet decision to withdraw the larger (and more expensive) space station and replace it with the Soyuz spacecraft ended plans to use the mission for beneficial science studies,"[92] science itself could no longer serve as a sound reason for the flight. It was necessary to find another rationale with which to convince the Congress and the president to support the joint manned space project. This could become a practice of rescue operations in space, which received a meaure of public support. An article published in *The Houston Chronicle* in September 1971 expressed the hope that "American and Russian cooperation in space could pay many dividends, most importantly by providing an increased safety factor."[93] Although, as described previously in this chapter, rescue missions using existing hardware seemed to be rather impractical. Thus the idea for the whole mission could have found itself in jeopardy if the legislature and the executive power had not found the safety factor to be a sufficiently good reason for conducting it.

The need to evaluate all the pros and cons of a mission involving a test of space rescue procedures caused the U.S. space officials to avoid reporting for a while about the possibility of the Apollo–Soyuz mission even to the U.S. government. In mid-April 1972, more than a week after the Soviets and Americans agreed "that a joint test rendezvous and docking mission would be planned for 1975, using specially modified Apollo and Soyuz type space-craft," and that the primary rationale for this kind of mission would be to "facilitate emergency assistance to astronauts in difficulty" (Ref. 53, p. 29), NASA Administrator James Fletcher told Congress that "there have been technical discussions of the possibility of an early experimental mission in which an Apollo spacecraft would rendezvous and dock with a Soviet Salyut-type space station. No decision has been made on conducting such a mission...."[94] Fletcher did not mention the possibility of the Apollo–Soyuz flight at all.

A potential switch to docking between the two spacecraft had not been mentioned by Soviet space officials either. At the end of April 1972, Keldysh made a statement about the likelihood of the U.S.–Soviet agreement on the plan to dock a U.S. Apollo spacecraft with a Soviet Salyut space station. Keldysh's words were considered by the American press as "the most positive official statement regarding the joint manned mission plan made by either country."[94]

It is highly unlikely that Keldysh was keeping any information concerning Soviet–American cooperation in space secret from the Soviet government. However, there is a good reason to assume that Keldysh was withholding some information deliberately on behalf of U.S. space officials who did not want the possibility of the Apollo–Soyuz flight to be released for a while. This kind of assumption seems to be a plausible explanation for Keldysh's statement, taking into consideration that by the spring of 1972 representatives of the two aerospace communities had established not only good working, but even trustful, relationships among themselves. In

December 1971, the Soviet space authorities requested that NASA with-hold some information about the Soviet space program from the American public, and the request was granted. As John P. Donnely, assistant for pub-lic affairs for NASA, explained, "our position is that it's kind of like some-one writes you a letter. If they want to make public what they write in the letter, that's fine, but it's their prerogative and not ours."[95]

FORMAL AGREEMENT FOR ASTP

Whatever doubts there had been about the formalization on the highest level of a possible Apollo–Soyuz mission, they had been dispersed in May 1972, when President Nixon and Chairman of the USSR Council of Ministers Alexei Kosygin signed a space agreement between the two coun-tries. The significance of the agreement consisted not only in giving U.S.–Soviet cooperation in space an official character, but also in demon-strating the difference between the interagency and intergovernmental approaches to the pooling of efforts of the two countries in space. A certain difference was shown in the attitude toward two issues: the significance of the joint manned project in the context of U.S.–Soviet space relationships and future perspectives for bilateral cooperation in space.

First, whereas NASA and the Soviet Academy of Sciences definitely con-sidered a rendezvous and docking of the Apollo command service module with Soyuz spacecraft as the most dramatic event in Soviet–American space cooperation,[96] the leaderships of the two countries maintained a more mod-erate position on this matter. The first issue in order of importance (Article 1 of the Agreement) was the development of "cooperation in the fields of space meteorology; study of the natural environment; exploration of near earth space, the moon and the planets; and space biology and medicine" (Ref. 53, p. 27). The parties agreed that it was especially important to "cooperate to take all appropriate measures to encourage and achieve the fulfillment of the Summary of Results of Discussion on Space Cooperation Between the U.S. National Aeronautics and Space Administration and the Academy of Sciences of the U.S.S.R. dated January 21, 1971." However, the Summary of the Results did not contain a single word about a joint manned flight.

Article 2 of the Agreement dealt with the ways "to carry out such cooper-ation," described in Article 1. And finally, Article 3 was about the U.S.–Soviet agreement "to carry out projects for developing compatible rendezvous and docking systems of United States and Soviet manned spacecraft and sta-tions,... envisaging the docking of a United States Apollo-type spacecraft and a Soviet Soyuz-type spacecraft with visits of astronauts in each other's space-craft" in 1975 (Ref. 53, p. 27). The leaderships of the two countries accepted a safety rationale for the mission by stressing that the compatible systems would be developed above all "to enhance the safety of manned flight in space."* The opportunity to conduct joint scientific experiments in the future was mentioned after the safety factor.

The reason why the leaderships of the two countries did not emphasize the critical role of the joint manned space mission in the broader context of Soviet–American cooperation in space was probably because the mission on such scale was technically and politically more prone to failure than any less-visible projects such as weather data or lunar sample exchanges. Thus, if this kind of mission failed, it could have given the impression of a general failure of U.S.–Soviet cooperation in space, which could have been considered by the public as a failure of the leaders who formalized the cooperative agreement.

As to future cooperation in space, the U.S. and Soviet space agencies envisioned more far-reaching perspectives than their respective governments. Just before the beginning of the April 1972 talks in Moscow, a reporter of *The New York Times*, John Noble Wilford, visited the Gagarin Cosmonauts' Training Center in Star City near Moscow. The visit was significant per se and implied a major change in Soviet policy toward space cooperation—no Western journalist had previously been permitted inside this key center of Soviet manned space facilities. During his visit, Soviet space officials made Wilford understand that they might be willing to explore the moon jointly with the United States.[98] Immediately after the April 1972 meeting in Moscow between the representatives of U.S. and Soviet space agencies, NASA Administrator James Fletcher said at a dinner he hosted for newsmen, "We certainly don't believe that the first mission of American and Russian astronauts together will be the last."[99] The U.S./USSR Rendezvous and Docking Agreement, made at the April meeting in Moscow, envisaged "future generations of manned spacecraft of both the United States and the Soviet Union" docking with each other (Ref. 52, p. 28). This could be a U.S. Space Shuttle, a potential Soviet moon vehicle,

---

*It is interesting to note that there was probably some kind of disagreement about what should be the primary rationale for the joint flight even within NASA leadership. This is the list of rationales (in the order of importance) presented by Dale Myers, assistant administrator–Manned Space Flight:

　1) To test the compatible rendezvous system in orbit.

　2) To test the androgynous docking assembly.

　3) To verify the techniques of transfer of astronauts and cosmonauts and then to perform the activities of the U.S. and USSR crews in docked flight in accordance with the program yet to be determined.

　4) To gain experience in conducting joint flights by U.S. and USSR spacecraft including rendering aid in emergency situations.

However, at the same press conference Glynn Lunney, special assistant to the Apollo program manager, stressed one more time that the primary purpose of the mission is rescue "and the secondary purpose then which would naturally accrue would be one of conducting planned joint cooperative exercises we have agreed upon ahead of time."[97]

or upgraded Salyut or Skylab space stations. This attitude enjoyed public support as expressed in an article in *The Washington Post*: "The Russians want to develop a space station, while the United States is going ahead with the reusable shuttle. If the two nations ever want to send men to Mars, they must marry the two techniques and share the burden of going to Mars together."[100]

The Nixon–Kosygin space agreement, like the agreement between NASA and the USSR Academy of Sciences, also suggested that U.S.–Soviet cooperation in space would go beyond the Apollo–Soyuz project. The project was considered the first experimental flight, implying that there would be other missions of this kind. The agreement was supposed to be in force for five years. However, the leaderships of the two countries avoided making any commitment to "future generations of manned spacecraft," having pledged to "determine other areas of cooperation in the exploration and use of outer space for peaceful purposes" (Article 5).[53]

Overall the meeting between President Nixon and Premier Kosygin, in the words of NASA Administrator Fletcher, "brought to fruition the most meaningful cooperation in space yet achieved by our two nations."[88] The Nixon–Kosygin space agreement was characterized by Fletcher at the press conference after the signing as "open ended," allowing "for new agreements, not just on docking, but other cooperative space agreements as time goes."[96]

AMERICAN POLITICAL COMMUNITY IMPRESSIONS

The U.S.–Soviet space accord was welcomed by the U.S. political community. Representative Hechler was "gripped by the historic significance" of the coming joint space mission; in his view it was significant "that we are not arguing emotionally over 'whether,' but we are talking about 'how'" (Ref. 101, p. 416). Senator Cook congratulated "NASA and the President on the decision for a new partnership in space exploration with the Soviet Union" and expressed his hope that the two nations "can begin to develop greater understanding and a more cooperative disposition." In his view "nations everywhere must begin to recognize that it is only through mutual interdependence that this world can exist peacefully for many tomorrows to come. Our goals must be to work for the benefit of man on earth."[102] Senator William Proxmire went even further by suggesting that the United States halt development of part of the post-Apollo Space Transportation System and instead use the Soviet Soyuz spacecraft and Salyut space station for its future space flight program. In his opinion "the cooperative manned space flight mission agreement is going to result in a great saving of money to this country on the space program."[92]

U.S. AND SOVIET PUBLIC REACTION

The agreement was welcomed by both the U.S. and Soviet public. *The New York Times* stressed the political significance of the agreement:

> A corollary of all this is the growing likelihood that the chief dividends from space programs will be political gains here on earth. Born in the mad competition for status characteristic of the cold war, manned and unmanned space research has taught both sides how puny are man's resources in facing the mystery and challenge of the universe. As that lesson has sunk in, both sides have come to understand the advantages of cooperation as against useless and wasteful rivalry.[103]

The Wall Street Journal and *The Washington Post* emphasized the significance of the agreement for the advancement of space exploration, both in scientific and economic terms. On one hand, they expressed the hope of American officials that "the docking program, which requires module modification so that [Apollo and Soyuz] can be joined together in space, will lead to more ambitious projects," and will also change the Communist Party and Soviet military bureaucrats' attitude toward nonrelease of technical data to the West.[104] On the other hand, "by curtailing both competition and duplication in space activities, scientists on both sides would be able to do in good time the things that ought to be done without straining the financial resources of their governments nearly as much in the process."[105]

SOVIET POLITICAL COMMUNITY IMPRESSIONS

As to the Soviet reaction, Academician Petrov stressed that it was "difficult to overestimate the significance of the agreement which has been concluded and has already come into force" and called it "an important new act in the development of international relations" in space domain.[106] However, besides this, the signing of the agreement marked an important point in the development of the Apollo–Soyuz program. The project went from the decision-making to the decision-implementation phase.

ASTP: THE ROAD TO LIFTOFF

A year after signing the agreement, the ASTP had already reached the point where American and Soviet technical directors and their staff were meeting almost monthly in Houston and Moscow to work out the details of compatible systems for rendezvous and docking (Ref. 107, S13480). This kind of cooperative activity was solid proof of the fact that ASTP was on its way to practical realization. However, the decision-implementation phase involved new technical–political problems. Although there were no more doubts about what kind of hardware to use for the mission, the other questions remained. Among them the most important were

1) What would be the best purpose of the mission, and does it justify its cost?
2) Would it be safe to fly the mission taking into consideration recent Soviet failures in space?
3) What kind and what volume of technology transfer would be acceptable in the course of the realization of the project?
4) Does Soviet secretiveness correspond with the spirit of cooperation in space?

SHRINKING NASA BUDGET

Such questions were originated by the interaction of a few U.S.- and Soviet-related factors. The U.S. factor was a continuing shrinkage of NASA's budget. The aeronautics and space budget plan for Fiscal Year (FY) 1973 was cut sharply, and Nixon put forward a 1974 budget far short of the constant level approved by Congress (Ref. 107, S13479). Such budgetary cuts raised concern about the successful development of the basic elements of a new space transportation system including the Space Shuttle. ASTP became one of the potential objects of the money-saving policy.* Representative Olin Teague, a chairman of the House Space Committee, said that his committee would review in depth the need for the joint ASTP:

> This cooperative effort we are trying to make with the Russians, which I have considerable doubt about, runs about $300 million. Whether it is better used that way, I am yet to be convinced.... It's strictly a political, psychological effort and maybe it's great, but we are sure going to hold some careful hearings on it.[109]

The reason behind this statement was rooted in the three Soviet-related factors, which became obvious after American specialists became more familiar with the Soviet space program and the Soviet approach toward cooperation with the United States in space.

LEVEL OF SOVIET SPACE TECHNOLOGY

The first Soviet-related factor was the relative backwardness of the Soviet space technology compared to that of the Americans. As it was stated by Representative Philip Crane,

---

*The total funding estimate of $250 million was broken down as follows: FY 1972, $6.9 million; FY 1973, $30.1 million; FY 1974, $92 million; FY 1975, $111 million; and FY 1976, $10 million.[108]

One purpose of this [Nixon–Kosygin] agreement was to open the way for technology sharing between the United States and the Soviet Union. It was expected that such an exchange would prove helpful to both countries, that each could contribute its share and that such mutual contribution would assist in bettering relations.

Now, after experts have had an opportunity to investigate this matter further, it seems as if the United States will be contributing almost all of the technology. The only thing the Soviet Union will be doing is "sharing" such information, which means, in effect, that we will simply be providing the Soviet Union with advanced technological data which it has been unable to develop for itself.[110]

Overall, American experts came to the conclusion that

the present level of Soviet space technology is about midway between American Mercury and Gemini technology; that is, about 11 years behind our current state of art. The Soyuz can in no way compare with the Apollo spacecraft, and it is not even a serious piece of engineering compared to the Space Shuttle....[110]

Aviation Week and Space Technology magazine went even further by reporting that the Soviet capsule's "capability is below that available in the Mercury spacecraft flown by American astronauts almost 13 years ago."[111] Although one could argue to what extent this conclusion about Soyuz reflects the actual quality of the spacecraft (more than 20 years after the Apollo–Soyuz flight, the modified version of Soyuz, Soyuz-TM, was chosen as a lifeboat for the ISS, which speaks of the sound design and reliability of the spacecraft), at that time it was necessary to define clearly a sound rationale for ASTP.

ORIGINAL RATIONALE FOR THE JOINT FLIGHT. The initial motivation for the ASTP was to develop rendezvous and docking procedures together with the necessary hardware for rescue operations in space. However, ASTP critics pointed out that this rationale was no longer valid because ASTP would be the last Apollo flight.[112] Responding to this challenge to the project, Wernher von Braun pointed out that

[s]tandardized coupling devices are not only a necessity for mutual assistance and rescue operations in space. They are an absolute "must" for any major joint international space venture—for which, it is hoped, the Apollo–Soyuz Test Project will lay the groundworks. Perhaps, the most significant of all the 1975 mission's experiments may be its first trial in space history of a prototype of the compatible docking mechanism of the future.[113]

Initially NASA officials tried to substantiate the rescue rationale by putting it in the context of the upcoming Space Shuttle program. According to Fletcher, after 1975 all American spacecraft should have been equipped with the same kind of docking adaptor needed for rescue operations, as used by Apollo–Soyuz. However, even in this case ASTP did not have any continuity, which considerably devalued its significance. The internal contradiction of the Apollo–Soyuz project as a rehearsal of a possible rescue mission was acknowledged by Fletcher himself when he called ASTP "the first in a series of programs with the later flights not designed."[96] As it was explained by NASA Deputy Administrator Low at the press conference on 13 July 1972, if the Apollo–Soyuz test "is successful, our plans are to equip the Space Shuttle with this kind of a docking system. We don't think that we will need a follow-on test but instead we will be flying operationally with compatible systems beyond this point." Later, Glynn Lunney, technical director for ASTP flight, explained the rationale for the joint flight even in broader terms at the press conference in February 1973: "This project represents, then, a test of those ideas, those systems as to how well they're going to work in the real world. But probably in a larger sense, it's really a test of our ability to cooperate in what is really a fairly difficult exercise."

Apparently, some politicians as well as some U.S. space specialists did not consider this support of the joint space flight to be sound. The situation was aggravated even more by concerns expressed by Representative Robert Price, who said:

> It is my conclusion that they [the Soviets] have not ever lived up to any agreement they have ever made in history. And I think as long as they can they will use us to develop their equipment in every way they can, and then they will abandon us when the time comes. This idea of compatibility is a wonderful thing, and peace, all this sort of thing. But I don't think we want to rush into this thing blindly, not knowing that they technically will bleed us of everything they can bleed us of (Ref. 101, p. 416).

Concern about the absence of a reasonable technological gain from cooperation with the Soviets, and insecurity about Soviet commitment to ASTP, made it necessary to find a rationale that could sustain the Apollo flight on its own merits, without the rendezvous and docking mission. Senator Moss mentioned specifically "the need for prompt decision on meaningful experiments for the U.S. flight," taking into consideration that in command service module (CSM) there was a "space for 400 pounds of experiments not now committed" (Ref. 107, S13479). The chief investigator behind the ASTP hearings in the House Manned Space Flight Subcommittee, congressman and former astronaut, Jack Swigert, felt that the United States had "no business flying ASTP unless it could also stand on its own feet without the rendezvous and docking mission." His suggestion was that NASA "should take out the added reaction control system propellant unit (for back-up deorbit) to open up an additional weight capability so that we can carry more experiments."[116]

NEW RATIONALE: A PROGRAM OF EXPERIMENTS FOR ASTP. In response to Swigert's concern, Deputy Administrator Low repeated that the basic purpose of the mission "is an experiment in rendezvous and docking so that in the future we could work together in space or rescue each other." However, NASA had enhanced the mission with an additional $10 million worth of experiments. This sum "was arbitrarily picked by NASA management as a balance between enhancing ASTP and not cutting too deeply into any other NASA work." Low specifically mentioned that "the basic constraint in not adding additional experiments is money and not weight or volume."[116]

The issue of rationale was finally solved by the development of the program of experiments for ASTP. The program received high praise from von Braun, who believed that "even without taking into account the momentous political and long-range aspects of the Apollo–Soyuz Test Project (ASTP), its program of experiments will make it worth every penny of its $250 million price tag" (Ref. 113, p. 40). ASTP Program Director Chester Lee expressed the firm belief that with the additional efforts NASA had expended on experiments, the Apollo mission could justify itself without a rendezvous with the Soviet Union (Ref. 101, p. 421). Overall 15 experiments were planned to be performed during the flight, 5 of which were joint U.S.–Soviet experiments.[117] Moreover, the goal of ASTP as a promoter of cooperative efforts in space was extended beyond the limits of bilateral Soviet–American cooperation. Two experiments were provided by the Federal Republic of Germany and one experiment was cooperatively supported by a U.S. industrial firm. As was expressed by Senator Moss, "these experiments, therefore, provide a model for conducting future cooperative experiments. This will be most important not only in developing the Space Shuttle payloads at reduced cost to the United States, but could provide a new era of international scientific and medical cooperation."[118]

CONTINUING SOVIET SECRECY

The second Soviet-related factor that could potentially hamper ASTP was the continuing Soviet policy of secrecy, which remained "the main roadblock to further U.S.–Soviet cooperation."[119] Secrecy permeated even the organizational aspects of the cooperative efforts. Most of the representatives of the Soviet space industry were introduced to their American colleagues as Intercosmos employees. Intercosmos, headed by Academician Boris Petrov, was the Soviet umbrella organization for international cooperation in space. Other Soviet engineers and space officials were introduced as employees of Mission Control Center, or Gagarin Cosmonauts' Training Center. According to the Soviets themselves, "this was not the best solution. Americans knew very well who was who, but these were the rules of the game" (Ref. 90, p. 198). This situation continued through 1980. Intercosmos coordinated joint activities, took care of the legal aspects of cooperation, provided space for joint work, and financed hosting U.S. delegations in the Soviet Union (Ref. 90, p. 198).

Soviet secrecy manifested itself in different ways. Among members of the Soviet delegations involved in the Apollo–Soyuz project there were intelligence officers pretending to be interpreters or engineers. Every Soviet manager of the project had to coordinate his activities with the security services. The main Soyuz spacecraft technical data were classified and needed to be cleared by security services for transfer to the Americans. However, some information was secret and could not be transferred at all. In that case special government bodies were required to issue a permission to release data to U.S. specialists. Among the data that required a special clearance was information about Soyuz radio frequencies and coordinates of mission control ground stations.

U.S. DISSATISFACTION WITH SOCIET SECRECY

Even though Soviet and American "technical specialists demonstrated understanding in dealing with inconvenience caused by secrecy, were patient and tried to ask for as little [information] as possible" (Ref. 90, p. 201), this secrecy nevertheless had a twofold negative impact on the attitude of the U.S. public and politicians to ASTP. On one hand, the Soviet nondisclosure policy increased concern about unilateral benefits the Soviets would achieve from this flight. As was expressed by Representative Teague,

> The only question in my mind is the definition of success in this endeavor. It is intended to be an exchange of scientific and technological knowledge. It seems that at this time the editorialist is correct when he writes: "The U.S.–Soviet test project promises no technological benefits to this country."[120]

According to Teague, the very simple reason why this kind of statement was true was because of "the Soviet's shroud of secrecy around their space program."[120] Representative Robert Huber presented the following fact as an example of inequality in Soviet–American space cooperation: "there are no less than 75 Soviet engineers and other technicians now [30 April 1974] in Houston gathering information. The number of U.S. people that will be in the Soviet Union is 10, and you can expect their access to Soviet information be severely limited."[121]

Hubert's expectations turned out to be true. The Soviets refused to allow NASA ASTP crews to conduct a familiarization training session with the flight Soyuz for the mission while it was being built and prepared in the factory. The Soviets also declined a proposal that U.S. specialists should participate in the installation of U.S.-built communication equipment or be available during the installation and test at the factory in case of difficulties.[122] Finally, the Soviet Union rejected requests that U.S. newsmen be permitted to come to Baikonur to view the launching of Soviet cosmonauts during ASTP. Soviet newsmen, however, were permitted to view the liftoff of the American astronauts from Cape Canaveral.[123]

The U.S. public began voicing growing dissatisfaction with ASTP, as expressed by an article in *The Los Angeles Times*: "The Soviet Union will gain much more than will the United States in this technological exchange. And hoped-for political benefits are being limited by the Russian's obsession with secrecy." According to the article, this kind of Soviet policy significantly undermined U.S. public support toward the future of U.S.–Soviet cooperation in space, which should not take place "unless the flow of technology in each direction will be substantially equal, and unless the Kremlin is prepared to allow normal news coverage of what, after all is supposed to be a demonstration of peaceful cooperation."[124] The Soviet refusal to release space-related information raised suspicions concerning the real motives behind their cooperation with Americans. Representative Huber had "no doubt that this space cooperation will enable them [the Soviets] to catch up even as the transfer recently of some of our computer technology enabled the USSR to speed up its schedule for MIRV [Multiple Independently Targeted Reentry Vehicles] missiles."[121] Some Americans believed that what the Soviets "are learning about inertial guidance alone is going to increase enormously the accuracy, and thus the deadliness, of their huge ICBMs armed with atomic warheads and aimed at all major American cities."[125]

Soviet secrecy, besides arousing suspicions about the Soviet's taking unilateral advantage, raised doubts about the safety of the Soviet hardware scheduled to be used for ASTP and, as a result, about the safety of U.S. astronauts when flying in docked configuration with their Soviet counterparts. These concerns were augmented by the number of Soviet failures in space, including the death of the Soyuz-11 crew in June 1971 and consequent failures of Salyut-2 and another Salyut (whose true identity was hidden behind the generic name Kosmos 557) in April and May 1973 and also by the unsuccessful attempt of Soyuz-15 to dock to the Salyut-3 space station in August 1974 (Ref. 89, pp. 136, 137, 146). "Two months ago, I would have said the Apollo–Soyuz mission was as solid as anything on our books," said one U.S. space official soon after Salyut failures in 1973, "I can't say that right now."[126] The Soviets were reluctant to release information about the causes of the failures, which made the U.S. side think that they were trying to avoid major embarrassment by demonstrating the true level of reliability and technological sophistication of their flight hardware and mission control facilities. For example, Konstantin Bushuyev, ASTP technical director on the Soviet side, refused to release any data about the exact location of the Soviet mission control center and the true reasons for the Salyut (Kosmos 557) failure.[127] Senator Proxmire, once a strong advocate of the Soviet–American linkup, became one of its harshest critics. He estimated the failure rate throughout the Soyuz mission profile as 28% and based on that had expressed his "deep concern that the upcoming joint Apollo–Soyuz experiment may be dangerous to American astronauts."[128] When the Soviets made public their plans to conduct two space missions simultaneously: the linkup of a Soyuz spacecraft with the Apollo vehicle in a flight to be launched 15 July 1975 and the continuation of the Salyut space station venture, Senator Proxmire asked NASA to postpone the joint space flight until the Salyut experiment was finished. He cited a report by Carl Duckett, deputy CIA director for science and technology, who said in effect that the Soviet Union could not handle the two missions at once.* Based on these facts, an article published in the *Chicago Tribune* suggested that the Soviet

> effort to upstage NASA may not be so harmless after all. If their command facilities are as limited as Mr. Duckett believes, the Apollo crew conceivably could be endangered by the Russian proneness to run risks. NASA is presumably in close touch with its Soviet counterpart, and it should examine closely the capabilities of the Star City space flight center. If this isn't possible, or if NASA isn't thoroughly satisfied by such an examination, then it should postpone the mission, as Sen. Proxmire suggests.[129]

This kind of combination of ASTP pros and cons resulted in the complicated attitude of NASA officials toward the upcoming joint mission. ASTP Program Director Chester Lee believed that

> [i]n essence, then, it is much more than blasting off from the launch pad and meeting another spacecraft in space. We are not establishing with this mission a rescue capability as such, although we could bring back the cosmonauts in our spacecraft. We are, however, verifying the

---

*It is interesting to note that a number of Soviet engineers were also against conducting the two missions at once (among them were NPO Energia General Designer Valentin Glushko and chief manager of ASTP on the Soviet side Konstantin Bushuyev). This opinion was rebutted by the future General Designer of NPO Energia Yuri Semyonov, Soviet EVA and pressurized suits Chief Designer Gai Severin, and cosmonaut Konstantin Feoktistov. The latter group gained the support of the Minister of General Machine Building Sergei Afanasiev. As a result, ASTP was conducted at the same time as the Salyut-4–Soyuz-19 flight (crew: Pyotr Klimuk and Vitaly Sevastyanov; Ref. 90, p. 205).

system and technique and gaining valuable experience for meeting and joining up with the spacecraft of another country in space. This is fundamental to any future rescue or joint space operations.... ASTP has also provided a window—however small—in a part of the Soviet society. With this window and resulting knowledge and understanding there is some contribution to détente.[130]

Overall, when Lee was asked to quantify what percentage of cooperation America was getting, assuming that full cooperation was 100%, Lee estimated it in the 90% range (Ref. 101, p. 420).

However, according to *The Economist*:

Most officials of America's National Aeronautics and Space Administration regard the link-up as a dead end in every sense save the political. The docking is not intended to last for more than two days. It has no object beyond a we'll-have-drinks-at-your-place-today-and-you'll-come-to-us-tomorrow exchange between crews....[131]

SOVIET ATTITUDE TOWARD ASTP

Concerning the attitude of Soviet space officials toward ASTP during the realization of the project, it is more difficult to judge than the American view. Representatives of Soviet space-related organizations could not voice their concerns about the project in the same way as their American colleagues. They were supposed to follow the party line, which was clearly favoring ASTP for the sake of "strengthening peace on the planet."[132] However, there are some indirect indications of the lack of unanimity within the Soviet space community with relation to cooperation with Americans. Although many U.S. space officials and politicians noticed that the Soviets did not commit themselves to a second joint manned mission,[133] Chester Lee in his memorandum to James Fletcher, written after the July 1973 Working Group Meeting, mentioned that "it is...evident from casual conversation with some of the Soviets that they are thinking beyond the current joint effort."[134]

An apparent lack of consensus among Soviet space officials with regard to continuing to pool efforts with Americans in space was because of two factors. First, this cooperation did not meet Soviet expectations in terms of giving them experience in rendezvous and docking procedures with the American Skylab, which could grant them more access to U.S. space technology—something the Soviets were counting on to further the development of Salyut stations. The role that docking technology was supposed to play in future Soviet space projects was emphasized by the ASTP director on the Soviet side, Konstantin Bushuyev, in his article published in the Soviet *Nauka E Zhizn* [Science and Life] magazine:

The further development of spaceflights is inseparably linked with docking technology. Without it, for example, the creation in near space of large orbital complexes for the solution of various scientific and national-economic problems is unthinkable. The basis for these complexes will be multipurpose orbital stations consisting of modules with different assignments.[135]

Therefore it is not difficult to imagine the disappointment of the Soviets when they learned about the cancellation of the second Skylab. NASA Administrator Fletcher told the Senate Aeronautical and Space Sciences Committee that the Soviets, when informed of the U.S. plan not to develop a second Skylab mission, "found it incredible."[136]

The second factor was the unavoidable exposure of the weaknesses of the Soviet space program in the course of international cooperation, which was damaging to the image of "one of the most technologically advanced nations of the world," as the Soviet leaders claimed the Soviet Union was.

Nor can the real attitude of the Soviet public toward ASTP be judged. On one hand, the public virtually had no access to space data. They had no choice but to rely on official interpretation by the Soviet mass media of Soviet–American interaction in space and thus could not really evaluate all the pros and cons of space cooperation with Americans. On the other hand, there was no place for any public discontent in the totalitarian society.

SOVIET COMMITMENT TO ASTP COOPERATION

Despite some ambivalence in the attitude of Soviet space officials and specialists toward cooperation with Americans, Chester Lee was "convinced that the Soviets are fully committed to making this mission a success."[134] Such an observation could be confirmed by a number of facts. The first was that the Soviets put safety concerns even above technological progress in the area of spacecraft design. In 1972, Soviet spacecraft designers discussed with the leadership of the Ministry of General Machine Building the possibility of using a considerably modified Soyuz spacecraft (later called Soyuz-T). Many engineers and space officials wanted to use it for the joint space flight because Soyuz-T spacecraft had better performance than Soyuz. However, Soyuz-T was still at the experimental testing stage, and it was decided to use a "classic" Soyuz to ensure the safety of the Apollo–Soyuz flight.*

Even the classic Soyuz still needed to undergo certain modifications to meet requirements for ASTP. The development of the ASTP-suited Soyuz (later called Soyuz-M) began in the fall of 1972. At that time, the spacecraft was called 7K-

---

*Critics calling the Soviet spacecraft used for ASTP obsolete should keep in mind that the Soviets sacrificed technological innovation for the safety of the mission.

TM (item # 11F615A12). For many years this code remained hidden from U.S. specialists. On 15 December 1972, Mishin's design bureau finished the general design of Soyuz-M spacecraft and determined the requirements that it had to face. The Soyuz-M was heavier than Soyuz, and its life support systems would allow it to stay in orbit up to 15 days. (The ordinary Soyuz was designed for only three days of autonomous flight.)

The increased weight of Soyuz-M required a more powerful booster. In November 1972, the chief manager of ASTP on the Soviet side, Konstantin Bushuev, and Soviet rocket designer, Dmitry Kozlov, proposed using a modified Soyuz booster (item # 11A511U) that was in development for Soyuz-T. The new launch vehicle could lift an additional 200 kg. To assure the flawless performance of the new type of the booster, it was launched seven times with automatic spacecraft and one time with an unmanned version of Soyuz-M.

Among other measures aimed at increasing the safety of Soyuz-M spacecraft were the development of a new rescue system, a new power supply system that enabled Soyuz-M to stay in orbit for an unlimited period of time (the power system of a regular Soyuz would enable the spacecraft to fly autonomously for only three days), and a new fire protection system. Overall, the modifications to Soyuz-M added 35–40% to the R&D cost of a regular Soyuz spacecraft.

Test flights of hardware developed for ASTP were an important part of mission safety assurance. The first unmanned Soyuz-M spacecraft under the name Kosmos-638 was launched in April 1974.[*] The vessel passed a comprehensive in-flight testing; however, it unexpectedly did a ballistic instead of controlled reentry. The second unmanned flight of Soyuz-M prototype called Kosmos-672 flew in August 1974. The flight went flawlessly (Ref. 90, pp. 198–202). Finally, in December 1974, the Soviets flew a six-day-long Soyuz-16 mission, which was called by the Soviet press "a rehearsal for the joint flight of Soviet and American cosmonauts." It was piloted by the primary backup crew for the ASTP flight and was equipped with the same docking equipment to be used during the Apollo–Soyuz joint mission.[137] The flight was called "a total success" and the mission commander, Anatoly Filipchenko, reported after their flight that "it is possible to proceed with the docking with Apollo without any doubts."[132] After the 1971 Soyuz-11 tragedy, the Soviets also doubled up on pressure seals and valves, so that the Soyuz had the same type of safety margin enjoyed by U.S. spacecraft.[138] The overall conclusion made by NASA officials and astronauts was that "the Soyuz capsule has just as much safety built into it as does the Apollo."[139]

Another clear indication of Soviet determination to make the ASTP flight a success was their plan to assemble and count down two complete space

---

[*]All Soviet unmanned Soyuz-type and some other types of automated spacecraft were called Kosmos.

vehicles with cosmonauts onboard to ensure one would lift off on schedule on 15 July 1975. There were three reasons for this decision. In the beginning of technical discussions, the United States proposed to launch Apollo first and then rendezvous after Soyuz was in orbit. This proposal was dropped when Soviet specialists explained their A-2 booster (Soyuz launch vehicle) had less power than Saturn. As a result, Soyuz could not carry enough fuel to play the role of active partner in rendezvous and docking.[140] The second reason was that if anything went wrong with the booster, the Soviets would have known it before the Apollo launch. The Apollo liftoff would have been postponed until the Soviets prepared their second booster for the launch. Finally, if the Apollo launch was postponed for some reason, Soyuz, because of its limited turnaround time (approximately five days), could not wait for very long for its American counterpart. Thus, most likely the Soviets would need to launch another Soyuz to be able to accomplish rendezvous and docking with a delayed Apollo.

Two crews—a primary and a backup—were assigned to each Soyuz vehicle, totaling four crews and eight cosmonauts prepared for ASTP.* The Apollo main crew also had a backup.† However, because the Apollo crew consisted of three crewmembers, and there was only one Apollo vehicle prepared for rendezvous and docking with the Soviets, the total number of astronauts trained for ASTP was six.

---

*Soyuz main crew: 1) *Alexei Leonov* at the time of ASTP flight was a Lieutenant Colonel in the Soviet Air Force (he is currently a Major General, retired). He was born 30 May 1934, in Listvyanka, Altai Region. He graduated from the Chuguyev Higher Air Force School in 1957, and from Zhukovski Air Force Engineering Academy in 1968. He was the copilot of Voskhod-2 spacecraft in 1965 (a mission commander was Pavel Belyaev), and became the first man to perform extravehicular activity in space.

2) *Valery Kubasov* is a civilian. He was born 7 January 1935, in Vyazniki, Central Russia. He graduated as an engineer for aircraft design from the Moscow Aviation Institute in 1958. He was the back-up flight engineer for Soyuz-5 and flight engineer on Soyuz-6 in 1969.

†Apollo main crew: 1) *Thomas Stafford* at the time of the ASTP flight was a Brigadier General in the U.S. Air Force (he is currently a Lieutenant General, retired). He was born 17 September 1933, in Weatherford, Oklahoma, and has a B.S. from the United States Naval Academy (1952). He was selected in the second group of astronauts in 1962. Stafford was the backup pilot on Gemini 3, the pilot on Gemini 6, the command pilot on Gemini 9 upon the death of a prime crewmember, the backup spacecraft commander on Apollo 7, and the spacecraft commander on Apollo 10. He was also chief of the Astronaut Office from August 1969 to May 1971 and was subsequently named deputy director for Flight Crew Operations, NASA Johnson Space Center, a position that he held until February 1974.

2) *Donald Slayton* was a Major in the U.S. Air Force at the time of his selection by NASA as one of the original seven astronauts in 1959. He was born in Sparta, Wisconsin, on 1 March 1924 and received a B.S. (aeronautical engineering) from the University of Minnesota (1949). He was chosen as the command pilot for the Mercury-Atlas 7 flight but was removed due to the discovery of a heart murmur. He resigned his commission as an Air Force Major in November 1963, but continued as an active member of the astronaut team, becoming director of Flight Crew Operations, a position that he held until February 1974. Slayton was returned to flight status in March 1972.

3) *Vance Brand* is a civilian. He was born 9 May 1931 in Logmont, Colorado. He received a B.S. (business) in 1953 and a B.S. (aeronautical engineering) in 1960 from the University of Colorado. He received an M.B.A. from the University of California at Los Angeles (1964). He became an astronaut in 1966 with the fifth group and was backup command module pilot of Apollo 15, as well as backup commander for Skylab 3 and Skylab 4 (see Ref. 41, pp. 30–31; Ref. 141).

NIXON'S COMMITMENT TO THE SUCCESS OF ASTP

One of the factors that helped some U.S. space officials and politicians to overcome their skepticism toward ASTP was the strong presidential commitment to the success of the project. In January 1974, President Nixon in his State of the Union Address mentioned that Americans were "moving full speed ahead" in their plans "for a joint space venture with the Soviets in 1975."[142] In his Report on Aeronautics and Space Activities made in April 1974, Nixon noticed that "notable progress has...been made with the Soviet Union in preparing the Apollo–Soyuz Test Project scheduled for 1975."[143] Later President Ford confirmed the full support of the White House to ASTP. He said that the joint Apollo–Soyuz project was not only a "tremendous technological achievement, but is far broader in its implications as far as the world is concerned.... The broader we can make our relationship, the better it is for us in America and for our friends in the Soviet Union."[144]

The political significance of ASTP was recognized even by those U.S. politicians who had expressed certain doubts in the technological and scientific validity of the joint flight. Senator Moss, who was insisting on the development of the program of "meaningful experiments for U.S. flight," expressed his expectation that

> the interchange of scientific and technological information will continue to contribute to the general détente between Russia and the United States in political and cultural fields. Cooperation between the two powers, even in science, is a dramatic departure form the atmosphere of mutual distrust that has marked our relations during the past half-century.[107]

Representative Teague, one of the most arduous critics of ASTP, acknowledged that "the joint space flight must be viewed chiefly as a political effort—a demonstration of the potential of détente...."[120]

ASTP-RELATED CONCERNS EXAGGERATED

There was another factor that moderated the opposition of some of the U.S. public, space specialists, and politicians to the Apollo–Soyuz flight. As the project approached its realization, it became obvious that some of the ASTP-related concerns were exaggerated.

TECHNOLOGY TRANSFER

The first concern was technology transfer. As was explained in a letter written by Gerald Griffin, assistant administrator for Legislative Affairs, to Senator Bob Packwood,

The Soviets are not given details of our space program. The ASTP is carefully structured to avoid the transfer of technology. Both parties are independently building their own systems to meet design considerations which are being jointly developed. The Soviets have made substantial inputs to common designs, particularly of the docking mechanism. While we exchange such information about the characteristics of our systems as is necessary for successful conduct of the flight test, these limited and well balanced exchanges exclude any transfer of manufacturing know how.[145]

One of the examples of a nontransfer of technology was the American RCA system—a "built-in insurance against the failure of the rendezvous."[146] U.S. officials insisted that the system that worked so well in the lunar landing and Skylab programs be used in the joint mission. Three sets of the $300,000 passive radar systems that were supposed to be installed in Soyuz had been shipped to the Soviet Union. Officials were not worried about giving away any secrets: the active part of the system that did most of the work had been installed on the Apollo.[146] Probably the only two visible effects that cooperation with Americans had on the Soviet space program and facilities were the reconfiguration of "the physical appearance of their mission control center to almost duplicate the...NASA mission control in Houston,"[147] and the use by the Soviets of a neutral buoyancy water tank to train their cosmonauts for space walks "after having observed this to be an effective method at the Marshall Space Flight Center in the early 1970's" (Ref. 148, p. 77). Summarizing a long discussion about whether the Soviets really grabbed U.S. technology or not, Deke Slayton, one of the Apollo crewmembers said the following:

> Let's get rid of this crap once and for all that the Russians were stuffing our technological secrets under their sweaters. We did not transfer carloads of technical data to them. You had to understand that the engineering of their Soyuz was 20 years old. But you also had to keep in mind that Apollo was hardly new; hell, it was a 10-year-old ship by then. We had already developed entirely new technologies for our Space Shuttle. What the Russians learned from us—if they learned anything at all—it was our system of management. That's where we really shone, and they recognized our way of doing things.[149]

COST OF THE MISSION

Another concern was the cost of the mission. Some U.S. space officials believed that "if the mission covered in the agreement [Nixon–Kosygin] is any indication of future cooperation, the U.S. will be upping its space expenditures." The reason for this sort of conclusion was that "not only is the U.S. to conduct the longest and most difficult part of the mission, but it is also committed to developing the docking collar for the mission as part of its lion's share expenditure...."[150] However, modifications introduced by the Soviets in the design of Soyuz spacecraft to make it able to dock with Apollo and to enable crew transfers, "along with two unmanned tests of the Soyuz craft and the complete rehearsal flight of Soyuz 16, convinced American space officials that the Soviet spending on the project at least equaled the American of $250 million."[151]

SOVIET SECRECY

One of the factors initially contributing to U.S. skepticism about ASTP was Soviet secrecy. This secrecy raised doubts about the reliability of the Soviet space hardware, and it failed to correspond with the U.S. ideal of détente and cooperation. However, as the Americans worked with the Soviets, it became obvious to them that much of the Soviet secrecy could be blamed not on the unwillingness of the Soviet officials and specialists to release the information to their American counterparts but on the bureaucracy of the Soviet Union. Konstantin Bushuyev frankly admitted that because of the Soviet inertia he did "have a problem in meeting commitments on documentation and providing replies to specific questions and requests for amplifying information...."[134]

Ultimately the Americans understood the magnitude of the bureaucratic problems in the Soviet Union. When Representatives Larry Winn and Olin Teague took a trip to the Soviet Union in August 1972, they asked to see Star City, the cosmonaut training center, and some of the tracking stations in the Soviet Union. Their request was denied. However, later Teague acknowledged, "It wasn't quite fair how we did it. We went over without telling a soul. I am sure if we had gone through the right channels, we would have been shown everything." Space Subcommittee Staff Director Jim Wilson, who accompanied Teague and Winn to Moscow and made the advanced arrangements, indicated that if Ambassador Anatoly Dobrynin had been in Washington, D.C., at the time, "he is sure that the Soviets would have arranged to show the committee members almost everything they wanted to see" (Ref. 101, p. 416). When Congressmen Camp and Winn went to the Soviet Union approximately a year after Teague and Winn's trip, they noted the difference in cooperative spirit among the Soviets, in contrast to Teague and Winn's experience in 1972. Referring to Alexei Leonov, the Soviet commander of Soyuz, Camp and Winn said the following:

Mr. Camp. When we were in the Soviet Union, they were very, very cooperative. We spent a day with Leonov, who briefed us, and you could not ask for a better association. Would you agree, Mr. Winn?

Mr. Winn. No doubt about it, compared to about a year before that (Ref. 101, p. 422).

The importance of the issue of openness in ASTP consisted not only in the need to ensure the safety of the mission but to promote détente as well. This issue became an example of a really close working relationship between representatives of the U.S. and Soviet space communities, who were willing to learn each other's cultural and political differences to successfully accomplish their joint project. Glynn Lunney, ASTP manager at NASA Johnson Space Center, said that it was not unusual for the Soviets not to inform the U.S. space agency about space failures. When NASA ran into trouble with Skylab, Lunney pointed out that the American space officials didn't "pick up the phone and call Russia."[152] NASA also declined to get involved in proposing a project for Soviet and U.S. youths to exchange correspondence on the ASTP and visits to the launch sites in the United States and the Soviet Union at the time of the joint missions. The refusal was made on the ground that it "would fuel Soviet suspicions that we are involved in ulterior political and public affairs purposes."[153]

NASA's search for mutual understanding and trust with its Soviet counterparts became especially evident in the light of public criticism against the agency's policy not to ask the Soviets about their space hardware problems "unless it concerned a major hang-up in one of the Soviet systems." This criticism was expressed by an article in *The Huntsville Times*:

> If the U.S. space agency wants the support of American tax-payers— and obviously it cannot long operate without that support—then it should follow a consistent, unvarying policy about a full and free flow of information. A science project is not, you know, an affiliate of CIA— or at least we hope, it isn't.[154]

However, in the end, those who expected ASTP to make both the Soviet space program and Soviet society more open had good reason to be satisfied with the results. An example demonstrating that the Soviets were "keeping NASA fully informed about any matters concerning Soyuz relative to the safety of the mission" was that the Soviets had revealed that the automatic system in the docking phase of the Soyuz-15 with the Salyut station had been deviant. The Soyuz crew could not dock with the Salyut automatically and no manual docking was attempted because, according to the Soviets, it was not relevant to the mission. With respect to the failure of Soyuz-11 that resulted in the death of its crew, the Soviets "have been fully and frankly responsive to the request from NASA on the causes of that space accident and what had been done to correct these."[155]

RESULTING OPENNESS FROM ASTP

Two significant breakthroughs in furthering the spirit of openness and cooperation happened during preparation for the ASTP flight. The first was the agreement that representatives from each country should be present in the other's mission control center during the flight. The presence of U.S. personnel at the Soviet launch site at the time of launch had also been agreed to (Ref. 41, p. 24). The second was the massive advance mass media coverage of the flight, which was especially striking with the usual secrecy of Soviet space shots. This was the first time that Moscow had announced a shot in advance, and the first time that a Soviet launch had ever been shown live on Soviet television. According to an article in *The Christian Science Monitor*, ASTP could definitely take credit for this, taking into consideration that the

> unaccustomed abundance of information is a necessary price the Soviets had to pay to cooperate with the U.S. in the venture. Since details would be made public by the American side in any case, Moscow preferred to announce the plans itself rather than have its citizens find them out from the Voice of America.[156]

As for allowing access to the U.S. mass media to access ASTP-related information, the Soviets for the first time opened a press center where American newsmen were supposed to be registered to cover a USSR space mission. The Soviets also for the first time agreed to release a launch video in real time for viewing in the West, and to release in the West real-time air-to-ground transmissions between their control center and spacecraft.[157]

Although by the time of the Apollo and Soyuz liftoff, ASTP had already brought some promising political results, but its instrumental significance would have been significantly devalued if the project had ended in failure. Very few people would have believed, in that case, that the project had not been just a technologically meaningless political game, involving the risk of human lives. The ultimate test of ASTP took place on 15 July 1975, when Alexei Leonov (mission commander) and Valeri Kubasov (flight engineer) in the Soyuz and Thomas Stafford (mission commander), Donald Slayton (the docking module pilot), and Vance Brand (command module pilot) in the Apollo were launched into orbit. On 17 July the two spacecraft docked. They flew together in docking configuration for four days. During this period the crews performed joint experiments and transferred from one vehicle to another. While in orbit, they were greeted by General Secretary Brezhnev and President Ford. Both leaders stressed the importance of the flight for détente and expressed their hopes that ASTP would be a forerunner of future cooperative projects between the two countries. On 21 July the American and Soviet spacecraft separated, and Soyuz made its landing on the same day. The crew of Apollo performed experiments in space for three more days and then made its splashdown on 24 July—this was the end of ASTP.

POST-ASTP: RAISED AND FALLEN EXPECTATIONS

Having reviewed Soviet–American space relations during 1976–1980, the authors of a 1976–1980 report on the Soviet space program suggested three generalizations: "namely, the primacy of politics in determining the limits of cooperation; the prevailing sense of realism wherein advantages of cooperation are not foreclosed; and the demonstrated willingness of the Soviets to cooperate in some areas" (Ref. 158, p. 6).

Despite the fact that scientists regarded the Apollo–Soyuz mission as "a real sacrilege" that "had nothing to do with science" (Ref. 7, p. 174). ASTP played an important role in the promotion of Soviet–American cooperation in space in addition to its contribution to détente. It proved a practicability of pooling the efforts of the two countries in the most ambitious and complicated field of space exploration—manned space flights. It also gave Soviet–American space relations "the momentum" that U.S. and Soviet space scientists and officials tried to maintain.[159] They clearly indicated their willingness to continue, and eventually to broaden, the cooperation. Major General Vladimir Shatalov, chief of Gagarin Cosmonaut Training Center (The Star City), expressed his hope that Apollo–Soyuz would not be the last joint experiment and that Soyuz backup crews trained for the ASTP mission would prove useful later.[160] Wernher von Braun went even further by envisaging at least four main Soviet–American joint ventures that could result from ASTP:

> An international space station, jointly manned by astronauts and cosmonauts, to succeed our Skylab and the Russians' Salyuts;
> A manned base on the moon including an astronomical observatory and a complex of long habitable structures, to be established by the U.S., the USSR, and Europeans. Our coming Space Shuttle could serve as a launch vehicle;
> An unmanned mission to Mars to return samples. The Russians have successfully done it on the moon. The attempt might possibly involve remote-controlled roving vehicles, also successfully demonstrated on the moon by Soviet Lunokhods. NASA has studied a conceptual design and operating characteristics for future Viking borne Mars rovers;
> A manned expedition to Mars. It may be too costly for any single nation, but entirely feasible in the future as an international super-adventure.[161]

Preliminary talks about another joint flight started even before the ASTP blasted off. Senator Moss believed that "the fact that post-ASTP plans are

already being discussed would seem to indicate that this spirit of cooperation will be a part of future joint missions."[162] However, this time Soviet
and American space officials were more concerned about the tangible technological and scientific results of the cooperation. "We said we saw no
need to repeat Apollo–Soyuz and they said they could not fly an
Apollo–Salyut in any reasonable length of time," said George Low.[163] Soon
after the ASTP flight, Boris Petrov, head of Intercosmos, the Soviet space
cooperation agency, expressed the true reason for the Soviet refusal to consider Apollo–Salyut. Petrov rejected any possibility of having an American
astronaut on a Soviet space station. Any such agreement must be recipro-
cally beneficial, and with no American manned missions scheduled, there
could be no reciprocal exchange for a Soviet cosmonaut.[164] Overall, both
sides appeared "to believe that whatever happens next is some years away,
and that agreeing on the right mission is more important than agreeing on
something quickly...."[162] Nevertheless, the project discussed most was
building compatible docking equipment into the Space Shuttle and the
Soviet orbiting space station, Salyut. The Shuttle clearly presented a less
expensive way for the Soviet Union to deliver men and material to their
space station.

In addition to the momentum developed during ASTP, there were three
more factors contributing to the maintenance of U.S.–Soviet space cooperation in 1975–1980. The first two—"the continued Soviet progress in space in
contrast with an apparent lagging U.S. effort" and "retrenchment in U.S.
space activities after the conclusion of the Apollo program"[158]—increased
U.S. interest in cooperation with the Soviets. The use of Soviet Kosmos biological satellites, for example, was the only U.S. life sciences flight opportunity between ASTP and the Shuttle.

The third factor was related to the Soviet foreign policy goals. After
ASTP, the party leadership, heads of the military–industrial complex, and the
space industry realized that manned flights could bring considerable international benefits (Ref. 7, p. 228). At some point Soviet space officials were
almost pushing their American colleagues to make a commitment for another joint manned flight even though the latter intended "to pursue a slower,
more cautious course in these discussions."[165] Brezhnev was portrayed as the
political architect of the ASTP mission. His words about the planet being big
enough to live on but too small for wars, and his phrase "détente in space is
irreversible," had been repeated tirelessly.[164] The strong political support of
the Soviet leadership to Soviet–American cooperation in space was one of
the reasons why space negotiations proceeded successfully despite setbacks
in arms negotiations.[166] On the 16th anniversary of Gagarin's flight,
Krasnaya Zvezda [The Red Star] newspaper, an official press organ of the
USSR Ministry of Defense, reported in 1977:

Consistently implementing the program of further struggle for peace, international cooperation, freedom and the independence of peoples advanced by the 25th CPSU Congress, the Soviet Union is seeking to achieve increasingly fruitful results from peaceful and business-like cooperation between states on a mutually advantageous basis. Space research presents particularly broad opportunities in this respect. It is capable of uniting the efforts of different countries for the sake of solving the large tasks facing the whole of mankind to their mutual benefit. This has been graphically demonstrated by the successful implementation of the joint Soviet–American Soyuz-Apollo flight and by cooperation with scientists of France, Sweden, India and other countries.[167]

Talks on future joint space ventures to follow up the Apollo–Soyuz manned orbital mission began in 1975. They resulted in the agreement between NASA and the USSR Academy of Sciences, signed 11 May 1977, and designed "to provide continuity of the joint technical, scientific and operational capability developed through the highly successful Apollo–Soyuz rendezvous and docking mission conducted in July 1975." Two joint working groups (JWGs) were set up to "define projects which might benefit from the flexible delivery capability and large capacity of the American Space Shuttle and the capability for longer stay-time in orbit represented by the Soviet Salyut."[168] In their studies, the JWGs were supposed to proceed on the assumption "that the first flight would occur in 1981" (Ref. 158, p. 393).

Overall, Soviet and American space professionals came up with three variants for performing joint research in space using the Salyut space station and the Space Shuttle. The first one was docking the Shuttle to Salyut to perform joint research in the docked configuration. The second was to have the Shuttle drop off items for experiments to be conducted onboard the Soviet space station and then returned to Earth. The third variant envisaged the Shuttle and Salyut performing joint research but without docking to each other at all.[169]

The NASA–USSR agreement was codified at a high political level two weeks later, on 18 May 1977 in Geneva, when Soviet Foreign Minister Andrei Gromyko and Secretary of State Cyrus Vance renewed a bilateral U.S.–Soviet agreement on space cooperation. The agreement was to become effective on 24 May 1977 and remain in force for five years. It established the basis for Soviet–American space cooperation through the early 1980s. It was also "a very important political instrument because it insured continuity in Soviet–American space relations" (Ref. 158, p. 217).

A trend toward the search for more concrete results in cooperation, which was indicated after the successful realization of ASTP, became even more obvious after the signing of the agreements. U.S. and Soviet space specialists acknowledged that "any joint program must be based on a sound science and technology base." The JWGs were "looking to a program approximating 50-50 split of resources" to "make recommendations to the NASA Administrator or the Soviet Academy as to whether there is a useful program" (Ref. 158, p. 219).

The groups were supposed to make similar recommendations by October 1979. However, this never happened for the reason anticipated by *The Washington Post* correspondent Thomas O'Toole in 1972, soon after the signing of the Nixon–Kosygin agreement. He wrote, "In the long run the biggest barrier is the relations between the two governments."[170] Maybe this is why the Gromyko–Vance agreement incorporated all of the major points of the NASA–Academy of Sciences agreement but one—it did not specify a possible date of the first joint flight. Probably politicians more than scientists were aware of the coming problems in Soviet–American relations, which could delay or even cancel the realization of the project.

SUBORDINATION OF JOINT SPACE EXPLORATION TO POLITICS

In the late 1970s several political factors arrested "the momentum toward final agreement on the Shuttle/Salyut mission." Among these factors were human rights issues and the U.S. establishment of relations with China. The Soviet Union retaliated by delaying momentarily the conclusion of the SALT II agreement. However, final U.S. approval of the treaty, after the formal signing in June 1979, became doubtful with the publication of official evidence of a Soviet brigade in Cuba (Ref. 158, p. 219).

From the U.S. perspective, certain Soviet policies violated the spirit of détente as well: the resumption of antisatellite (ASAT) testing in 1976, the tightening of the Soviet policy on Jewish emigration, the stationing of SS-20s missiles in Europe, an alleged violation of the Biological Weapons Convention in Sverdlovsk in 1979, and finally the Soviet invasion of Afghanistan in December 1979 (Ref. 171, p. 277).

All of these factors had a negative impact on U.S.–Soviet relations in the field of science and technology. In early 1980 the U.S. National Academy of Sciences canceled workshops, seminars, and symposiums with the Soviet Union for at least six months in protest of the banishment of Nobel laureate Andrei Sakharov.[172] Then the Soviet government called for a halt in the development of the American Space Shuttle—a spacecraft that was supposed to dock with the Soviet Salyut station. The Soviets viewed the Shuttle as an antisatellite weapon, which because of its maneuverable arm should have been banned within the framework of an antisatellite treaty.[173] Following the Soviet invasion of Afghanistan, the United States deferred or put off all major space undertakings with the Soviet Union, including joint working groups' meetings and large-scale projects.[174] Finally, in May 1982, President

Ronald Reagan's decision to retaliate for the declaration of martial law in Poland by refusing to renew the 10-year-old agreement on Soviet–American space cooperation effectively doomed, for at least the time being, any hopes for the resumption of the mutually beneficial interaction in space.

DEVELOPMENTS LEADING TO THE RENEWAL OF SPACE COOPERATION

The next five-year period in the history of Soviet–American space relations, which coincided with Reagan's presidency, finally ended in the renaissance of Soviet–American space cooperation in the second half of the 1980s. This period was distinguished by nine developments.

RETURN TO THE COLD WAR

The first was a return to the Cold War, which did not favor any kind of cooperation, particularly in space. The worsening political climate had especially severe effects in the area of manned programs and space applications, largely because of the greater operational involvement required by both sides for a meaningful program. By 1985, there were no ongoing bilateral projects in this field (Ref. 171, p. 286).

REAGAN'S ATTITUDE TOWARD SPACE EXPLORATION

The second development was the president's attitude toward space exploration. He regarded it basically as a means of achieving political goals. "A president who normally touted large budget cuts, supported a relative increase for NASA—especially for its keys problems—because this continued to send a message to the Soviets about American superiority in space" (Ref. 175, pp. 244–245).

The subordination of space exploration to politics encouraged Soviet–American cooperation in space when Reagan shifted his approach toward the Soviet Union and decided to use space as a means of rapprochement between the two countries. This happened toward the end of his first presidential term. The change was due in part to a tight fiscal environment that made it difficult to compete against the Soviets in both military and civilian space programs.

U.S. POLITICIANS ENDORSE COOPERATION

The third development was the lawmakers' attitude toward U.S–Soviet space relations. Because "at the root of public officials' choices about the space program were political calculations about what they could support on behalf of American taxpayers,"174 the new attitude of the president toward cooperation in space with the Soviets was endorsed by Congress. This would have helped the legislature reach two goals. The members of Congress "supported cuts for NASA in favor of funding domestic programs" (Ref. 175, p. 245), and "renewed cooperation in space could save billions of dollars as space science and exploration lead...into a new age of awesome potential."176 However, Congress was concerned about the possible extension of the arms race into space and was considering space cooperation as a substitute for this race. Representative George Brown, a member of the House Space Subcommittee, urged NASA in February 1984 to invite the Soviet Union to participate in the development of the new U.S. space station, which he said would be an important confidence-building measure.177 Eleven members of the House joined later with Representative George Brown and Senator Paul Tsongas in participating in the new Coalition for the Peaceful Uses of Space, which was seeking a test moratorium on ASATs and a negotiated ban on ASATs and a funding cap on the Department of Defense's 1985 fiscal year strategic defense programs aimed at or based in space.178 In May 1984 Representative Mel Levine told the International Security and Scientific Affairs Subcommittee of the House Foreign Affairs Committee that the world was standing at the critical point of either proceeding with an arms race in space or cooperating in space for peaceful purposes and if no steps were to be made, "we could pass the point of no return."179 From the view of the members of Congress, the potential U.S.–Soviet joint projects could have involved an unmanned mission to Venus, an unmanned mission to the moon, manned and unmanned missions to Mars, a defense against a possible asteroid collision (Project Spacewatch), Shuttle–Salyut flights, and a manned lunar base. On 10 October 1984, the Senate proceeded to consider "the joint resolution (S.J. Res. 236) relating to cooperative East–West ventures in space as an alternative to a space arms race," which had been reported from the Committee on Foreign Relations, with amendments as follows:

> Resolved by the Senate and House of Representatives of the United States of America in Congress assembled, that the president should—
> 1) Endeavor, at the earliest practicable date, to renew the 1972–1977 agreement between the United States and the Soviet Union on space cooperation for peaceful purposes;
> 2) Continue energetically to gain Soviet agreement to the recent United States proposal for a joint simulated rescue mission;
> 3) Seek to initiate talks with the Government of the Soviet Union, and with other governments interested in space activities, to explore [the] further opportunities for cooperative East–West ventures in space [as an alternative to an arms race in space,] including cooperative ventures in such areas as space medicine and space biology, [space rescue,] planetary science, manned and unmanned space exploration....[180]

The Senate Joint Resolution 236 was signed into law by Reagan on 30 October 1984. Although he found "portions of the language contained in the preamble to the Joint resolution very speculative," he repeated his desire to increase contacts with the Soviets and confirmed the readiness of the United States "to work with the Soviets on cooperation in space in programs which are mutually beneficial and productive."[181]

However, even earlier in January 1984, Reagan made the first step toward resumption of more formal and high-profile space cooperation with the Soviets. Just a few days before his State of the Union address, in which he invited U.S. friends and allies to participate in the construction of a space station, the United States privately proposed to the Soviets the idea of a simulated space-rescue mission involving the U.S. Space Shuttle and the Salyut-7 space station. In the proposed space-rescue demonstration, the American Shuttle was supposed to rendezvous with the Soviet station. One or more U.S. astronauts would have flown to the Salyut wearing the jet-powered backpacks used on some of the Shuttle missions.[182] The next step would have been retrieving a Soviet cosmonaut and flying him back to the Shuttle. Publicly and privately, the Soviets did not express any enthusiasm about the idea, perhaps because of the perceived asymmetry of a mission in which the American spacecraft would simulate a rescue of cosmonauts from the Soviet station.[183]

SPACE OFFICIALS BACK COOPERATION

The fourth development was the highly politicized attitude of U.S. public officials toward the issues of space exploration and space cooperation and the influence this had on U.S. space officials. At the root of choices for the space officials were "political calculations about how best to sell technologically complex projects, the immediate benefits of which were not always obvious" (Ref. 175 p. 244). In spring 1986, the National Commission on Space made several recommendations to the White House, in particular to colonize Mars by 2027. A manned expedition to Mars, as the Commission advocated, was not in itself sufficient; automated spacecraft could explore Mars better. However, "a mission like a joint American–Soviet journey to Mars could help foster trust and divert the superpowers' rivalry to peaceful ends."[185]

IMPACT OF STRATEGIC DEFENSE INITIATIVE

Reagan's intention to adopt the Strategic Defense Initiative (SDI) was the fifth development. Together with the Cold War, SDI encouraged the Soviets to improve the general context of bilateral relations and prevent the militarization of outer space. *Sovetskaya Rossiya*, one of the official papers of the Communist Party and the Soviet government, mentioned in April 1983, a month after Reagan's SDI announcement, the experience of the seventies, when equal and mutually advantageous cooperation between the USSR and the United States in space and in other spheres of science and technology was bringing real benefits both to those directly involved in it and to all mankind.... Friendly handshakes in space between representatives of different countries and peoples must become landmarks in mankind's endless ascent to the stars.[186]

As for practical steps, in August 1983 the General Secretary of the Central Committee of the Communist Party, Yuri Andropov, announced a unilateral Soviet moratorium on launching ASATs into space. Although this move had nothing to do with a joint peaceful exploration of space, it created ultimately a climate of trust in which the renewal of Soviet–American cooperation in space became possible.

INTENSIFICATION OF SOVIET COOPERATION WITH WESTERN COUNTRIES

The sixth development was the considerable intensification by the Soviets in the second half of the 1970s and early 1980s of international cooperation in space, involving not only their Eastern European satellites, but also a number of Western countries. One of the reasons for this policy was explained by Nancy Lubin, a project director from the Office of Technology Assessment:

Soviet political leaders have consistently used their own space program not only to enhance cooperation, but to pursue other foreign policy objectives which are more competitive and more confrontational in nature; such as using space to enhance their own prestige and influence, portraying the United States as a threat to international peace, and deflecting attention from the military character of the Soviet space program onto that of the United States (Ref. 148, p. 44).

The Soviets' efforts generated fear in the United States of possible American isolation from international efforts in space. According to a panel of scientists convened by the Office of Technology Assessment in 1984, "the danger may lie in not cooperating." The panel members pointed out that the Soviets "have been improving their capabilities so rapidly—and have been pursuing non-U.S. partners so vigorously—that the Americans might one day find themselves isolated in space science."[187] The same situation could have been observed in the area of manned space flights. By the mid-1980s the Soviets had put nearly a dozen non-Soviets into orbit, whereas the United States took only one non-American into space.[188]

ADVANTAGES OF COOPERATION ARE CLEAR

Seventh, although U.S.–Soviet cooperation in space was significantly reduced, especially if compared with the mid-1970s, it was still maintained at a level high enough to show the advantages of cooperation. Besides "lunar and planetary studies, where one nation's data has helped the other to plan its subsequent spacecraft mission,"[187] and a mission to Halley's comet, the United States and the Soviet Union, together with Canada and France, had participated since 1979 in a highly successful satellite-aided research and rescue project, known as Cospas–Sarsat. This humanitarian program helped to save 475 lives by the mid-1980s, "vividly demonstrating the tangible benefits of international space cooperation" (Ref. 148, p. 24). However, one of the most productive and visible examples of this cooperation was the Cosmos unmanned biosatellite program. In 1974, the Americans were invited to fly on one of the Soviet series of Cosmos automatic spacecraft. The U.S. scientists participated initially in 1975 on Cosmos-782. Similar invitations were subsequently extended to the Americans each time that a Soviet biosatellite was planned. By 1985, a total of 36 U.S. investigations from 33 different universities and research institutes from 15 states had participated in these missions (Ref. 148, p. 102). Moreover, it turned out that the United States was benefiting more from Soviet–American space cooperation than the Soviets. According to Gerald Soffen, NASA's director of life sciences, the scientific information in the late 1970s was flowing only in the American direction, but Soffen hoped that the United States "will be able to return more with the Shuttle."[189] This assessment of U.S.–Soviet cooperation in space was later confirmed by Senator Spark Matsunaga:

> Especially in space medicine and biology the United States has learned a great deal from cooperation with the Soviets and could learn a lot more.... American scientists piggybacked into a Soviet biosatellite which proved so successful that our Government allowed American participation to continue even after the space agreement was terminated (Ref. 175, S 1276).

STRENGTHENING SOVIET CAPACITIES

The eighth development was that the strengthening of Soviet capacities in space helped eliminate concern on the U.S. side about the lack of scientific value of Soviet–American space cooperation and thus increased the interest of U.S. space scientists and officials to pool efforts in space with the Soviets. The report prepared by the Office of Technology Assessment in 1985 stressed that "Soviet technical capabilities are now excellent, especially in planetary and space science areas, and this can help stimulate cooperation."[190]

EFFECT OF THE ASTP EXPERIENCE

Finally, the overall positive experience that the Soviets and Americans had with ASTP became a real bridge for the two countries' joint future in space. According to Jim Nise, manager of the Russian programs office in the International Space Station program, the original plan of the reopening of Soviet–American cooperation in space was built around what the Soviet Union and the United States had established in the late 1960s and 1970s and what finally became ASTP. Moreover, in 1978 a NASA commission identified nine areas in which the Americans and the Soviets could go forward together, and that ultimately provided the foundation for the research program conducted during the Shuttle–Mir program in the mid-1990s.[169]

Thus, by the mid-1980s, a number of political and technological preconditions already existed for the renewal of broadscale space cooperation between the two countries. However, this kind of cooperation needed a final boost to become a reality. This boost was expressed in the words of NASA Administrator James Beggs in 1985:

> Of course, the political relationship between the United States and the Soviet Union will remain the most important factor determining the level and scope of cooperative space activities between our two nations. A key reason why U.S.-USSR space cooperation has not always yielded its full potential has been aggressive Soviet behavior which has undermined the political basis so necessary for fruitful partnership. During the past decade, there have been a number of Soviet political and military actions which have directly and adversely affected progress in U.S.-USSR space cooperation (Ref. 148, p. 30).

In March 1985, Mikhail Gorbachev was elected General Secretary of the Communist Party. His election ultimately marked the end of the Cold War and consequently, according to Beggs' predication, the resumption of meaningful cooperation between the Soviet Union and the United States.

CONCLUSION

At the end of the 1960s and the beginning of the 1970s, Soviet–American space cooperation finally started taking shape, which ultimately resulted in ASTP. This renaissance was the result of a number of factors that could be divided into two categories. The first includes a variety of U.S. and USSR domestic and foreign developments that finally brought the Cold War to its end and created a general cooperative environment in the bilateral Soviet–American relationship.

The second category consisted of factors related to changes in the space policies and space industries of the two countries. The development that underlay most of these changes was the end of the moon race. On the Soviet side, the weakening of the competitive drive behind the USSR space activities resulted in attempts to make up for the defeat in the moon race by seeking cooperation with the Americans in projects where both could be presented as equal partners. In addition, the Soviets hoped to gain docking experience from their American colleagues, in which the latter had advanced considerably further than the Soviets. The shift of balance in the leadership of the Soviet space program from those who actively supported competition with the United States in space to those who, at most, did not mind cooperating with them also contributed to space rapprochement between the two countries. Finally, General Secretary Brezhnev was portrayed as an architect of a joint mission that was designed to play an important role in Soviet plans to promote détente.

On the American side, there were also a few factors that ultimately contributed to the blossoming of U.S.–Soviet space relations. First, the American people had a decreasing interest in continued competition in space with the Soviets after the successful Apollo landings. Second, Nixon, unlike Kennedy or Johnson, was not a pro-space-race person and eventually wanted to save on space exploration. Third, Nixon realized that space cooperation could have a healing impact on the U.S.–Soviet relationship and started promoting it as an alternative to an arms race.

Although high-level political support in the United States to the joint project ultimately played an important role in the realization of ASTP, the first move toward pooling efforts with the Soviets in space was made by American space officials. The end of the moon race could have had a more destructive effect on the U.S. program than on that of the Soviets because the former was dependent on public support considerably more than the Soviet program. The flight with the Soviets could have again sparked the interest of the American people to the space venture. The joint mission had also reasonably high chances of enjoying support from the U.S. Congress, which could view the project as one of the areas of common interest with the Soviet Union. Besides, a Soviet–American mission could have used the remaining Apollo hardware, filled a gap in U.S. manned space flights, provided NASA with the time to define more fully its requirements for Shuttle-era subsystems, and built mutual confidence and trust to ensure future joint operations in space, including the creation of permanently manned space stations in the Earth's orbit and possibly a journey to Mars.

The successful development of U.S.–Soviet cooperation in space, driven by the mutual interests of the respective space communities, facilitated its formalization by the top leaderships of the two countries in the form of a Nixon–Kosygin space agreement signed in 1972. The agreement determined the final profile of the mission—a joint flight performed by Apollo and Soyuz spacecraft.

The decision-making part of ASTP prior to signing the Nixon–Kosygin agreement involved two kinds of questions: 1) what kind of mission to fly and 2) what kind of hardware to use. The decision implementation part of

ASTP, involving joint work on the realization of a concrete project resulted in unavoidable exposure of some of the previously hidden weaknesses of the Soviet space program to U.S. space officials and specialists, such as the relative backwardness and not very high safety record of Soviet space hardware. Also, contrary to U.S. expectations, the Soviets did not reciprocate the same degree of openness of the U.S. space program to Soviet specialists. Thus, the American politicians and space officials raised the following questions: Would the U.S. gains from the joint flight with the Soviets be considerable enough to justify the cost of the mission? Would it be safe to fly with the Soviets? What kind and volume of technology transfer would be acceptable? Does Soviet secretiveness correspond with the spirit of détente?

On the Soviet side, there was also a feeling of disappointment about cooperation with the Americans. First, the public comparison of U.S. and Soviet space technologies stressed the superiority of the former, undermining further the prestige of one of the greatest Soviet achievements. Second, the Soviets expected that ASTP would be followed by a joint flight of Soyuz with the space station Skylab, thus giving them more access to U.S. space technology and experience. However, the Americans refused to implement this plan, which diminished Soviet interest in the coming mission even further.

Despite all the questions and doubts on both the U.S. and Soviet sides concerning ASTP, the project ended in success for a number of reasons. First, it was the strong political commitment of the Soviet and U.S. leaderships to accomplishing the mission. Second, ASTP contributed to détente in terms of lifting, partly, the veil of secrecy over the Soviet space program and society in general. And finally, technology transfer and mission cost concerns on the U.S. side turned out to be exaggerated.

The post-ASTP period in U.S.–Soviet space cooperation was marked by several trends. First, Apollo–Soyuz proved practicability of joining efforts of the two countries in manned space flights and gave momentum to U.S.–Soviet space cooperation. Second, continued Soviet progress in space, in contrast with the apparent lag in U.S. efforts and retrenchment in U.S. space activities after the conclusion of the Apollo program increased the interest of U.S. space officials and specialists to cooperate with the Soviets. These tendencies strengthened technical and scientific rationales for the bilateral cooperation in space. Finally, Soviet leadership used international cooperation in space, with the United States in particular, to promote its foreign policy goals.

Negative political developments in the late 1970s and early 1980s in the overall U.S.–Soviet bilateral relationship significantly reduced U.S.–Soviet cooperation in space. However, the positive experience that ASTP provided the Soviet Union and the United States, and the intentions on both sides to explore possible future joint manned missions along with the continuing although limited U.S.–Soviet space cooperation, brought results tangible enough to prove beneficial. Among the benefits were the strengthened Soviet space capabilities and the attempts of Soviet and American leadership to use cooperation in space as a means of warming U.S.–Soviet relations. Finally, the U.S. fear of becoming isolated in space exploration, while the Soviet Union was increasingly developing its space ties with other countries, created preconditions for the blossoming of cooperation as soon as the overall context of U.S.–Soviet relations improved.

REFERENCES

[1]*Nashe Otechestvo*, Tom II [Our Motherland, Vol. 2], Terra, Moscow, 1991, p. 503.

[2]Gartoff, R. L., *Détente and Confrontation: American–Soviet Relations from Nixon to Reagan*, Brookings Inst., Washington, DC, 1994, pp. 9–29.

[3]Nixon, R., "I Shall Consecrate," *Washington Post*, 20 Jan. 1969. NASA Historical Reference Collection, NASA History Office, NASA Headquarters, Washington, DC.

[4]Gaddis, J. L., *Russia, the Soviet Union and the United States: An Interpretive History*, McGraw–Hill, New York, 1990, pp. 268–269.

[5]Oberg, J. E., *Red Star in Orbit: The Inside Story of Soviet Failures and Triumphs in Space*, Random House, New York, 1981.

[6]O'Toole, T., "Visiting Cosmonauts Assert Space Is for Science and Not for War," *Washington Post*, 24 Oct. 1969. NASA Historical Reference Collection, NASA History Office, NASA Headquarters, Washington, DC.

[7]Sagdeev, R., *The Making of a Soviet Scientist*, Wiley, New York, 1994.

[8]Stanford, N., "Space Cooperation? U.S. and Soviet Technical Exchanges Fan Hope," *Christian Science Monitor*, 23 Oct. 1970. NASA Historical Reference Collection, NASA History Office, NASA Headquarters, Washington, DC.

[9]Clarity, J. F., "American Space Experts Arrive in Soviet," *The New York Times*, 25 Oct. 1970. NASA Historical Reference Collection, NASA History Office, NASA Headquarters, Washington, DC.

[10]Clarity, J. F., "Soviet to Weigh U.S. Space Offer," *The New York Times*, 10 July 1970. NASA Historical Reference Collection, NASA History Office, NASA Headquarters, Washington, DC.

[11]Gwertzman, B., "U.S. and Russians Reach Moon Pact," *The New York Times*, 22 Jan. 1971. NASA Historical Reference Collection, NASA History Office, NASA Headquarters, Washington, DC.

[12]Mishin, V., "Pochemu My Ne Sletali Na Lunu" [Why We Did Not Fly to the Moon], *Cosmonautics/Astronomy*, Dec. 1990, pp. 27–28.

[13]Cassutt, M., *Who's Who in Space*, Macmillan, New York, 1993, pp. 260–261.

[14]Daniloff, N., *The Kremlin and the Cosmos*, Knopf, New York, 1972, pp. 166–167.

[15]O'Toole, T., "U.S.–Soviet Space Links Not Seen for Some Time," *Washington Post*, 30 Oct. 1970. NASA Historical Reference Collection, NASA History Office, NASA Headquarters, Washington, DC.

[16]Congress, Senate, Committee on Aeronautical and Space Sciences, *Statements by the Presidents of the United States on International Cooperation in Space: A Chronology, October 1957–August 1971*, No. 92-40, 92nd Congress, 1st sess., 1971.

[17]Hoff, J., "The Presidency, Congress, and the Deceleration of the U.S. Space Program in the 1970s," *Presidential Leadership and the Development of the U.S. Space Program*, edited by R. D. Launis and H. E. McCurdy, NASA Center for AeroSpace Information, Linthicum Heights, 1964.

[18]Krug, L. T., *Presidential Perspective on Space Exploration: Guiding Metaphors from Eisenhower to Bush*, Praeger, New York, 1991, p. 48.

[19]Leavitt, W., "Space: Now the Nixon Years," *Space Digest*, Dec. 1968, p. 73.

[20]Paine, T. O., "Meeting with the President," *Memorandum for the Record*, 22 Jan. 1970. NASA Historical Reference Collection, NASA History Office, NASA Headquarters, Washington, DC.

[21]"Priorities in Space," *Wall Street Journal*, 3 Dec. 1968. NASA Historical Reference Collection, NASA History Office, NASA Headquarters, Washington, DC.

[22]Logsdon, J. M., "Leadership, Preeminence, and U.S. Space Policy," Keynote Talk at NASA–American Univ. Symposium Presidential Leadership, Congress, and the U.S. Space Program, 25 March 1993.

[23]*Aeronautics and Space Report of the President, Fiscal Year 1991 Activities*, NASA, Washington, DC, 1991, p. 180.

[24]*U.S. Department of State Bulletin*, Vol. LXI, 11 Aug. 1969, pp. 111–112.

[25]*Public Papers of the Presidents of the United States: Richard M. Nixon*, U.S. Government Printing Office, Washington, DC, 1969.

[26]*Weekly Compilation of Presidential Documents*, April 1970.

[27]James A. Lovell, John L. Swigert, Fred W. Haise, Private Communication to Konstantin Feoktistov, 11 May 1970, ANSER/CIAC Library, Arlington, VA.

[28]*Defense/Space Daily*, 19 Sept. 1974, p. 88.

[29]Republican Coordinating Committee, *Choice for America: Republican Answers to the Challenge of Now*, March 1968, p. 424.

[30]*Space Business Daily*, 16 Dec. 1968, p. 198.

[31]Khrushchev, S., *Nikita Khrushchev: Krisisy e Rakety* [Nikita Khrushchev: Crises and Missiles], Novosti, Moscow, 1994, Vol. 1, p. 445.

[32]Leavitt, W., "On the President-Elect's Agenda—A Balanced Space Program for the 1970s," *Space Digest*, Nov. 1968.

[33]Hines, W., "Nixon Should Reveal Russian Space Data," *Philadelphia Inquirer*, 30 Dec. 1968. NASA Historical Reference Collection, NASA History Office, NASA Headquarters, Washington, DC.

[34]Gallup, G. H., *The Gallup Poll, Public Opinion, 1935–1971*, Random House, New York, 1972.

[35]Harvey, B., *Race into Space: The Soviet Space Programme*, Wiley, New York, 1988, p. 208.

[36]"Are U.S., USSR Inching Toward Space Cooperation?" *Space/Aeronautics*, June 1970, p. 17.

[37]Wilford, J. N., "U.S. and Soviet Space Trusts: A Contrast," *New York Times*, 14 April 1972. NASA Historical Reference Collection, NASA History Office, NASA Headquarters, Washington, DC.

[38]O'Toole, T., "U.S.–Soviet Joint Efforts in Space Seen," *Washington Post*, 6 April 1972. NASA Historical Reference Collection, NASA History Office, NASA Headquarters, Washington, DC.

[39]Oberg, J., "U.S.–Soviet Space Flight," *Los Angeles Times*, 30 Sept. 1973. NASA Historical Reference Collection, NASA History Office, NASA Headquarters, Washington, DC.

[40]Sullivan, W., "U.S. and Russia May Combine Space Rescues," *New York Times*, 9 Oct. 1970. NASA Historical Reference Collection, NASA History Office, NASA Headquarters, Washington, DC.

[41]Zegel, V. A., *Background and Policy Issues in the Apollo–Soyuz Test Project*, Congressional Research Service, Washington, DC, 1975, p. 4.

[42]Ezell, E. C., and Ezell, L. N., *The Partnership: A History of the Apollo–Soyuz Test Project*, NASA, Washington, DC, 1978.

[43]*Space Business Daily*, 6 July 1972, p. 18.

[44]*Space Business Daily*, 19 Sept. 1969, pp. 82–83.

[45]Frutkin, A. W., *Memorandum for A/Dr. T. O. Paine, Attachment B*, 31 July 1970, NASA History Office, Washington, DC.

[46]Lannan, J., "Shift in Space Goals Is Urged," *Washington Star*, 1 June 1969.

[47]Califano, J., "For the President's Night Reading," Memorandum, 2 Dec. 1968. NASA Historical Reference Collection, NASA History Office, NASA Headquarters, Washington, DC.

[48]*Space Business Daily*, 16 May 1968, p. 80.

[49]Paine, T., "Letter to the President," 7 Nov. 1969. NASA Historical Reference Collection, NASA History Office, NASA Headquarters, Washington, DC.

[50]Paine, T., "Letter to the President," 9 Jan. 1970. NASA Historical Reference Collection, NASA History Office, NASA Headquarters,

Washington, DC.

[51]Grigorovich-Barsky, S., "Poyezdka s Kosmonavtami Nikolayevym e Sevastyanovym 18–28 Oktyabrya 1970 [Travel with Cosmonauts Nikolaev and Sevastyanov]," *Memorandum*, Oct. 1970.

[52]*Agreement Between the National Aeronautics and Space Administration and the Academy of Sciences of the USSR, 29 October, 1970*, NASA Release No. 70-210, 9 Dec. 1970. NASA Historical Reference Collection, NASA History Office, NASA Headquarters, Washington, DC.

[53]Congress, House, Committee on Science and Astronautics, *U.S.–U.S.S.R. Cooperative Agreements: Hearing Before the Subcommittee on International Cooperation in Science and Space*, 92nd Cong., 2nd sess., 13–21 June, 1972.

[54]Dobrynin, A., interview by author, 28 Jan. 1996.

[55]"Cooperation in Space," *New York Times*, 2 Nov. 1970.

[56]Low, G. M., "International Aspects of Our Space Program," Speech at the National Space Club Luncheon, 26 Jan. 1971. NASA Historical Reference Collection, NASA History Office, NASA Headquarters, Washington, DC.

[57]*Space Business Daily*, 5 Oct. 1970, p. 145.

[58]O'Toole, T., "Space Cooperation Talks Start Today," *Washington Post*, 26 Oct. 1970. NASA Historical Reference Collection, NASA History Office, NASA Headquarters, Washington, DC.

[59]*Space Business Daily*, 11 May 1970.

[60]Kirkman, D., "Russians Deny Moon-Walk Plans," *Washington-Daily News*, 29 Oct. 1970. NASA Historical Reference Collection, NASA History Office, NASA Headquarters, Washington, DC.

[61]"Moscow Agrees to Space Talks," *Baltimore Sun*, 9 Oct. 1970. NASA Historical Reference Collection, NASA History Office, NASA Headquarters, Washington, DC.

[62]Sullivan, W., "Russians to Join Space Rescue Panel," *New York Times*, 9 Oct. 1970. NASA Historical Reference Collection, NASA History Office, NASA Headquarters, Washington, DC.

[63]Wilford, J. N., "5 NASA Officials to Visit Moscow," *New York Times*, 13 Oct. 1970. NASA Historical Reference Collection, NASA History Office, NASA Headquarters, Washington, DC.

[64]Mills, D., "U.S., Soviet End 1st Space Talks," *Baltimore Sun*, 29 Oct. 1970. NASA Historical Reference Collection, NASA History Office, NASA Headquarters, Washington, DC.

[65]*Summary of Results of Discussions on Space Cooperation Between the Academy of Sciences of the USSR and the U.S. National Aeronautics and Space Administration*, Moscow, 18–21 January 1971. NASA Historical Reference Collection, NASA History Office, NASA Headquarters, Wash-

ington, DC.

[66]"U.S.–USSR Space Meeting," NASA Release No. 71-9, 21 Jan. 1971.

[67]O'Toole, T., "U.S., Russia to Devise Single Space Linkup," *Washington Post*, 18 June 1971. NASA Historical Reference Collection, NASA History Office, NASA Headquarters, Washington, DC.

[68]Mills, D., "Space Docking Accord Expected," Baltimore Sun, 4 Dec. 1971. NASA Historical Reference Collection, NASA History Office, NASA Headquarters, Washington, DC.

[69]"Soviet–U.S. Parley on Space Canceled," *New York Times*, 23 May 1971. NASA Historical Reference Collection, NASA History Office, NASA Headquarters, Washington, DC.

[70]*Space Business Daily*, 23 June 1971.

[71]Maloney, J., "U.S., Russ Talk Joint Space Effort," Houston Post, 24 June 1971. NASA Historical Reference Collection, NASA History Office, NASA Headquarters, Washington, DC.

[72]*Space Business Daily*, 8 Feb. 1972.

[73]Strickland, Z., "Apollo-Salyut, Soyuz-Skylab Docking Eyed," *Aviation Week and Space Technology*, 13 Sept. 1971.

[74]Sedlazek, M. F., "Review of US/USSR Joint Rendezvous and Docking Meeting at MSC—July 2 1971," *Memorandum for the Record*, NASA History Office, Washington, DC.

[75]Sehlstedt, A., Jr., "U.S., Soviet Agree on 90 Percent of Compatible Space Plans," *Baltimore Sun*, 1 Sept. 1971. NASA Historical Reference Collection, NASA History Office, NASA Headquarters, Washington, DC.

[76]*Congressional Record*, 4 May 1972, E4690-1.

[77]"Space Talks Should Pay Dividends," Houston Chronicle, 6 Sept. 1971. NASA Historical Reference Collection, NASA History Office, NASA Headquarters, Washington, DC.

[78]Lyons, R. D., "U.S. Astronauts Expected to Use Salyut," *New York Times*, 1 Sept. 1971. NASA Historical Reference Collection, NASA History Office, NASA Headquarters, Washington, DC.

[79]NASA Release No. 71-244, 7 Dec. 1971.

[80]Press Conference of Vladimir Kirillin, Chairman, Committee for Science and Technology and Academician Boris Petrov on Agreement Concerning Cooperation In the Exploration and Use of Outer Space for Peaceful Purposes, 24 May, 1972. NASA Historical Reference Collection, NASA History Office, NASA Headquarters, Washington, DC.

[81]*Soviet Aerospace*, 5 March 1973.

[82]"Moscow and the Moon," *New York Times*, 23 March 1972. NASA Historical Reference Collection, NASA History Office, NASA Headquarters, Washington, DC.

[83]Harris, L., "The Harris Survey. Nixon Trip to Moscow Is Backed by

Public," *Washington Post*, 17 April 1972. NASA Historical Reference Collection, NASA History Office, NASA Headquarters, Washington, DC.

[84]"Soviets Urged Space Cooperation with U.S.," *Washington News*, 12 April 1971. NASA Historical Reference Collection, NASA History Office, NASA Headquarters, Washington, DC.

[85]"Burzhe—Den Kosmonavtov" [Bourgue Is the Day of Cosmonauts], *Pravda*, 6 June 1971.

[86]*Space Business Daily*, 9 Feb. 1972.

[87]*Space Business Daily*, 4 April 1972.

[88]Davis, P. O., "Keynote of the 1970s. Joint Ventures Into Space," *Air University Review*, Sept.–Oct. 1973, p. 18.

[89]*TRW Space Log 1957–1991*, Space and Technology Group, Redondo Beach, CA, 1991.

[90]Semyonov, Y., (ed.), *Raketno-Kosmecheskaya Korporatsiya "Energia" Imeny S.P. Korolyova* [Korolev Rocket-Space Corporation], RKK Energia, Moscow, 1996.

[91]"Friendly, Yes, But Trying to Be First," *New York Times*, 26 March 1972. NASA Historical Reference Collection, NASA History Office, NASA Headquarters, Washington, DC.

[92]*Space Business Daily*, 2 June 1972, p. 176.

[93]"Space Talks Should Pay Dividends," *Houston Chronicle*, 6 Sept. 1971. NASA Historical Reference Collection, NASA History Office, NASA Headquarters, Washington, DC.

[94]*Soviet Aerospace*, 24 April 1972.

[95]"Data from Soviet Withheld by U.S.," *New York Times*, 16 Dec. 1971. NASA Historical Reference Collection, NASA History Office, NASA Headquarters, Washington, DC.

[96]Press Conference of the Vice President; Dr. James C. Fletcher, Administrator, NASA; Glynn S. Lunney, Assistant to the Manager for Operational, Experiment and Government Furnished Equipment, NASA; and Dr. Edward E. David, Jr., Science Advisor to the President, 24 May, 1972. NASA Historical Reference Collection, NASA History Office, NASA Headquarters, Washington, DC.

[97]"News Conference on U.S./USSR Rendezvous and Docking Agreement," *NASA Press Release*, 24 May 1972. NASA Historical Reference Collection, NASA History Office, NASA Headquarters, Washington, DC.

[98]"Moscow and the Moon," *New York Times*, 23 March 1972. NASA Historical Reference Collection, NASA History Office, NASA Headquarters, Washington, DC.

[99]O'Toole, T., "U.S.–Soviet Efforts in Space Seen," *Washington Post*, 6 April 1972. NASA Historical Reference Collection, NASA History Office,

NASA Headquarters, Washington, DC.

[100]O'Toole, T., "Summit in Space: June 15, 1975," *Washington Post*, 7 May 1972. NASA Historical Reference Collection, NASA History Office, NASA Headquarters, Washington, DC.

[101]Congress, House, Committee on Science and Technology, *Toward the Endless Frontier: History of the Committee on Science and Technology, 1959–79*, U.S. Government Printing Office, Washington, DC, 1980.

[102]Congress, Senate, Senator Cook Speaking for the United States–Soviet Space Agreement, *Congressional Record*, 5 June 1972, S 8782.

[103]"Unity in Space," *New York Times*, 25 May 1972. NASA Historical Reference Collection, NASA History Office, NASA Headquarters, Washington, DC.

[104]Keatley, R., "U.S., Russia Sign Space, Science Accords; Talks Adjourn to Brezhnev Country Home," *Wall Street Journal*, 25 May 1972. NASA Historical Reference Collection, NASA History Office, NASA Headquarters, Washington, DC.

[105]"The Promise in Moscow," *Washington Post*, 26 May 1972. NASA Historical Reference Collection, NASA History Office, NASA Headquarters, Washington, DC.

[106]Moscow TASS International Service in English, 1110 GMT, 26 May 72. NASA Historical Reference Collection, NASA History Office, NASA Headquarters, Washington, DC.

[107]Congress, Senate, Keynote Address by Senator Moss to AIAA/ASME/SAE/Space Mission Planning and Executive Meeting, *Congressional Record*, 14 July 1973.

[108]*Business Space Daily*, 13 Sept. 1972.

[109]*Space Business Daily*, 9 Jan. 1973.

[110]Congress, House, Representative Crane Speaking for the Need to Reconsider the Apollo–Soyuz project, *Congressional Record*, 11 Dec. 1974, E 7070.

[111]Fialka, J., "Prestige and Propaganda are Stowaways on Space Shot," *Washington Star*, 25 March 1975. NASA Historical Reference Collection, NASA History Office, NASA Headquarters, Washington, DC.

[112]Pipp, E. G., "Space Spectacular: U.S.–Soviet Rendezvous," *Sunday News*, 1 Dec. 1974. NASA Historical Reference Collection, NASA History Office, NASA Headquarters, Washington, DC.

[113]von Braun, W., "We Get Set for Astronaut-Cosmonaut Space Linkup," *Popular Science*, Jan. 1975, p. 41.

[114]Press Conference of George M. Low, NASA Deputy Administrator, Academician Boris N. Petrov, Chairman of USSR Intercosmos Council, Alexis Tatistcheff, Interpreter, Jack King, PAO, 13 July 1972. NASA Historical Reference Collection, NASA History Office, NASA Head-

quarters, Washington, DC.

[115]Prime Crew Press Conference, Brigadier General Thomas P. Stafford, Vance D. Brand, Donald K. Slayton, Dr. Glynn S. Lunney, Technical Director, ASTP, 1 February, 1973. NASA Historical Reference Collection, NASA History Office, NASA Headquarters, Washington, DC.

[116]George M. Low to James C. Fletcher, *Memorandum*, 3 Oct. 1973. NASA Historical Reference Collection, NASA History Office, NASA Headquarters, Washington, DC.

[117]Congress, House, Committee on Science and Astronautics, *Status Report on Space Shuttle, Space Tug, Apollo–Soyuz Test Project*, 93rd Cong., 2nd. sess., Feb. 1974, p. 460.

[118]Congress, Senate, Senator Moss speaking on Apollo–Soyuz Project, *Congressional Record*, 5 March 1975, S 3168.

[119]Perry, A., "Doubt on Joint Space Flight...and on the Russian's Secrecy Fetish," *Los Angeles Times*, 30 June 1974. NASA Historical Reference Collection, NASA History Office, NASA Headquarters, Washington, DC.

[120]Congress, House, Representative Teague speaking on the Soviets keeping their space data, *Congressional Record*, 22 July 1974, E 4900.

[121]Congress, House, Representative Huber speaking on Soviet-United States cooperation turning into massive Soviet technology grab. *Congressional Record*, 30 April 1974, E 2619.

[122]*Space Business Daily*, 4 Dec. 1974.

[123]"Soviet Refuses to Lift Its Ban on U.S. Newsmen at Launching," *New York Times*, 12 Oct. 1974. NASA Historical Reference Collection, NASA History Office, NASA Headquarters, Washington, DC.

[124]"A Wheat Deal in Sky?" *Los Angeles Times*, 6 Nov. 1974. NASA Historical Reference Collection, NASA History Office, NASA Headquarters, Washington, DC.

[125]Rusher, W., "Just 'How Crazy Can We Get?" *San Francisco Chronicle*, 31 Oct. 1974. NASA Historical Reference Collection, NASA History Office, NASA Headquarters, Washington, DC.

[126]O'Toole, T., "2d Russian Space Shot Fails," *Washington Post*, 4 May 1973. NASA Historical Reference Collection, NASA History Office, NASA Headquarters, Washington, DC.

[127]Press Conference, ASTP Personnel, USSR, U.S., Johnson Space Center, Houston, Texas, 20 July 1973. NASA Historical Reference Collection, NASA History Office, NASA Headquarters, Washington, DC.

[128]Congress, Senate, Senator Proxmire speaking on the possible reevaluation of Apollo–Soyuz test mission, Congressional Record–Extension of Remarks, 9 April 1975, S 5527.

[129]"Russian Roulette' in Orbit?" *Chicago Tribune*, 8 July 1975. NASA Historical Reference Collection, NASA History Office, NASA Head-

quarters, Washington, DC.
[130]"NASA's Defense of ASTP," *Aviation Week and Space Technology*, 11 Nov. 1974, p. 9.
[131]"Thought We Were Partners," *Economist*, 13 July 1974. NASA Historical Reference Collection, NASA History Office, NASA Headquarters, Washington, DC.
[132]Konovalov, B., "Prazdnik v Zvyozdnom [Holiday in Zvyozdny]," *Izvestiya*, 11 Dec. 1974.
[133]*Soviet Aerospace*, 5 March 1973.
[134]Chester Lee to James Fletcher, *Memorandum*, 24 July 1973, NASA History Office, Washington, DC.
[135]Bushuyev, K., "Vzaimopomoshch v Kosmose [Mutual Aid in Space]," *Nauka E Zhizn* [Science and Life], April 1973, p. 6.
[136]*Soviet Aerospace*, 12 March 1973.
[137]Chriss, N. C., "Russ Plan 'Rehearsal' of Apollo–Soyuz Flight," *Los Angeles Times*, 13 September, 1974. NASA Historical Reference Collection, NASA History Office, NASA Headquarters, Washington, DC.
[138]Sallsbury, D. F., "U.S.–Soviet Space Plans Still Intact, NASA Says," *Christian Science Monitor*, 29 Aug. 1974. NASA Historical Reference Collection, NASA History Office, NASA Headquarters, Washington, DC.
[139]Salisbury, D., "Joint Flight Project Still Has Its Critics," *Christian Science Monitor*, 15 July 1975. NASA Historical Reference Collection, NASA History Office, NASA Headquarters, Washington, DC.
[140]Harris, G., "Apollo–Soyuz Summit in the Sky...Or, as They Say in Russia...," *Centennial Star—Florida Magazine*, 23 Feb. 1975. NASA Historical Reference Collection, NASA History Office, NASA Headquarters, Washington, DC.
[141]Hawthorne, D. B., *Men and Women of Space*, Univel, Inc., San Diego, CA, 1992.
[142]Congress, House, The State of the Union—Address by the President of the United States, *Congressional Record*, 30 Jan. 1974, H 377.
[143]Congress, Senate, Report on Aeronautics and Space Activities— Message from the President, *Congressional Record*, 8 April 1974, S 5346.
[144]"2 Russians Taken to Picnic by Ford," *New York Times*, 8 Sept. 1974. NASA Historical Reference Collection, NASA History Office, NASA Headquarters, Washington, DC.
[145]Gerald Griffin to Bob Packwood, 29 March 1974. NASA Historical Reference Collection, NASA History Office, NASA Headquarters, Washington, DC.
[146]Congress, House, Representative Huber speaking on the continuing of the great space technology grab, *Congressional Record-Extension of Remarks*, 8 Oct. 1974, E 6327.

[147]Congress, Senate, Committee on Commerce, Science, and Transportation, *Soviet Space Programs: 1976–1980 (with Supplementary Data Through 1983), Manned Space Programs and Space Life Sciences*, Rept. prepared at the request of Bob Packwood, Part 2, 98th Cong., 2d sess., U.S. Government Printing Office, Washington, DC, 1984, p. 456.

[148]Congress, House, Committee on Science and Technology, *Assess Potential Gains and Drawbacks of Civilian Space Cooperation with the Soviets: Hearings Before the Subcommittee on Space Science and Applications*, 99th Cong., 1st sess., 30–31 July 1985.

[149]Shepard, A., and Slayton, D., *Moon Shot: The Inside Story of America's Race to the Moon*, Turner Publishing, Atlanta, GA, p. 340.

[150]*Business Space Daily*, 2 June 1972.

[151]McElheny, V. K., "Behind the Flight: 5 Years of Joint Effort," *New York Times*, 13 July 1975. NASA Historical Reference Collection, NASA History Office, NASA Headquarters, Washington, DC.

[152]Casebolt, B., "Joint Flight Still on, Top Officials Insist," *Huntsville Times*, 30 Aug. 1974. NASA Historical Reference Collection, NASA History Office, NASA Headquarters, Washington, DC.

[153]Oscar Anderson to Arnold Frutkin and John Donnelly, 19 Sept. 1973. NASA Historical Reference Collection, NASA History Office, NASA Headquarters, Washington, DC.

[154]"Now, Tell Us Again," *Huntsville Times*, 31 Aug. 1974. NASA Historical Reference Collection, NASA History Office, NASA Headquarters, Washington, DC.

[155]Levy, L., *Down to Earth—A Woman's View of Space*, Rept. (Nov. 1974). NASA Historical Reference Collection, NASA History Office, NASA Headquarters, Washington, DC.

[156]Pond, E., "Soviets Hail Space Cooperation," *Christian Science Monitor*, 15 July 1975. NASA Historical Reference Collection, NASA History Office, NASA Headquarters, Washington, DC.

[157]Donnely, J. P., "ASTP Coverage," *Aviation Week and Space Technology*, 4 Nov. 1974, p. 62.

[158]Congress, Senate, Committee on Commerce, Science, and Transportation, *Soviet Space Programs: 1976–1980. Supporting Vehicles and Launch Vehicles, Political Goals and Purposes, International Cooperation in Space, Administration, Resource Burden, Future Outlook*, Rept. prepared at the request of Bob Packwood, 97th Cong., 2nd sess.

[159]Congress, House, Committee on Science and Technology, *U.S.–U.S.S.R. Cooperative Agreements in Science and Technology: Hearing Before the Subcommittee on Domestic and International Scientific Planning and Analysis*, 94th Cong., 1st sess., 18–20 November 1975, 280.

[160]*New Times*, No. 1, Jan. 1975, p. 22.

[161]von Braun, W., "What Could Apollo–Soyuz Lead To?" *Popular Science*, Jan. 1975, NASA Historical Reference Collection, NASA History Office, NASA Headquarters, Washington, DC.

[162]Congress, Senate, Senator Moss speaking on background and policy issues in the Apollo–Soyuz Test Project, *Congressional Record*, 5 March 1975, S 3174.

[163]O'Toole, T., "Détente's Space Spectacular: Apollo–Soyuz Linkup Thursday," *Washington Post*, 13 July 1975. NASA Historical Reference Collection, NASA History Office, NASA Headquarters, Washington, DC.

[164]Piper, H., "Fling with Détente Not Likely to Displace Space Rivalry," *Baltimore Sun*, 28 July 1975. NASA Historical Reference Collection, NASA History Office, NASA Headquarters, Washington, DC.

[165]"Soviets Urge Joint Mission Decision," *Aviation Week and Space Technology*, 5 Dec. 1977, p. 49.

[166]Ball, J., "Space Agreement Near: NASA," *Sentinel Star*, 12 April 1977. NASA Historical Reference Collection, NASA History Office, NASA Headquarters, Washington, DC.

[167]*Krasnaya Zvezda* [The Red Star], 12 April 1977.

[168]NASA, Soviets Agree On Further Manned Space Cooperation, NASA Release No. 77-98, 17 May 1977.

[169]Nise, T., "Space Station and the International Human Exploration and Development of Space," 1998 AAS National Conference and 45th Annual Meeting, 17 Nov. 1998.

[170]O'Toole, T., "Summit in Space: June 15, 1975," *Washington Post*, 7 May 1972. NASA Historical Reference Collection, NASA History Office, NASA Headquarters, Washington, DC.

[171]Moltz, J. C., "Managing International Rivalry on High Technology Frontiers: U.S.–Soviet Competition and Cooperation in Space," Ph.D. dissertation, Univ. of California, Berkeley, 1989.

[172]O'Toole, T., "U.S. Scientific Exchanges with Soviets Are Canceled," *Washington Post*, 26 Feb. 1980. NASA Historical Reference Collection, NASA History Office, NASA Headquarters, Washington, DC.

[173]Burt, R., "Soviet Said to Ask Space Shuttle Halt," *New York Times*, 1 June 1979. NASA Historical Reference Collection, NASA History Office, NASA Headquarters, Washington, DC.

[174]*Defense/Space Daily*, 3 March 1980.

[175]Ragsdale, L., "Politics Not Science: The U.S. Space Program in the Reagan and Bush Years," *Presidential Leadership and the Development of the U.S. Space Program*, edited by R. D. Launis and H. E. McCurdy, NASA Center for AeroSpace Information, Linthicum Heights, 1964.

[176]Congress, Senate, Senator Matsunaga speaking for East-West cooperation in space as an alternative to a space arms race. (S.J. Res. 236),

Congressional Record, 9 Feb. 1984, S 1277.

[177]*Defense Daily*, 3 Feb. 1984.

[178]*Defense Daily*, 3 April 1984.

[179]*Defense Daily*, 4 May 1984.

[180]Congress, Senate, Consideration of the joint resolution (S.J. Res. 236) relating to cooperative East–West ventures in space..., *Congressional Record, Attachment C*, 10 Oct. 1984, S 14018.

[181]Statement on Signing S.J. Res. 236 "Cooperative East–West Ventures in Space" into Law, *Administration of Ronald Reagan*, 30 Oct. 1984, 1691.

[182]O'Toole, T., "U.S., Soviets Plan '85 Space Mission To Demonstrate Astronaut 'Rescue,'" *Washington Post*, 16 Nov. 1984. NASA Historical Reference Collection, NASA History Office, NASA Headquarters, Washington, DC.

[183]U.S. Congress, Office of Technology Assessment, *U.S.–Russian Cooperation in Space*, U.S. Government Printing Office, Washington, DC, April 1995, OTA-ISS-618, p. 45.

[184]"NASA's Future in Space," *New York Times*, 4 April 1986.

[185]"How to Regain Face in Space," *New York Times*, 28 May 1986. NASA Historical Reference Collection, NASA History Office, NASA Headquarters, Washington, DC.

[186]Khozin, G., "Cause of All Men," *Sovetskaya Rossiya* [Soviet Russia], 22 April 1983.

[187]Waldrop, M., "OTA Studies U.S.–Soviet Space Cooperation," *Science*, Vol. 225, 21 Sept. 1984.

[188]Interview with James Beggs, NASA Administrator, *U.S. News and World Report*, 23 April 1984, p. 49.

[189]Weisman, D., "Red Scientists Elated over U.S. Cooperation," *Daily Citizen*, 27 Oct. 1979. NASA Historical Reference Collection, NASA History Office, NASA Headquarters, Washington, DC.

[190]Covault, C., "Reagan Renews Plan for Joint Space Mission with Soviets," *Aviation Week and Space Technology*, 15 July 1985, p. 17.

Chapter 5

COOPERATION IN THE INTERNATIONAL
SPACE STATION PROJECT

PRELUDE TO COOPERATION

The next period (1985–1998) in the history of Soviet/Russian–American cooperation in space, which began soon after Gorbachev was elected General Secretary of the Soviet Union's Communist Party Central Committee and *perestroika* was launched, resulted in almost 100 U.S.–Russian joint space projects involving the public and private sectors of the two countries. However, for a number of reasons, this chapter will concentrate on U.S.–Russian cooperation on the ISS Alpha project. First, U.S.–Russian cooperation on the ISS is "the largest and most visible sign of the new relationship in space activities."[1] Second, ISS more than any other joint space project, was politically controversial. On one hand, cooperation in human space flight is more prone to political fluctuations in bilateral relationship, and on the other, "the space station's high visibility in the overall political relationship and the Clinton Administration's strong commitment to the project could help insulate it from transitory political strains" (Ref. 1, p. 17). Third, if U.S.–Russian cooperation on ISS—the biggest cooperative space project between the two countries to date—fails, it may negatively affect future space ventures between the two countries. Fourth, the author's personal experience as a candidate for space flight results in a natural inclination toward using manned space flights, including the Shuttle–Mir and ISS projects, as the best example of U.S.–Russian cooperation in space.

CHANGE IN THE SOVIET UNION

U.S.–Russian cooperation on the ISS was preceded by a number of interrelated developments in the domestic politics of the Soviet Union, Soviet–American relations, and U.S. and Soviet space activities. Gorbachev wanted to transform the Soviet Union into a "civil society." His foreign policy activity was designed to play an important role in this transformation by incorporating the Soviet Union into "the civilized world." However, this incorporation was inseparably linked to the end of Cold War

155

confrontation, which in turn required demilitarization of external relations and of the internal economic and social system, as well as the abolition of the "image of the enemy." The latter was in fact a depiction of an external threat based on ideological grounds that was used to justify the repressive regime inside the country, militarization, and external hostility. In the final analysis, "only a Soviet leader could have ended the Cold War because it rested on the Marxist–Leninist assumption of a struggle to the end of two social–economic political systems, the capitalist world and the socialist (communist world)."[2] The changes in Soviet foreign and domestic politics as well as in Soviet–American relations had a direct impact on Soviet space activities. The end of U.S.–Soviet global confrontation greatly diminished the significance of the Soviet space program as a tool in the struggle for Soviet global leadership. As a result, Soviet leadership started losing interest in supporting the program, which became especially obvious during Gorbachev's visit to the Baikonur launch site in May 1987. Everybody who listened to his speech was very disappointed by the "emptiness of his promises to support the space program, or to preserve a scientific–technological and industrial potential of the country" (Ref. 3, p. 441).

REDUCED SUPPORT FOR THE STATE INDUSTRY

The demilitarization of foreign relations resulted in reduced state support to the military–industrial complex in general and to the space program, which was part of the Soviet Military–Industrial Complex (MIC), in particular. However, unlike the other branches of the MIC, the space industry came under especially heavy public criticism. During the entire period of its existence, the Soviet space program was shrouded in secrecy. In 1990, one of the leading Soviet journalists, the chief of the science and technology department for *Pravda*, Vladimir Gubarev, publicly acknowledged that the Soviet mass media was lying to the Soviet people about official idols of Soviet society—cosmonauts (Ref. 4, pp. 225–226). The space program's real significance for the national economy and defense, its achievements as well as its cost and failures, had never been openly discussed. Thus, it is not surprising that the program was perceived by the Soviet people as a burden on the country's economy. People were sending letters to the papers proposing to kill the space program altogether. Soviets clearly demonstrated their attitude toward national space efforts during the realization of the "Journalist in Space" project, which became the first openly announced and publicly discussed recruitment of potential cosmonauts in the history of the Soviet space program.

The idea of sending a Soviet journalist into space was initially advanced by the Soviet/Russian "Number One" space journalist and writer Yaroslav

Golovanov in 1964 during his conversation with Soviet rocket Chief Designer Sergei Korolev. Korolev liked the idea and promised to support it on one condition—a future space journalist should have a background in engineering. On 21 January 1965, Golovanov wrote an official letter to Korolev requesting to be included in the group of cosmonauts candidates. On 12 February 1965, Korolev granted Golovanov such approval. Two more journalists were admitted to the cosmonauts' training at that time—Mikhail Rebrov from the *Krasnaya Zvezda* newspaper and Yuri Letunov (from Gosteleradio state company).

After Korolev's death in 1966, the idea of sending a journalist into space lost its popularity among the Soviet elite, although in 1967, soon after Vladimir Komarov's death, Yuri Gagarin expressed his confidence that a Soviet journalist would fly in space. In 1968, Golovanov wrote a letter to the Central Committee of the Soviet Communist Party asking to continue preparing the journalists for a space flight. However, he was told that there was no reason for such a mission. The party bureaucrats explained to Golovanov that because the Soyuz spacecraft was a two-seat vehicle, it had places only for a mission commander and a flight engineer, but no room for a passenger. The space officials indicated, however, that when new multiseat spacecraft would become operational, the project might be put back on track.

Golovanov brought up this issue in his letter to the Central Committee two more times, in February 1979 and January 1981. His letter to the Central Committee in January 1981 was encouraged because the Soviet Union had started operating a three-seat version of the Soyuz spacecraft (Soyuz-T), which could potentially accommodate a passenger. The Central Committee never gave a straight "no" to Golovanov, but it never gave a go ahead to a practical realization of the project. The idea of a journalist in space was silently buried in the Soviet Union until it was resurrected in 1989 as a reaction to the intention of the Japanese broadcasting corporation to send its reporter to the Soviet space station in a Soviet spacecraft.

On 27 March 1989, Glavkosmos (at that time the prime managing body of the Soviet space program) signed a contract with the Japanese TV company TBS, which decided to celebrate its 40th anniversary by sending a journalist into space. The Soviet journalists felt insulted that the first journalist in space would become a Japanese reporter flying in a Soviet Soyuz spacecraft. *Pravda* announced a contest on 6 April, a week after the information about the Glavkosmos–TBS deal was published in the Soviet press.

According to the letters sent by the Soviet people to different publications, their attitudes toward the project fell into two main categories: those who favored the flight of a Soviet journalist, particularly before the Japanese one, and those who were against it entirely. Although the former considered the flight of a Soviet journalist as another small achievement in a space race,* the latter believed that this flight would not be worthwhile because it was undertaken for the purpose of prestige.† They favored the flight of a Japanese journalist as a way to inject hard currency into the Soviet economy. The supporters of the second approach, supported by NPO Energia, finally won and a Soviet journalist never flew in space.

A number of populist politicians, among them Boris Yeltsin, proposed to postpone the realization of a number of space projects for five to seven years to "considerably improve the living standards of the Soviet people."[5] Later, Yeltsin's attitude toward the space program was hardened by the fact that Oleg Baklanov, the chief of the Military–Industrial Commission that was in charge of the program, was one of the organizers of the August 1991 coup d'etat in Moscow, which was aimed particularly at toppling Yeltsin's government.

COMMERCIALIZATION OF THE SOVIET SPACE PROGRAM

Such attitudes on the part of the general public and the politicians resulted in the further shrinking of the Soviet space budget. To survive, Soviet space enterprises started looking for national and foreign customers to whom they could sell their hardware and launch services. Commercialization of the space program also corresponded with Gorbachev's effort to introduce elements of the market economy into Soviet society.

The opening of the Soviet space program to Soviet society and to the West, as well as the willingness of the Soviet authorities to pursue cooperation in space with the Western countries more vigorously, had important political significance. It was one of the clearest indications of *glasnost* and of the fall of the Iron Curtain. Furthermore, the new Soviet leadership was

---

*Vladimir Gubarev, expressing such views, said: "The flight of a journalist is not just a flight of Ivanov, Petrov or Sidorov [common Russian names]. Fifty years from now people will say: 'The first satellite was a Soviet sputnik, the first man in space was Yuri Gagarin, the first man to walk in space was Alexei Leonov, and the first journalist in space was also a Soviet.' This is a real benefit of the flight of a Soviet journalist."

Another well-known Soviet science and technology journalist Andrei Tarasov viewed the flight of a Japanese journalist as a starting point of the sale of the Soviet space program to foreigners. He told one of the mid-level managers of NPO Energia, Flight Director and former cosmonaut Vladimir Solovyov, "Volodya [a nickname for Vladimir], remember, this is how we will end up by manning all our spacecraft by the foreigners. When you will be released from your position as a Flight Director because somebody at the top will tell you: 'We want Mr. Hopkins to become a Flight Director. He will pay us millions for this,' you will ask: 'Are you nuts?!' And then you will recall the day, when we started selling our space program by pieces. This is where we might finally end up" (Ref. 4, p. 229).

†The idea of sending a Soviet journalist into space was supported even by the Minister of Chemical Industry of the Soviet Union Yuri Bespalov, who offered a sponsorship to the project. He presented three rationales for his offer. First, by helping the Japanese reporter to become the first journalist in space, the Soviet Union would trade in its prestige for money. Second, Bespalov wanted to help Soviet journalists write an objective story about the Soviet space program from inside. Finally, the minister believed that his ministry could increase sales of its products by advertising them from space.

considering space cooperation as a means of alleviating tensions between the Soviet Union and the United States.

The first sign of the political significance the new Soviet leadership attached to space cooperation came during the celebration of the 10th anniversary of the Apollo and Soyuz linkup in 1985. A number of Soviet papers, still controlled by the Soviet government and thus expressing its official attitude, called for a renewal of détente and new joint superpower space projects. *Pravda*, the press organ of the Communist Party, considered these two goals interrelated. According to this paper, it was crucial to "revive and consolidate détente" to realize the Apollo–Soyuz kind of projects. *Komsomolskaya Pravda*, the Communist Party youth newspaper, voiced the view that the importance of joint space projects lay "not so much in the desire to seek the answers to technical questions but in the political and, if you like, human significance of the flight." Diplomats viewed such articles as further evidence of Gorbachev's "détente offensive," which was aimed at improving U.S.–Soviet relations.[6]

The first practical step aimed at integrating Russia into the global space community was the signing of a "memorandum of understanding" between the Soviet Union and Intelsat—the world's largest telecommunications carrier. Although neither Soviet nor Intelsat authorities indicated that the understanding would lead to full-fledged membership for the country, which was the largest nonmember-user of the satellite network, the memorandum nevertheless laid "the groundwork for increased use of Intelsat's network for global transmission of Soviet voice, data, and television transmissions."[7]

In March 1986, as the Soviet spacecraft Vega I and Vega II flew by Halley's comet, the Space Research Institute, the leading managing organization of the Vega project, was opened up to foreign visitors, providing access that many Western scientists believed was unprecedented. In April, a manned space liftoff was shown live on national television for the first time in the history of the Soviet space program. Later the same month, foreign reporters were admitted for the first time to the classified facility of the Soviet mission control center in the Moscow suburban city Kaliningrad to watch a televised interview with two cosmonauts—Leonid Kizim and Vladimir Solovyov—orbiting the Earth in the Mir space station. In April, reporters were allowed into the normally closed Center for Cosmonauts' Training (Star City) to attend a press conference held in honor of the 25th anniversary of Yuri Gagarin's flight.[8] Finally in June, the Soviet Union announced that it was prepared to launch foreign satellites, indicating that the Kremlin was ready to open a commercial space program. Soviet leadership also proposed the establishment of a World Space Organization, affiliated with the United Nations, "to oversee joint international space projects and to police space to ensure their peaceful use."[9]

POSITIVE U.S. REACTION TO CHANGES IN THE SOVIET SPACE PROGRAM

The changes in Soviet politics and in the Soviet space program provoked a positive reaction from U.S. politicians. As in the Soviet Union, the initial U.S. space cooperative overtures toward the Soviet Union were driven by political considerations. In the fall of 1985, Congressman Bill Nelson proposed a "political show in space"—U.S.–Soviet communication via a Space Shuttle–Salyut signal exchange. An agreement for such an exchange was accomplished through discussion with both the managers of the Soviet space program in the fall of 1985 and with NASA's Associate Administrator for Space Flight, Jesse Moore. The plan was never realized because of the sudden return to Earth of the Soviet Salyut 7 crew in November 1985 and the subsequent lack of appeal in contacting an empty space station.[10] In 1986, Senator Spark Matsunaga, in his book *The Mars Project: Journeys Beyond the Cold War*, called for the resumption of U.S.–Soviet cooperation in space. He viewed the benefits of cooperation as primarily political. He believed that existing tensions between the Soviet Union and the United States could be overridden by cooperative activity targeting the "emerging scientific and technical elite, the most sophisticated and cosmopolitan segment of Soviet society, who enjoy the highest status, upon whom the totalitarian power structure depends for running the system." This cooperation "must be a long-term undertaking that locks [the Soviets] into an expanding program of joint activity. Soviet policy-makers must find it sufficiently appealing to stay in, although it places their reactionary systems of internal controls increasingly at risk."[11] In June 1986, then-Senator Albert Gore said that he "would like to see a manned mission to Mars in concert with the Soviet Union—a joint U.S.–Soviet mission to put human beings on the surface of Mars."[12]

The idea of renewed U.S.–Soviet cooperation in space was also supported by representatives of the U.S. academic and business communities. The demonstration that Soviet and American scientists could set aside political differences and work together in space exploration was called by *The New York Times* "one of the most notable achievements of the Soviet Union's Vega spacecraft."[13] Authors of a report on the American space effort issued in 1986 by the Business–Higher Education Forum, whose members include the leaders of some of the most prestigious businesses and universities in the United States, expressed regret about the dichotomy in the U.S. position regarding relations with other countries on space matters:

> On the one hand, the feeling persists that the United States should strengthen and promote its competitive posture and prevent technology transfer. The other attitude characterizes space as an instrument of international policy and diplomacy, which should be used to promote cooperation and synergism. The fact that these potentially conflicting posi-

tions have not been reconciled is retarding progress in many aspects of space development (Ref. 14, p. 56).

The forum encouraged the U.S. administration and Congress to accelerate its efforts to explore mechanisms for international collaboration on the peaceful uses of space, to review policies affecting international relations, and to seek ways for the United States to cooperate with other space nations (Ref. 14, p. 5). Finally, the idea of Soviet–American cooperation in space was shared by some American grassroots organizations, which proposed to give a ride to a Soviet journalist in the Space Shuttle and to his American colleague in the Mir space station in 1992 (Ref. 4, p. 231).

NASA's Attitude Toward Renewed Cooperation with the Soviets

NASA had a slightly different attitude from U.S. politicians, businessmen, and scientists toward cooperation with the Soviets. The increased cost of major space projects contributed to the desirability of greater space cooperation. A mission to Halley's comet, involving collaboration between the Soviet Union and the United States, in particular, "was a prime example of cost sharing in an endeavor that might have seemed prohibitively expensive if pursued by any single nation alone."[15]

Another important factor that indirectly contributed to U.S.–Soviet space cooperation in the late 1980s was that unlike the situation in the 1960s, pooling efforts together with the Soviets would not have undermined the rationale for a vigorous and competitive U.S. space program.

However, unlike the case in the early 1970s, cooperation with the Soviets did not serve as one justification for the continuation of the U.S. space effort that had been significantly relaxed after the American victory in the moon race. According to Dick Kohrs, who was a manager of the system engineering of the Space Shuttle in the mid-1980s, public interest in the Space Shuttle program declined after four or five Shuttle flights. However, this decline was not as dramatic as after the completion of successful U.S. moon landings. The agency did not object to cooperation with the Soviets, but it preferred this cooperation to be confined to some particular projects that would not require a complex and labor-consuming, large-scale bridging of U.S. and Soviet space programs. Also, in the mid-1980s NASA was flying a significant number of Department of Defense missions that prevented the agency from cooperating with the Soviets, who were still viewed as a Cold War rival.[16]

Besides the wish to avoid complexity and the potential leak of military secrets that could follow cooperation with the Soviets in space, NASA had one more reason not to seek broadscale collaboration with them. Until the late 1980s, NASA officials and U.S. politicians were still thinking of

U.S.–Soviet space relations in terms of a race. When Thomas Moser, who was a Space Station program director from 1986 through 1989, tried to convince Congress to allocate money for Space Station Freedom in fiscal year 1987, he presented three rationales for building it.

1) The need to have a microgravity facility in orbit.
2) The need to fulfill U.S. commitments to Europe, Canada, and Japan, who were part of the Freedom program.
3) The need "to have the Station because the Russians have it, and if we don't put the Station into orbit, the Russians will surpass us by a cumulative number of man-hours." [17]

Despite the still-race-dominated mentality, NASA took some practical steps aimed at rapprochement with its Soviet counterparts. In July 1985, NASA Administrator James Beggs confirmed the readiness of the United States, as expressed in 1984 by President Reagan, to resume cooperation with the Soviet Union on civilian space missions. Beggs was referring to simulated space rescue missions, which could "demonstrate our capabilities to work peacefully together in space for humanitarian purposes." [18]

The improvements in bilateral relations facilitated professional contacts between Russian and U.S. space scientists. In September 1986, a 10-member U.S. space team, headed by Jet Propulsion Laboratory Director Lew Allen, discussed possible joint Mars studies with the Soviets. The meeting was initiated by the U.S. side, which proposed the visit after the Soviet Union indicated it would avoid linking U.S. abandonment of SDI as a precondition to such discussions. [19]

U.S.–SOVIET SPACE RAPPROCHEMENT

On 15 April 1987, U.S. Secretary of State George Shultz and Soviet Foreign Minister Eduard Shevardnadze signed an Agreement Between the United States of America and the Union of Soviet Socialist Republics Concerning Cooperation in the Exploration and Use of Outer Space for Peaceful Purposes. The parties agreed to cooperate in "such fields of space science as solar system exploration, space astronomy and astrophysics, earth sciences, solar-terrestrial physics, and space biology and medicine (Ref. 20, pp. 217–220). The agreement did not envisage broadscale cooperation in human space flight, and overall was rather limited in scope. Its scope corresponded at that time both with the real interest of the U.S. and Soviet space communities toward cooperation with each other and with the state of Soviet–American relations. However, the real importance of the agreement was that it contributed to the further bridging of U.S. and Soviet space efforts.

IMPROVEMENT NEEDED IN U.S.–SOVIET POLITICAL RELATIONS

Several more factors contributed to U.S.–Soviet space rapprochement in the late 1980s. Among them were the gradual improvement in U.S.–Soviet relations after the Reykjavik, Iceland, and Washington, D.C., summits of 1986 and 1987, respectively; the indication by the Soviets that they would avoid demanding the U.S. abandonment of SDI as a precondition for full-fledged bilateral space cooperation; and the increased interest of U.S. and Soviet space communities, or at least of their "space flyers'" sectors toward cooperation with each other. The latter was demonstrated by the activity of the Association of Space Explorers (ASE). In 1987, ASE met in Mexico City and advanced the proposal that the United States and the Soviet Union find ways to cooperate in the future to move forward with space exploration. In 1989, at its annual meeting in Al Riyadh, Saudi Arabia, ASE went even further by proposing that Soviet cosmonauts fly on the Space Shuttle and that American astronauts fly on space station Mir. The rationale behind the proposal was that the Soviets would be able to gain some expertise flying in the Shuttle, while Americans would have the advantage of flying their astronauts on long-duration missions. Such experience would have been an important asset both for the Soviets in developing their Buran spacecraft (an equivalent of the Space Shuttle) and for the Americans working on Space Station Freedom. This decision made by ASE was reported to the Bush and Gorbachev administrations.[21]

Although there is no indication that the recommendation made by ASE had a direct impact on the U.S. and Soviet decision to consider cooperation in human space, in June 1990 Gorbachev and Vice President Dan Quayle tentatively agreed during their meeting in Washington, D.C., to send an American astronaut to the Mir space station and a cosmonaut on a NASA Space Shuttle mission. Gorbachev also agreed to discuss Soviet participation in an international lunar base and Mars missions with the United States. However, the meeting between Gorbachev and Quayle did not bring about any substantive progress on human space flight-related accords because everything was "on hold because of the arms control and trade agreements."[22]

PROFESSIONAL INTERACTIONS BETWEEN THE TWO SPACE COMMUNITIES

Despite the lack of progress in negotiations at the top political level regarding cooperation in the area of human space exploration, professional interactions between space communities of the two countries became more frequent and productive. On the U.S. side, it was primarily because of the changes in NASA leadership. In 1989, Admiral (Ret.) Richard Truly, a man who was deeply involved in ASTP and had remained a firm supporter of

Soviet–American cooperation in space, became NASA Administrator. As Truly described,

> the actions and initiatives that I took with my Soviet colleagues while NASA Administrator in 1989–92 had their roots as far back as 1974–75 when I worked for Tom Stafford on the support crew for the Apollo-Soyuz (ASTP) mission. I had been the ascent/rendezvous/docking capcom ("capsule communicator") for all 3 of the Skylab missions, and Tom asked me and Bob Crippen (Skylab entry capcom) to join the ASTP support crew late in the game. Both Bobko and Bob Overmyer had been on the support crew for some time. It was a fascinating experience, and I worked hard at it. It demonstrated how unbelievably difficult it was to accomplish even a relatively straightforward technological mission when language, culture and initial mistrust stood in the way. I went to Russia on the final training trip, lived in Star City, went to Baikanur, etc., etc. But by far the most important part of it was to get to know the cosmonauts on the prime, backup and support crews that we worked with. Although some were very well known to the West at the time (Alexei Leonov, etc.), others were totally unknown (Dhzanibekov, Romanenko, etc.). I consider Dhzani one of my closest friends.
>
> It's my personal belief that the maintenance of these astronaut/cosmonaut friendships in the intervening years played a big role in making it possible to begin some U.S./Russian cooperations in the late 80s/early 90s, and space was just one of them (although it was the one I was involved in).[23]

For Truly, who considered U.S. and Soviet space programs as "the most successful on Earth," cooperation in space between the two countries "simply seemed an opportune thing to explore." Potential political and economic advantages of such cooperation were taken for granted by him. As an ASTP student, Richard Truly learned one more lesson from Apollo–Soyuz. He wanted to have a meaningful cooperation in space between the Soviet Union and the United States. He was not interested in "showboat demos that weren't mutually beneficial."[23]

On the Russian side, one of the main reasons for the Soviet–American space rapprochement was the professional interest of the Soviet space industry. In the late 1970s and during the 1980s, the Soviet Union was realizing the Buran space program. Buran was in fact the Soviet Space Shuttle, designed to provide continuous and relatively inexpensive delivery of crews, cargo, and supplies to the Mir space station. However, in the late 1980s it became clear that the Buran program was heading nowhere. After the first and only successful automated flight of Buran (without a crew onboard) in November 1988, the program was virtually terminated. Soviet aerospace officials and engineers felt very frustrated about it because Buran was integrated in the plans of the long-term operation of the Mir outpost. For this reason, at the end of 1980s, Soviet space specialists started looking for a replacement for Buran. The most natural choice would be a U.S. Space Shuttle.[24]

Shortly after becoming NASA Administrator in July 1989, Truly was invited to visit the Soviet Union to discuss potential cooperation. Not everybody in the top U.S. political leadership shared his ideas regarding resumption of large-scale U.S.–Soviet cooperation in space at this time. His trip was stopped by the National Security Council. The invitation, however, remained opened and in October 1990, Truly visited Soviet space facilities, including design bureaus, the Baikonur launch pad, and the Mission Control Center. According to Truly,

> All of my hosts were friendly, cooperative and encouraging about finding ways that the programs might work together across the spectrum of science, technology, flight programs, etc. From my point of view, the "cold war foe in space" was behind us, although both sides were extremely interested to learn anything we could about past programs.[23]

In June 1991, Minister of General Machine Building (*Minobshchemash* one of the primary space managing bodies of the former Soviet Union) Oleg Shishkin and NPO Energia Chief Designer Yuri Semyonov visited the United States, where they toured different U.S. space facilities, including the NASA Johnson Space Center. It was the first visit of *Minobshchemash* representatives to the United States in 15 years. After the tour they met with Vice President Quayle. Both sides agreed that 50% of the Soviet's space technology was five to six years ahead of the United States, whereas the United States was ahead of the Soviet Union in the other 50%. They also agreed that the two countries could complement each other in space.[24]

Truly was a bit more specific in his assessment of the differences between U.S. and Soviet space technologies. In his view,

> [Soviet] technology was very "brute force" compared to ours, but in other cases, I was extremely impressed with their advancement over us. A good example was in several metallurgy areas that enabled their engines to work, and their exquisite large-scale automated welding machines that I was shown in one or two of their booster plants.

Meeting with the Soviets made Truly even more convinced of the benefits of cooperating with them in space. Surprisingly, he found that "our own National Space Council was a bigger difficulty than my Soviet friends. With the Soviets, language was a hindrance, but finding a way to get started (a program that was affordable and politically acceptable) was what we spent the most time working on."

Eventually, Truly decided that

> the easiest way to start was a swap of crew flights (a cosmonaut flight aboard Shuttle, and a long-duration astronaut flight on MIR). I worked this with Oleg [Shishkin], and before I proposed it to the National Space Council, I brought the JSC Director (Aaron Cohen) and key astronauts (Dan Brandenstein, etc.) to Washington to make sure they were on board. Incidentally, it was me that continually insisted on the "long duration" aspect of the MIR flight, because to me, that was the one thing that we could get out of the swap.[23]

At the same time they prepared for cooperation in the area of human space flight, the Soviet Union and the United States continued cooperating in space science. On 15 August 1991, the Soviet Union launched its atmospheric monitoring satellite Meteor 3, which carried NASA's Total Ozone Mapping Spectrometer. It became the first Soviet satellite to have American instruments onboard.[25] During the same period of time, the U.S.–Soviet Earth Science Working Group was established. In August 1991, the Soviet Union offered the United States access to its three biggest radio telescopes as a part of the Search for Extra-Terrestrial Intelligence program, and in October 1991, Department of Defense and NASA officials traveled to the Soviet Union to consider expanded use of Soviet launch vehicles (Ref. 26, p. 99).

ECONOMIC PROBLEMS IN SOVIET SPACE PROGRAM

The continuing disintegration of the socialist economy and the transition to market conditions aggravated the economic problems of the Soviet space program. Overall, evaluating the Soviet space budget is a very complicated task because the space complex management had an administrative, not an economic, nature. Financing the space program under such a management regime was contingent not on the need to obtain funds and effectively use them, but on the timely adoption of relevant space policy decisions by the top Soviet leadership. After those decisions had been made, it was relatively easy to coordinate activities of the Ministry of Finances, Gosplan (the USSR State Planning Committee), and Gossnab (the USSR State Supplying Committee). These bodies were responsible for the overall planning, financing, and material support of the different branches of Soviet industry and of the space industry in particular.

Minobshchemash, which often initiated appropriate space policy decisions, was also responsible for following through with these decisions. A Military–Industrial Commission (VPK) issued direct orders to ministries and organizations involved in the realization of the *Minobshchemash* project, coordinated their activity, and ensured fulfillment of the program developed by the ministry.*

The complexity of such a system made it virtually impossible to use budget figures as an indicator of space activity. Furthermore, the budgets of individual space programs and projects were not even calculated—they were incorporated into a single line item with strategic missiles under *Minobshchemash*.[27] However, some analysts believe that the 1990 budget for Soviet space activities—both military and civilian—fell from the 1989 level of 6.9 billion rubles to a reported 6.3 billion rubles, a decline of 8.7%.[28]

TRANSITION: SOVIET–AMERICAN TO RUSSIAN–AMERICAN COOPERATION IN SPACE

The disintegration of the Soviet Union, the emergence of Russia as the major space power among the former Soviet republics, and the rise of new space powers, such as Europe and Japan, introduced new elements into U.S.–Russian space cooperation.

First, unlike its Soviet predecessor, the Russian space program did not receive consistent budgetary funding. The economic health of the program was aggravated by a generally worsened economic situation in Russia, which had a negative impact on all branches of the Russian economy. In general, Russian industrial production levels in January 1992 were 82% of their January 1990 levels, and in January 1993, that number had fallen to 69%. Inflation, which became a very serious problem, struck the Russian space industry as well. The cost of a Proton launch rose from 17.1 million rubles in 1991 to 4 billion rubles in 1993, a hundredfold increase. The monetary situation was complicated even more by uneven price inflation on different materials and components, which prevented industry from handling the inflation by "acting as before" (Ref. 29, p. 5). This resulted in the further economic loss of the Russian aerospace sector. Cooperation with the West came to be seen as one of the major sources of income for the Russian space industry.

Second, the low monthly salaries of the workers and specialists employed in the space industry (at the end of 1992, space companies paid their personnel only 4,500 rubles per month, while the national average was 7,200 rubles per month; Ref. 29, p. 6), as well as a loss of prestige in working for the space program, resulted in a labor and brain drain from the industry. This factor, coupled with the fall of the Iron Curtain and the significant weakening of KGB control over Russian society, encouraged some

---

*The difficulty of coordinating the activities of numerous organizations, design bureaus, and factories involved in the realization of a space project could be illustrated by one example: the Buran (Soviet Space Shuttle) was realized by 74 ministries and 1124 enterprises.

rocket and missile specialists to look for job opportunities abroad. Among countries that expressed a willingness to hire Russian specialists as well as to buy Russian rocket and missile technologies were those who wanted to use these assets to create weapons for use in international terrorism. Such developments alarmed the international community. U.S. Secretary of State James Baker expressed his concern "that the creative talents of scientists and engineers in the former Soviet Union not be used there or elsewhere to enhance the military capabilities of potentially hostile regimes." [30] One of the ways for the United States to prevent this from happening was to create job opportunities for Russian rocket and missile specialists at home by involving Russia in cooperative bilateral and international space projects.*

Third, with the final disappearance of one of the main rivals in the Cold War, it was formally over. This lessened fears about a possible technological exchange between former adversaries and also contributed indirectly to the development of Russian–American space cooperation.

Fourth, the new Russian government clearly considered space one of the means to promote rapprochement between the two countries. One of the first steps of the Yeltsin administration in the field of Russian–American relations was to join the celebration of the signature of the "Charter of Russian–American Partnership and Friendship" with "Space Flight Europe–America 500." On 15 November 1992, a Soyuz spacecraft with space capsule Resource 500 was launched from a military launch pad (Plesetsk). The capsule was in orbit for one week and landed 200 miles off the coast of Washington State. It was filled with messages of peace from Russian President Yeltsin and leaders of the European community as well as samples of products from Russian and European companies. This flight celebrated the 500th anniversary of the discovery of America by Columbus. The goal of the mission was "to highlight the future of normal business relations with Russia." [31]

Fifth, the demise of the Soviet Union also resulted in the disintegration of the Soviet space complex, with Russia left with about 85% of it. Thus Russia emerged as a major player in potential cooperative space projects between Western countries and the former Soviet states. Although almost all of the former Soviet republics, with the exception of the Baltic states, signed an agreement envisaging continuation of the space program as a cooperative enterprise and even created an "interstate space council," this cooperation

---

*There are different opinions regarding the possible impact on the U.S. and international security of keeping Russian rocket and missile specialists at home. Director of Space Policy Institute at George Washington University, John Logsdon, said that if U.S.–Russian relations return to the Cold War, Russian missile specialists might be used by a hostile Russian government to create new weapons aimed at the United States.

turned out to be limited to interactions between Russia, Ukraine, and Kazakhstan.

Sixth, the disintegration of an established framework for space activities in Russia resulted in a loss of clarity in Russia's goals in space. Because of the end of global confrontation/competition, there was no longer a race in space, and because of the disastrous state of the Russian space industry, a manned expedition to Mars—the potential goal of Russian long-term manned flights—was out of the question. The absence of clear-cut goals introduced an element of chaos into the Russian space program, but it also encouraged program managers to consider large-scale coordination of space activities between Russia and the West.

Seventh, the disintegration of the Soviet space complex resulted in the collapse of the established management system of the Soviet space program. Some changes were introduced in the control of particular parts of the program as far back as the late 1980s. In 1988, Valentin Glushko, a chief designer of NPO Energia (since 1994, RKK Energia)* died. Workers and specialists from Energia sent a letter to the Central Committee of the Soviet Union's Communist Party and to the Military–Industrial Commission of the Council of Ministers, asking for the appointment of Yuri Semyonov to general designer of the enterprise. The request was granted.[32] The appointment of Semyonov meant not only the change of general designer, but a change of the managing structure of NPO. In the past, the enterprise had been run virtually by two CEOs—the general director and the general designer. Semyonov changed this system by leaving only one CEO, the general designer, while lowering the position of the general director to the level of deputy general designer (Ref. 3, pp. 434–436). This was the beginning of the creation of Semyonov's monopoly and the loss of centralized state control over the Soviet space program. This loss of control was clearly demonstrated by the aforementioned Journalist in Space project. During his meeting with *Pravda* editors on 26 October 1989, Mikhail Gorbachev said that the decision to launch a Soviet journalist into space was made. Moreover, according to Gorbachev, the Soviet journalist was supposed to blast off before the Japanese reporter, whose flight was scheduled for late 1990.[33] However, the general secretary's decision was never implemented. NPO Energia said that it was impossible to prepare a Soviet journalist for the mission in the time remaining before the Japanese flight (Ref. 4, pp. 223–225), although the true treason was a lack of interest of NPO in a noncommercial

---

*NPO stands for *Nauchno-Proizvodstvennoye Ob'edineniye* (Scientific-Research Enterprise). RKK stands for *Raketno-Kosmicheskaya Korporatsiya* (Rocket-Space Corporation). Energia has developed a majority of Soviet manned spacecraft, including Vostok, Voskhod, and Soyuz spacecraft; some Salyut space stations; the Mir space station; the Buran shuttle; and the Energia heavy booster.

flight.* As a result, a Soviet journalist never flew in space, although six journalists were admitted to the General Cosmonauts' Training in Star City and even completed it more or less successfully.

After the ministries that had managed the space complex were closed by political decree in September 1991, leaving the space program without government direction or coordination (Ref. 4, p. 7), Semyonov's power was consolidated even further. In early 1992, Semyonov assumed control of the development and operations of the Mir space station and later of a Mission Control Center. This complicated the decision-making process in the Russian space sector and later had a noticeable impact on Russian–American cooperation in space.

Eighth, the uncontrolled privatization of space enterprises, such as NPO Energia or the Khrunichev State Research and Production Space Center,† had a twofold impact on their attitude toward international space cooperation. On one hand, privatization coupled with desperate economic conditions forced space enterprises to seek commercial cooperative projects promising immediate or short-term returns. As a result, these enterprises lost a significant amount of interest in state noncommercial projects, including those that were part of Russia's commitments to international partners. On the other hand, enterprises like Energia or Khrunichev had difficulties responding to normal market pressures because of a large overlap between scientific R&D and production. This overlap occurred because investments in R&D take longer to create economic returns, and quick profits on capital investment are therefore almost impossible to realize. The overhead expenses of space enterprises, burdened by costly test facilities and research laboratories, was skyrocketing between 1992 and 1994 up to 900%. Such trends did not let these enterprises manufacture competitively priced products. It meant that the principal client of such enterprises remained the state and that their main financial source continued to be the state budget.

Ninth, in February 1992, Russia created a state body to manage civil space activities—*Rossiiskoye Kosmicheskoye Agentstvo* (RKA), or the Russian Space Agency (RSA), which was supposed to become the Russian equiva-

---

*An observation about the lack of interest of NPO Energia in the flight of a Soviet journalist as a main reason for the unwillingness of NPO to launch a Soviet reporter could be confirmed by NPO employees themselves. Vladimir Solovyov, a mid-level manager of Energia, confirmed that NPO had to follow orders in the past regarding accelerated training of foreign cosmonauts who had to be launched for political reasons. One such case was the flight of an Afghan cosmonaut in August 1988 (see Ref. 4, p. 224). After the liftoff, Afghan President Mohammad Najibullah declared a unilateral armistice in the Afghan civil war for the duration of the flight.

†Khrunichev State Research and Production Space Center is a manufacturer of the Russian Proton heavy booster, which is a primary launch vehicle for the delivery of space station modules into orbit. It is often addressed as just Khrunichev (like Energia), Khrunichev Space Center, Khrunichev plant, factory, or enterprise.

lent of NASA. Since its creation, the RSA has become a powerful entity—in some aspects even more powerful than the Ministry of General Machine-Building, which did not have the same degree of access to the USSR's foreign cooperative partners and was subject to dual supervision by the Communist Party and the Military–Industrial Commission.[34] RSA's access to foreign partners laid the groundwork for the increased involvement of Russia in international cooperative space projects, especially because one of the officially declared goals of the agency was "to interact with the relevant bodies of CIS states and other foreign countries in the area of space exploration, including the use of ground space infrastructure."[35]

Tenth, by the end of the Cold War, the United States was more likely to see its traditional partners as "friends and competitors" rather than as "friends and allies." This factor, coupled with the easy access of foreign partners to relatively inexpensive and highly reliable Russian space hardware, raised considerable concern on the U.S. side. In May 1992, Congress noted that the collapse of the former Soviet Union

> has already enabled foreign competitors of the United States industry to obtain unique advanced technology from the former Soviet Union's military, research and industrial organizations for a tiny fraction of their development cost. It is in the national interest that the National Aeronautics and Space Administration aggressively identify, examine, and where appropriate import unique space hardware, technologies, and services available to the United States from the former Soviet republics' design bureaus, scientific production associations, and research institutes.[36]

Such unique hardware existed in the former Soviet Union. When the Subcommittee on VA-HUD-Independent Agencies asked NASA in early 1982 about the agency's potential interest in space hardware other than Soyuz spacecraft, in the former Soviet Union, NASA Administrator Richard Truly mentioned several other areas in which Soviet space technology could represent attractive opportunities for NASA. Among them was an automated rendezvous and docking system. The United States did not have a similar capability. Another area of immediate interest was the possible acquisition of tracking time on ground antennas belonging to the former Soviet Union's deep space network. The third area was space nuclear reactors power systems, and the fourth, Russian rocket engines, such as the RD-170.[30]

Finally, after the disintegration of the Soviet Union, Russia managed to achieve some success in creating a legal framework for its space activities and articulating its official space policy. In the past, such a framework did not even exist. All space activities were controlled by the party, which set goals for the Soviet Union in space and determined ways to achieve them. A number of documents adopted by Russian authorities created a legal and institutional framework for the Russian space program and codified international cooperation in space between Russia and other countries.

The first ever parliamentary hearings on space took place in Russia on 10 November 1992. These hearings were held by the Commission on Transport, Communications, Informatics and Space of the Supreme Soviet. The hearings resulted in two documents: a draft statement made by the Supreme Soviet, "On the Priorities of Space Policy for the Russian Federation"[37] and a draft resolution, "On Measures to Stabilize the Situation in Space Science and Industry."[38] These documents were adopted in April 1993. The statement "On the Priorities of Space Policy..." emphasized among other things the need "to deepen international cooperation and integration in space exploration on a mutually advantageous basis, and ensure the fulfillment of Russia's obligations under international agreements."[37]

The next step was the adoption of the "Law of the Russian Federation on Space Activities," which came into effect on 6 October 1993. The law stressed the need for the Russian Federation to "ensure the fulfillment of the obligations it has assumed in the field of space activity" and "to promote international cooperation in the field of space activity."[39] The law also laid the groundwork for the Russian "Federal Space Program up to the Year 2000," which was adopted on 11 December 1993. Among the major goals for the Russian space effort, the Federal Space Program included the promotion of international cooperation in science and technology for social, environmental, and other global concerns.[40]

FROM ASSURED CREW RETURN VEHICLE SUBCONTRACTOR TO FULL-FLEDGED PARTNER

The changes on the U.S. side, which increased NASA interest in cooperation with the Russians in the early 1990s and finally made possible the merging of U.S. and Russian efforts in the ISS Alpha program, did not have the same dramatic character as on the Russian side. These changes were caused mostly by problems and difficulties encountered by the Space Station Freedom (SSF), Alpha's predecessor. In 1986, when the funds for the ISS were budgeted, the managers of the program, to keep cost down, did not include the development of the Assured Crew Return Vehicle (ACRV), or lifeboat, into the budget of the station. This did not mean, however, that the managers considered the ACRV unnecessary. Designers and engineers were working on the vehicle, but the work was not officially included in the SSF program. The ACRV was supposed to fulfill the following tasks: to return one disabled space station crewmember during medical emergencies; to return space station crew in the event of accidents or failure of the SSF; or to return space station crew during an interruption of Space Shuttle launches.

In 1989, two companies, Lockheed Martin and McDonnell Douglas, were each awarded $1 million contracts to continue to study and refine the ACRV concept and definition. As a result, the cost of ACRV development and manufacture was determined to exceed $1 billion. This would have meant further augmenting the SSF budget, which had already been increased by $6.7 billion since being initially defined at $8 billion in 1986, to a total of $14.7 billion. In 1990, Bill Lenoir, then NASA associate administrator, Office of Space Flight, brought the ACRV program under the general SSF program, thus making further work on the vehicle part of SSF development.

But Lenoir's move did not persuade Congress to pass the ACRV budget. After this, NASA had only two choices. The first was to build the station without the ACRV, thus exposing the crewmembers to unnecessary risk by depriving them of the possibility of immediate return to the Earth in case of emergency. This was a totally unacceptable option especially in view of the recent *Challenger* accident. The second choice was to look for already existing space hardware. The most natural choice would be the workhorse of the Soviet space program, the Soyuz spacecraft. As a matter of fact, NASA was even encouraged to look to Russia by the politicians who considered the recent collapse of the Berlin Wall as the end of both the Cold War and the space race. An indirect indication of the latter was that Dick Kohrs, who was a Space Station program director from 1989 through 1993, unlike his predecessor Thomas Moser, did not have to use a "beat the Russians in space" argument as a rationale for building SSF.[41] The idea of using a Soyuz spacecraft as the ACRV was welcomed by the Soviets. In October 1991, during his meeting with Boeing representatives, Yuri Semyonov offered Soyuz spacecraft as a lifeboat for the SSF (Ref. 3, p. 517).

U.S. BUDGETARY CONSTRAINTS INFLUENCE COOPERATION WITH THE SOVIETS

The end of the space race and continuing improvement of Soviet–American relations had a twofold impact on the U.S. space program, including its biggest project—Space Station Freedom. On one hand, these developments undercut one of the budgetary rationales for the program, and on the other, they made its managers more susceptible to the idea of pooling efforts in space together with the Soviets. In 1990, authors of the *Report of the Advisory Committee on the Future of the U.S. Space Program*, known also as the *Augustine Report*, acknowledged that "the United States's civil space program was rather hurriedly formulated some three decades ago on the heels of the successful launch of the Soviet Sputnik," and that NASA had been intensively criticized particularly for its inability to estimate and control the costs of missions. One of the suggestions of the authors of the report, aimed at improving the current situation, was to redesign SSF "to lessen complexity and reduce cost, taking whatever time may be required to do this thoroughly and innovatively."

Although the authors of the *Augustine Report* stressed that "international cooperation should and will become an increasingly important aspect of future space activities," they did not emphasize the particular significance of U.S.–Soviet cooperation in space. This was done by another report called *America's Space Exploration Initiative* (*ASEI*), prepared a year later by a "Synthesis Group" under the leadership of a former U.S. (and Apollo–Soyuz) astronaut, Lt. Gen. Thomas Stafford. This report described a "recent increase in joint U.S. and Soviet cooperation in the life sciences" as "an encouraging development." The authors of *ASEI* put special emphasis on the unique Soviet orbital facility, the Mir space station, and on the cooperative opportunities it presented:

> Mir represents the only extended duration spacecraft in operation, and access to Soviet crews for joint medical studies represents a tremendous windfall that adds to the existing knowledge base. This resource represents a timely start on key medical and physiological concerns and should be aggressively pursued.[43]

The increased budgetary constraints on the U.S. space program, according to the authors of a Task Group Report called *A Post Cold War Assessment of U.S. Space Policy* (January 1992), heightened American "interest in benefitting from the capabilities and resources of other countries in achieving objectives in space."[44] NASA's budget was $13.9 billion in 1991, $14.0 billion in 1992, and $14.3 billion in 1993. However, adjusting for inflation, the same figures were $12.2 billion, $11.7 billion, and $11.6 billion in 1987 dollars (Ref. 26, p. 106; Ref. 45). Russia was described by the Task Group Report as "the country with perhaps the most to offer as a cooperative partner" to the United States.[44] The report issued a practical recommendation to the United States:

> to employ the existing space assets and capabilities of the former Soviet Union on a selective basis when they offer unique programmatic benefits, and should encourage collaboration between U.S. industry and the privatizing space organizations of the former Soviet Union in developing future space capabilities (Ref. 44, p. 43).

OPPOSING VIEWS ON PURCHASING SOVIET SPACECRAFT

One such asset clearly was the Soyuz spacecraft. In the fall of 1991, Yuri Semyonov and NASA Administrator Richard Truly attended hearings of the Senate Appropriations Committee on Science and Space, during which Senator Barbara Mikulski asked them to explore possibilities of using the Soyuz-TM spacecraft as an ACRV.[46] This hearing became the starting point of Russian involvement in the SSF project. Also in the fall, Congress, for the first time since the beginning of work on the ACRV, approved fiscal year 1992 money ($6 million) for the study of the concept of the vehicle. In March 1992, NASA Associate Administrator Arnold Aldrich visited Russia to become more familiar with the Soviet spacecraft. Based on what he saw and learned about this spacecraft, he concluded that if Soyuz could be adapted properly, it could be a bargain, whereas the development of a U.S. crew escape vehicle would cost from $1 billion to $2 billion. The low cost of Soyuz could have been partly explained by the fact that the Soyuz systems had already been certified for flights.[47]

Not everybody within the U.S. aerospace community viewed the reliability and relative simplicity of the Soviet/Russian space technology as an asset that could be used to the advantage of the United States space program. Whereas Aldrich gave a positive assessment to the Soyuz spacecraft, Dan Warrensford, an aerospace engineer, called Soviet equipment

> relatively durable, and boring, circling near Earth....Relying on hardware of decades-old design, whether ours or someone else's, won't provide us with vehicles for intra- and extra-solar operations. It will retard, rather than promote, evolution in exotic processes and materials. The state of the art in space technology will be caught in a time wrap.[48]

The idea of purchasing Soviet space technology provoked a complicated reaction within the U.S. ruling elite and highlighted once again the difference between Republican and Democrat attitudes toward cooperation in space with Russia. According to "federal and industry officials," the Republican Bush administration "quietly blocked the purchase of missiles, rocket engines, satellites, space reactors, spacecraft, and other aerospace technology from the former Soviet Union." The federal officials said that "their opposition to the purchases is part of an administration policy intended to force the Russian space and military industry into such a decline that it poses no future threat to the United States."[49]

The potential purchase of Soviet space hardware was generally welcomed by a Democratic Congress. The legislatures supported this idea for a number of reasons. The first was expressed by the Chairman of the Subcommittee on Space of the U.S. House of Representatives, Ralph Hall, who said that the U.S. economy in 1992 and "election year budget making" gave NASA "a very stiff challenge to live within some shrinking national means." Hall believed that this development was bringing Russia and the United States together and that possible U.S.–Russian space cooperation was "not just about saving or capitalizing on the former Soviet space program," but was about the U.S., European, and Japanese space programs, "and all the space programs that we are going to have in the future" (Ref. 50, p. 1). Such an attitude was shared by Representative James Sensenbrenner, who agreed that there were "some unlimited opportunities where the Russians and the Americans can complement each other, with each other's strengths in the area of space" (Ref. 50, p. 5).

The second factor underlying Congress' approach toward cooperation in space with the Russians was the consideration of space exploration as an element "of the new era of global diplomacy," which was supposed to contribute to further rapprochement between Russia and the United States and help remove the residue of Cold War tensions. Space cooperation was supposed to realize a twofold goal: to inject money into the Russian economy and to preserve Russian pride. The logic was clear: the Russians would get dollars not simply as "foreign aid," but "through increased trade...[in the area] where both nations can benefit and where the recipient of the good will of the American people can be accepted by the Russian people as a trade" (Ref. 50, pp. 2, 19).

The third factor that underlay the positive attitude of U.S. politicians toward space cooperation with Russia was their intention to help the former Soviet Union, which "devoted a much larger proportion of its assets, of its national wealth, of its intellectual potential to the defense sector than...[the United States] did," to solve "an enormous problem in converting that sector to constructive and civilian uses." According to former U.S. Ambassador to the Soviet Union, James Matlock, "bilateral space cooperation should ideally be part of a larger program of cooperation including areas which were part and still are of the military-industrial complex" (Ref. 50, p. 20).

Congress approved $15 million for fiscal year 1993 for the continuation of ACRV-related activities. This decision was made despite some concern about possible "job dislocation" in the U.S. aerospace industry through the purchase of Soviet hardware; possible dual-use technology transfer to the Russian military–industrial complex, a decision that gave Russia control over the critical path of the station through ACRV development, and the only partial technical suitability of the Soyuz spacecraft for use as an ACRV. The work on adapting the Soyuz spacecraft for the use as a Freedom lifeboat was supposed to continue through fiscal year 1998, with the funding reaching a peak of $159 million in fiscal year 1996 for the actual building of ACRV (Ref. 50, pp. 28, 40–41, 95–97; Ref. 51).

NEW U.S. SPACE POLICY UNDER GOLDIN

The ACRV activities as they were planned in 1991–1992, however, did not extend beyond 1993. This happened for both space policy and general policy reasons. The space policy reasons included changes in NASA leadership and management philosophy. On 1 April 1992, a new NASA Administrator, Daniel S. Goldin, took office. He introduced a new management style into NASA headquarters. This style was expressed in the words "Faster, Better, Cheaper." One of the results of the practical application of the new management style was Goldin's intention to cut the NASA budget by 35%, the SSF program in particular. "We should be a $22 billion agency, and we're going to have to operate on $10 billion. It's not easy," he said (Ref. 52, p. 66).

One of the ways to save money was to increase and diversify the purchase of Russian space hardware, which was considerably less expensive than American hardware. According to Richard Beatty, who studied possible areas of cooperation between the United States and Russia in human space flight for the consulting firm Booz, Allen and Hamilton, the general impression of NASA officials at that time was that "you can buy Russian space hardware practically for pennies."[53] In May 1992, NASA gave Booz, Allen and Hamilton (BAH) the task to "assess a variety of options for combining aspects of the Space Station Freedom Program with the currently operating Russian Mir space station complex."[54] BAH came up with four options. This is how they were presented and explained in the report prepared by the company:

1) *"Buy it and put an American flag on it!"*
    Benchmark case of no U.S. development
    U.S. would operate Russian payloads and systems
    Platform operation contracted to Russian Space Agency
    Potential to jointly purchase with International Partners (IPs)

2) *U.S. integrates its own payload, on-orbit, into existing Mir complex*
    Treated like other Russian international participants
    Could buy as much rack/drawer/crew space as offered
    To involve its IPs into the project in similar fashion

3) *U.S. outfits, prior to launch, an existing Kvant-class module with pay loads and systems*
    Some supporting system enhancement possible
    Greater control over operations—"U.S. outpost" on Mir
    Partners could participate in a similar fashion

4) *U.S. supplies SSF elements and systems modules for attachment to Mir*
    Baseline SSF elements with minimal modification
    U.S. payloads supported by U.S. systems
    Ops management shared by Russia and U.S.

Potential growth path to all-U.S. configuration
Could include International Partners in lieu of, or in addition to, U.S.
elements and systems
Several options are conceivable (Ref. 54, pp. 7–8)

The first option was never considered in detail because the Russians did
not offer their space station for sale. Others, however, were not found to be
mutually exclusive and received close attention from NASA officials.

BEGINNING OF BROADSCALE RUSSIAN–U.S. COOPERATION IN SPACE

In mid-June 1992, around the same time the report was published, NASA
Administrator Goldin and RSA General Director Yuri Koptev met for the
first time and "quickly agreed that there were many opportunities for
enhanced cooperation, particularly in the area of human spaceflight" (Ref.
20, p. 220). The tendency toward rapprochement between the two space
agencies was supported by the top leadership of the two countries, who evidently attached great political significance to intensified bilateral space
cooperation. During a summit meeting between Russian President Boris
Yeltsin and U.S. President George Bush a few days after the Goldin–Koptev
meeting, Russia and the United States announced their intention to put
"space cooperation between the two countries on a new footing, reflecting
their new relationship." The two governments agreed to give consideration
to the following:

> flights of Russian cosmonauts aboard a Space Shuttle mission (STS
> 60), and U.S. astronauts aboard the Mir Space Station in 1993; and a
> rendezvous and docking mission between the Mir and Space Shuttle in
> 1994 or 1995 (Ref. 20, pp. 220–221).

Thus, at the dawn of broadscale Russian–American cooperation in space,
the cooperative activities between the two countries for a number of technical, political, and economic reasons tended to go beyond a customer–subcontractor relationship within the ACRV framework. On 5 October 1992,
this tendency resulted in a more detailed agreement ("Implementing
Agreement Between the National Aeronautics and Space Administration of
the United States of America and the Russian Space Agency of the Russian
Federation on Human Space Flight Cooperation"), signed by RSA General
Director Koptev and NASA Administrator Goldin, concerning "three interrelated projects: the flight of Russian cosmonauts on the U.S. Space Shuttle,
the flight of U.S. astronauts on the Mir Space Station, and a joint mission
involving the rendezvous and docking of the U.S. Space Shuttle with the Mir
Space Station" (Ref. 20, p. 223).

The potential for broadening cooperation was even built into a $1 million contract concerning the study of the Soyuz spacecraft as an interim ACRV, signed in June 1992, between NASA and NPO Energia.[55] The use of the Soviet spacecraft was supposed to enable an early permanent manned capability (PMC) in 1997. The goal of the study was to provide a detailed description of Soyuz as well as the cost and schedules for vehicles with the modifications necessary to interface the Soyuz with SSF and the Shuttle. This study was part of the plan

> to acquire the Soyuz TM in conjunction with American industry. By achieving the capability for early PMC in 1997, the Space Station program will be able to significantly expand SSF science accomplishments during the years prior to the presently planned target date for PMC in the year 2000.[51]

Besides the use of the Soyuz spacecraft as an ACRV, this study covered three other areas that were clearly related to the potential application of Soviet space technological and life science achievements to the creation of a permanently manned orbital outpost. One of these areas was the Soviet automated rendezvous and docking system, another was Mir biomedical and environmental monitoring, and the third was called "additional application of Russian vehicles and systems to the U.S. space program and missions."[56]

In August 1992, NPO Energia signed contracts both with Lockheed—concerning the manufacture of a full-scale Soyuz spacecraft mock-up—and with Rockwell, concerning information about the Soyuz-TM spacecraft landing system. In December 1992, NPO Energia and NASA signed an agreement that eventually could have led to the use of the Soyuz-TM spacecraft as a lifeboat. The agreement was made after a group of U.S. and Russian specialists completed two weeks of meetings at the NASA Johnson Space Center to discuss the feasibility of using the Soyuz-TM capsule as an interim ACRV for the space station crews. No technical problems were identified that would prevent the Soyuz from being adapted to dock at SSF during the early phase of PMC.[57] In March 1993 NASA and NPO Energia signed another contract concerning a more detailed study of Soyuz-TM. The goal of the contract was not just to determine the feasibility of Soyuz as an ACRV but also to develop a concrete plan for a Soyuz-TM-based lifeboat. The major problem to be solved was how to extend the on-orbital duration of Soyuz from six months to three years. In December 1993, the development of technical documentation and preliminary tests of hardware were under consideration. However, these plans were never realized. A Soyuz full-scale mock-up was built for Lockheed, but the company never bought it and had to pay compensation to NPO Energia. This all happened because political factors were added to space policy factors. Such a combination ultimately resulted in the redesign of the SSF and reconsideration of Russia's role in the redesigned space station.

CLINTON ORDERS REDESIGN OF THE SPACE STATION

The first political (or economic) factor that prompted radical redesign of the space station was the policy of the new U.S. President, Bill Clinton (elected in November 1992), to reduce the huge budget deficit. The order to redesign the space station was issued by Clinton on 9 March 1993. The task was supposed to be fulfilled within 90 days. The goal of the redesign was to reduce the cost of SSF. The president wanted to see three options, costing $5 billion, $7 billion, and $9 billion. Even the most expensive $9 billion version was less than half as much as the 1993 overall cost estimate of $19.4 billion (Ref. 52, p. 68).

The president's intention to stretch the budget was not the only reason for his recommending the redesign of the station. While campaigning for the presidency, Clinton dwelled on the new opportunities for the U.S. civilian space program offered by the end of the Cold War and expressed his support for the program, particularly for "its role in building new partnerships with other countries." He promised to seek "greater U.S.–Russian cooperation in space [which would] benefit both countries, combining the vast knowledge and resources that both countries had gathered since the launch of Sputnik in 1957." In Clinton's view, if "organized effectively," Space Station Freedom could "pave the way for future joint international ventures, both in space and on earth."[58] Thus, from the beginning of his presidency, Clinton was suggesting that joint space activities could bring Russia and the United States closer together.

President Clinton had good reason to consider the space program as one of his foreign policy tools, particularly in dealing with Russia. The beginning of Clinton's first term in office was marked by the end of the honeymoon in Russian–American relations because of the changes in Russian foreign policy.

CHANGES IN RUSSIA'S FOREIGN POLICY

From its inception, the foreign policy of independent Russia has been dominated by five schools of thought. The first group was the "Westerners" headed by the Russian Foreign Minister Andrei Kozyrev, and sometimes by Yeltsin himself. The urgent goal of this group was political and economic integration of Russia into the West. The second school of thought believed that Russia's interests and responsibilities should be focused, above all, in the new states of the "near abroad," i.e., the former Soviet republics. The third, the moderate nationalist and centrist school of thought, also considered Russia a great power, but one resting on an ethnically defined domestic base, with authority residing in a strong state. A fourth, the isolationist and Slavophile school of thought, saw Russia as being in the midst of spiritual rebirth. A fifth, the extreme-nationalist and neo-communist school, combined some elements of the third and fourth schools, but took them to absurd length by blaming all of Russia's current problems on foreigners, "masons," and other obscure forces.

The ideas of the first, pro-Western group dominated policy formulation and implementation in relations with the West on arms control and regional problems and on Russia's position in the United Nations and the Conference on Security and Cooperation in Europe from August 1991 to the middle of 1992. Then, Russian foreign policy made a leap from "a more self-effacing and passive role seeking entry into the international community to a more assertive pursuit of national interests." This was the result of three factors.

1) A need to pay more attention to the outside world after the "initial imperative concentration" on internal problems.
2) A tactical move by Yeltsin to defend himself against criticism by his political opponents that he was "making up" to the West and basically selling Russian national interests for symbolic economic aid.
3) The "shock therapy" that brought the whole Russian economy and social structure to the brink of collapse.

Such domestic developments had their impact on the Kremlin's foreign policy. The Westerners were pushed aside by nationalists and hardliners. Moreover, the military allied with the latter in formulating foreign policy.

Examples of the impact of the military and nationalists on Russian foreign policy in 1992–1993 included attempts by the Russian military to play a more independent role in Russia's policy toward members of the Commonwealth of Independent States, such as Moldova, Georgia, and Tajikistan; a twice-postponed Yeltsin visit to Japan because of the "nationalist–military" team's fear that Yeltsin would trade away the four southernmost islands in the Kurile chain; the more pro-Serbian character of Russian policy toward the former Yugoslavia; the continued selling of Russian submarines to Iran and weapons to China; Russia's insistence on lifting the embargo against Iran; and finally, Kozyrev's refusal to sign NATO's Partnership for Peace program (Ref. 2, p. 781; Refs. 59–61).

Although all of these tendencies showed an increasing nationalization of Russian foreign policy, they were not of immediate concern to the United States because all of them had a regional Eurasian character. However, there was one development that could have had implications for international security and thus alarmed the American government.

GLAVKOSMOS–ISRO ISSUE

On 19 November 1990, an article published in the Indian newspaper *Economic Times* detailed the negotiations between Glavkosmos and the Indian Space Research Organization (ISRO) concerning the possible selling of cryogenic engines (both hardware and technology) to India. Although India claimed that the engines were destined for an entirely legitimate, peaceful space program, the U.S. government regarded the deal as a violation of Missile Technology Control Regime (MTCR). Even though launch vehicles boosted by cryogenic engines are not considered practical in modern warfare because of the significant amount of time needed to fuel the rocket with liquid oxygen and hydrogen, cryogenic engines potentially could be used for building long-range and high-payload ballistic missiles.

In 1990–1991, the United States made a number of demarches regarding the Russian–Indian deal, but all of them remained unanswered in the light of the growing influence of political hardliners. By 1992, the Glavkosmos–ISRO issue became a theme of "almost daily exchange" between Moscow and Washington and was one of the most important topics of the U.S.–Russian diplomatic agenda for 1992–1993. In early January 1992, during the meeting between Russian Foreign Minister Kozyrev and U.S. Secretary of State Baker, the U.S. side brought up the possibility of trade sanctions against Glavkosmos. However, this did not persuade the Russians to terminate the deal with India.

As a result of Russia's intransigence, the United States imposed sanctions on Glavkosmos and ISRO. The sanctions included a "two-year ban on 1) sales to both organizations of any equipment, including parts, requiring an American export license; 2) contracts with Glavkosmos and ISRO by U.S. governmental agencies; and 3) import to the United States of any products of these organizations." Although the sanctions did not really affect managerial and production units of the Russian space program, other than Glavkosmos, the U.S. move strained Russian–American relations without having any real impact on the Russian–Indian deal. Russia continued to fulfill the contract and the official U.S.–Russian dialogue on this matter was suspended until the Clinton administration came into the White House in January 1993.

Soon after Bill Clinton took office, the cryogenic rocket engines issue became the subject of intense activity on the part of the new president's National Security Adviser Anthony Lake and his staff. The immediate topic to be discussed was whether, in the face of continued Russian and Indian implementation of the deal, the U.S. government should impose sanctions on bodies of the Russian space program other than Glavkosmos. The issue was of special concern to Vice President Albert Gore who, while a Senator, had been an author of the missile sanctions law. The White House was also pressed by Congress for resolute action on the matter, and the Clinton administration realized that its attitude toward the Glavkosmos–ISRO deal would be a test of the White House commitment to nonproliferation. Meanwhile, Russia continued on its course. In January 1993, President Yeltsin issued a decree, "Concerning Imposition of Control over Export from the Russian Federation of Equipment, Materials and Technology Used for the Creation of Rocket Weapons," and approved a list of goods that could be exported only by licenses (Ref. 62, p. 89). However, during his visit to India in late January 1993, Yeltsin also assured Indian Prime Minister Narasimha Rao that the Kremlin remained committed to the cryogenic engine and technology sale.

U.S. INCENTIVES TO PREVENT THE GLAVKOSMOS–ISRO CONTRACT

Although the initial intention of the Clinton administration was to strengthen sanctions against Glavkosmos and possibly impose them on the other Russian space organizations and enterprises, the cooling of Russian–American relations prevented the White House from choosing this course of action. Instead, the Clinton administration decided to place greater emphasis on the "carrot" rather than on the "stick" and to consider expanding space cooperation with Russia as a means of demonstrating support for President Yeltsin and encouraging Russia to stay closer to the U.S. position on MTCR. Moreover, the Clinton administration decided that it would not insist on complete cancellation of the Glavkosmos–ISRO contract, but would use its incentives strategy to attempt to make Russia reverse only the deal's most sensitive component, the transfer of cryogenic engine production technology.

To facilitate the implementation of this plan, the Clinton administration took three steps. The first was the establishment of a special interagency task force, called the Space Cooperation Working Group. This group consisted of members of the U.S. trade representative's office; the Departments of Defense, Commerce, and Transportation; NASA; and the National Oceanic and Atmospheric Administration. The group held meetings approximately once every two weeks to discuss ways of broadening international cooperation in space.*

The second step was taken during the first summit meeting between Clinton and Yeltsin in Vancouver on 3–4 April. Clinton proposed that if Russia formally agreed to follow the guidelines of the MTCR, and if it promised not to transfer the cryogenic engines production technology to India, the United

---

*The background information on the Glavkosmos–ISRO contract was taken from Ref. 63. pp. 23–27, 32–34, 37, 44–45.

States would not oppose the transfer of the cryogenic engines themselves, and would even sponsor Russia for a full membership in the MTCR. Specifically, Clinton indicated that the United States would be willing to enter into agreements to guarantee substantial Russian participation in the space station project and to permit greatly increased Russian participation in the commercial space-launch market. According to U.S. officials, Yeltsin accepted these proposals (Ref. 63, p. 46). The two sides also agreed to establish a special U.S.–Russian Joint Commission on Economic and Technological Cooperation in Energy and Space, to be chaired by Vice President Albert Gore and Russian Prime Minister Victor Chernomyrdin. The commission was supposed to meet semiannually, starting in late June 1993, to discuss energy and space issues. This board was later called the Gore–Chernomyrdin Commission.

The third step of the Clinton administration that contributed to Russian involvement in the space station program was the establishment, in April 1993, of the Advisory Committee on the Redesign of the Space Station, also known as the "Vest Committee" (after its chairman, Charles Vest). The goal of the committee was to evaluate NASA proposals concerning different variants of the redesigned space station (Ref. 52, p. 68).

Although "the costs of the redesigned options, and of Space Station Freedom, were the focus of much of the Committee's effort" (Ref. 64, p. 3), Charles Vest nevertheless wrote to the director of the Office of Science and Technology Policy and White House Science Adviser John H. Gibbons, "requesting a statement of the Administration's first-level objectives for the space station and its strategic goals for the civil space program." In his response, Gibbons indicated the following:

> International cooperation in space activities can help the international community move beyond the Cold War. Working with our existing partners in Europe, Japan, and Canada, and with Russia and other parts of the emerging democratic world, we can forge additional relationships and contribute to global peace and prosperity. International cooperation in space...can provide important lessons on how nations, working together, define challenges and solve problems that no one nation alone could accomplish (Ref. 64, p. 70).

The administration's approach to the possibility of broadening international participation in the space station project enabled the Vest Committee to consider the benefits of bringing the Russians into the space station program. The first such benefit was a reduced risk of dependence "upon the space shuttle as the sole launch vehicle." The committee suggested using a number of alternative launch vehicles, such as the Russian Proton.[64]

The second benefit was the previously considered use of the Soyuz spacecraft as an ACRV. The plan for using a Soyuz spacecraft as an ACRV, despite all of its advantages, had one drawback. At that time, the Soyuz capsules needed to be deorbited every year or two for refurbishment before returning to the station. If the station inclination was 28.8 deg, the Shuttle needed to be refitted to carry a modified Soyuz into orbit in the cargo bay. This would have increased the cost of using Soyuz, thus diminishing one of the advantages of using the Russian space hardware. However, the Vest Committee offered a different approach to the problem: changing the orbital inclination of the station to 51.6 deg so that its orbit would be reachable by Soyuz boosters from the Baikonur Cosmodrome. This would have enabled Soyuz to be launched as an ACRV "on its own tested rocket rather than going through a costly adaptation to shuttle."[64]

Besides preserving the relatively low cost of using Soyuz spacecraft as an ACRV, the Russian orbital inclination would have a few more advantages: It would enable the use of Progress cargo vehicles for the resupply of the station, it would have "most of the U.S. and Kazahkstan territory in which to make its normal land recovery" should astronauts need to suddenly return to Earth, and a 51.6-deg orbit would have been accessible to vehicles launched from Japanese and Chinese sites.[64]

The third benefit of bringing the Russians into the space station program would be gaining access to Russian risk management techniques and experience, while the fourth would be important Russian capabilities that could be advantageous to the redesigned space station, such as "automated rendezvous and docking hardware, environmental control and life support systems, Mir and other mechanical components." The authors of the report concluded:

> Russian automated rendezvous and docking hardware would potentially permit the use of a common docking capability for both the Soyuz-Progress vehicles and for NASA's shuttle. With almost 20 years of experience, the Russian environment control and life support systems would help minimize electrical power requirements for the station. Finally, Mir offers another potential for early joint-nation cooperative research opportunities (Ref. 64, p. 59).

The overall recommendation of the committee to NASA and the Clinton administration was to "further pursue opportunities for cooperation with the Russians as a means to enhance the capability of the station, reduce cost, provide alternative access to the station, and increase research opportunities" (Ref. 65, p. 59).

LACK OF FUNDING FOR THE RUSSIAN SPACE PROGRAM

Parallel to the activities of the U.S. and Russian political leadership, the goal of increased U.S.–Russian cooperation in space through joint participation in the space station program was pursued by the space communities of the two countries. The desperate economic situation in the Russian space program made cooperation with the West virtually a key condition for its survival.

In present-day Russia, even after the formal separation of military and civil space activities and the placing of the latter under the aegis of the RSA, it is still difficult to determine the actual budget of the civil space program. Unlike most countries, Russia does not specify allocations to any particular space project in its federal budget. The amount of financing is determined by the previously achieved level of financing and the general economic situation in the country. The RSA's budget request is considered by the Ministry of Economics (with respect to the main areas of spending), and by the Ministry of Finances (with respect to its influence on the general structure of the federal budget). Having been evaluated by these bodies, the budget is passed to the State Duma (the lower house of the Russian parliament) for consideration and adoption. RSA's budget request is usually not changed, either by the Duma or by the Federation Council (the upper house of the Russian parliament). The final approval of the budget is made by the president (Ref. 29, p. 8).

Even taking into consideration the preceding complexity in defining the actual size of the Russian space budget, funding of the Russian space program was estimated at 72 billion rubles in 1993. According to some researchers, this figure constituted only 1% of the previous level of financing.[65]

RUSSIA'S DECISION TO COOPERATE ON THE SPACE STATION

A decisive step toward cooperation with the United States in the space station program was made by the managers of the Russian space industry in March 1993. On 5 March 1993, a delegation from NPO Energia, headed by Yuri Semyonov, came to Seattle, Washington, to conduct negotiations with the representatives of Boeing. The negotiations took place between 6–13 March. Among the issues discussed was the creation of the ISS using elements from the Mir 2 and Freedom stations. The ISS was supposed to consist of the Russian base, service, and docking modules and the air lock and the spacecraft Soyuz-TM and Progress. The United States was supposed to contribute the power, laboratory, and logistic modules. Based on the results of negotiations, RSA General Director Koptev and NPO Energia General Designer Semyonov addressed NASA Administrator Goldin with a letter containing a formal proposal to join efforts in the creation of the ISS. The letter stressed in particular that Russia

> has gained a lot of experience in the field of design and operation of the manned multi-purpose orbital complexes (Salyut, Mir), and is currently considering the concept of the Mir 2, a Space Station of the new

generation. The initial elements of Mir 2 are supposed to be put into orbit in 1997. We are also familiar with efforts of the United States and its partners to build the space station Freedom.

Such labor-intensive projects are not just limited to the solution of complex scientific technological matters. They also require significant financial resources. This financial burden combined with the growing complexity of the orbital manned facility and the tasks they are to accomplish becomes a serious problem. This problem requires a constant search for different solutions. One possible solution could be uniting of the scientific and technological efforts of the United States and Russia in working towards a future joint Space Station....

The expert assessment showed that by uniting technical resources within the framework of the future international program can save up to a few billion dollars as compared to the current projected expenses of separate national space station programs.

We believe that [our offer] has an enormous potential for national and international interests. We are ready to discuss with you this very important issue at your convenience (Ref. 3, pp. 496–497).

One of the factors contributing to Russia's integration into the Space Sation program was a growing productive business relationship between the Russian and U.S. space professionals. The Americans continued to have a significant presence in Russia, and the Russians in the United States, working on the Shuttle–Mir program. Impetus developed to see if the Russians had anything to contribute to the U.S. redesign that "could make life easier and the Station more functional."[66]

Although no formal decision was made at that time about Russian participation in the redesigned space station, at the end of April, NASA invited a group of the Russian consultants to provide the U.S. redesigning team with their advice. About 30 Russians were brought to the United States. The group included the leading Russian space station designers and space program managers, such as Yuri Semyonov and Yuri Koptev. Also invited were representatives of the Salyut design bureau, the Central Scientific-Research Institute of Machine Building (TsNIIMash), and the Institute of Biomedical Problems. The group of Russian consultants was located in Crystal City, Virginia, next to the building where the redesign team worked. The meetings between the Russians and the U.S. redesign team took place from April to July and from October to November 1993.[66]

The Russian delegation proposed to make a station that would combine Freedom and Mir elements. The U.S. redesign team, however, did not accept the Russian proposal. It offered instead a redesigned station that would include only one Russian element, a Soyuz-TM lifeboat.

The proposal of the U.S. redesign team did not fully meet the expectations of the space leaderships of the two countries, the U.S. Congress, or the president's administration that international participation in the space station program could be broadened, particularly through partnership with Russia. At the end of May, Koptev and Goldin signed a confidential Memorandum of Understanding concerning U.S.–Russian cooperation in space. The memorandum stated that "the U.S. has initiated a redesign of its Space Station program. Part of this redesign activity was focused on providing interim capability utilizing the Mir Station and the Space Shuttle as well as utilizing Russian assets to support a redesigned U.S. Station." The provision of the interim capability consisted of two parts: current activities and possible interim activities. The former included Shuttle–Mir rendezvous and docking missions (one of each), a 90-day stay on Mir by a U.S. astronaut, and two flights for two Russian cosmonauts (one flight for each) on the Space Shuttles. Possible interim activities included:

1) The flight of one U.S. astronaut on MIR, twice each year from 1995 through the planned duration of the MIR Station.

2) Placement of U.S. experiments and equipment in the Spektr and for Priroda Modules with an emphasis on life sciences. Placement of the payload onto Mir would occur as early as 1995 with data collection through the planned duration of the MIR Station.

3) The U.S. and Russia will consider joint missions to extend the useful lifetime of the MIR station and enhance its capability to accomplish meaningful science and technology experiments.

4) The transport of the U.S. and for Russian crewmembers will be either by Shuttle or Soyuz.

5) Two Space Shuttle flights will be flown to Mir in 1995 and four flights in 1996 and in each subsequent year through the duration of the MIR Station. U.S. crew members will stay for durations up to 6 months or as required to perform their missions.[68]

These cooperative missions were supposed to "not only serve to perform meaningful science and technology experiments, but also serve to begin validation of joint interface concepts (life support, power, command, and telemetry, etc.) that could support a unified U.S.–Russian Space Station." Although at that time the sides were not talking about a "joint" but a "unified" station, consisting of U.S. and Russian elements that could form cores of separate space stations, they agreed that "the U.S. element must be large enough to accommodate Russian participation on a meaningful basis, as a part of the unified Space Station, if Russia should choose a more integrated approach." The overall conclusion of the memorandum was that "a unified U.S.–Russian Space Station is feasible, and could operate in a fully integrated fashion with both sides contributing to the unified Space Station."[68]

DISSATISFACTION WITH THE SPACE STATION REDESIGN

One sign of Congress' dissatisfaction with the results of the redesign team's activity was the questions, comments, and suggestions by the members of the Senate Subcommittee on Science, Technology, and Space during the July 1993 hearings, almost two months after the team finished its work. Senator John Rockefeller, chairman of the Committee, asked the question:

> I want to know why, for example, the administration and any other space station proponents believe that this is the place to spend that money, the tens of billions of dollars, and not on, for example, other space programs, other science efforts, other aeronautical pursuits that would come under NASA, and obviously, also, other needs of this country (Ref. 69, pp. 2–3).

Senator Conrad Burns was "anxious to hear from NASA why it is in the national interest to spend $10.5 billion over the next five years on the Space Station" (Ref. 69, p. 6). Senator Ernest Hollings, noting "the uncertain cost of the redesigned space station, and the decision not to send humans to Mars or the Moon," proposed to consider "other more practical and affordable solutions" (Ref. 69, p. 4).

The reason for the overall dissatisfaction with the final version of the redesigned space station was that "none of the fully-implemented phases of the three station redesign options meets the Administration's cost targets of $5 billion, $7 billion, and $9 billion for fiscal year 1994 through fiscal year 1998, nor does any option meet the annual funding target while simultaneously achieving the schedule milestones desired" (Ref. 69, p. 15). The station, according to Louis Friedman, executive director of the Planetary Society,

> is not yet worth the cost. It can be. The seeds are there, and the Vest committee recommendations provide for it. And the NASA Administrator and the Clinton administration have stated their support for some of the necessary additions to the plan.... In a postcold war world, the idea of international space station, excluding the Russians, seems contradictory (Ref. 69, pp. 35–36).

Chairman of the House Committee on Science, Space, and Technology George Brown believed that inviting Russia into the space station program would have economic benefits above all. According to him,

> NASA found after its redesign activities that it still could not fit the program into the budget that was mandated by OMB [Office of

Management and Budget]. NASA began to look to the Russian participation as a means of reducing cost (Ref. 70, p. 2).

However, the support provided by the White House and the NASA administration for another redesign of the station was clearly motivated by foreign
policy considerations. Clinton suggested that "one element of the Space
Station Program could itself help define the new era that will follow the Cold
War—international cooperation on the space frontier" (Ref. 70, p. 7). By
linking together international cooperation in the space station program and
the end of the Cold War, Clinton made it clear that he was considering inviting Russia to join the program. NASA Administrator Goldin, while outlining
six basic rationales for the United States to build the space station, called the
following rationale "a very key factor": "As we leave a century full of war,
where nations worked together on megaprojects that developed weapons, we
have an opportunity on the largest project in history, to bring nations together for peaceful cooperation, to signal a new era" (Ref. 70, p. 17).

POLITICAL SIGNIFICANCE OF COOPERATION WITH RUSSIA

The reason for the increased political significance attached by Clinton and
Goldin to the space station project by July 1993 were the developments in
U.S.–Russian relations occurring between April and July. In the weeks following the Vancouver summit in early April, Yeltsin failed to fulfill the agreement concerning implementation of the MTCR's guidelines and cancellation
of the transfer of cryogenic engines technology to India. This caused another
wave of dismay in the United States. While emphasizing the benefits of
U.S.–Russian cooperation in space that Russia might enjoy by restricting its
deal with the ISRO, the White House threatened to extend its sanctions
beyond Glavkosmos and to impose them on other parts of the Russian space
complex, such as the Salyut and Chemical Machine Building design bureaus
and the Khrunichev missile and rockets production plant. The sanctions were
expected to cover four major cooperative projects. The first one was to prohibit the use of Russian launch vehicles to launch the Inmarsat 3 satellite,
which contained U.S. components; the second was to freeze Russia's participation in the Motorola Iridium project; the third was the expulsion of Russia
from the ISS project; and the fourth was the cancellation of a number of other
smaller contracts. Because of the proposed santions, the overall estimated
cost of the losses exceeded $1 billion, of which approximately 50%
($400–$600 million) resulted from expulsion from the ISS (Ref. 63, pp.
49–50).

The sanctions, scheduled to start on 24 June, were postponed until 15 July.
The fledgling U.S.–Russian partnership in space was very fragile and could
easily have fallen apart. According to Richard Beatty, who was in Moscow

at that time with Arnold Nikogossian, director of the NASA Office of Life and Microgravity Sciences and Applications, to discuss the science program for the first Shuttle–Mir flight and the extension of the Shuttle–Mir program beyond the first docking, Nikogossian called Goldin every night to check on the status of U.S.–Russian cooperation.[53]

The cooperation, however, survived for a number of reasons. First, the contract with India proved to be more modest than expected; second, the deal boosted the international competitiveness of only a very limited number of Russian space enterprises, while doing almost nothing for others. Finally, the pro-American faction in Yeltsin's "inner circle," such as foreign minister Kozyrev, managed to convince him to give preference to U.S.–Russian cooperation in space (Ref. 62, p. 90). As a result, the United States and Russia worked out a compromise by 16 July. According to a "Memorandum on Mutual Understanding of the Issue of the Export of Missile Equipment and Technology," Russia pledged to abide by the "standards and criteria" of the MTCR. Russia also signed a "Memorandum on Mutual Understanding Regarding Existing Contracts," under which it was "allowed" to transfer to India the cryogenic rocket engines but not the production technologies. The United States agreed to lift the existing sanctions on Glavkosmos and not to impose new sanctions on other Russian space enterprises.

REACTION AMONG ORDINARY CITIZENS

The solution of the cryogenic engines crisis removed one of the obstacles to successful development of Russian–American cooperation in space and was by this time welcomed not only by politicians and space professionals but also by ordinary citizens of the two countries. For example, on 15 July 1993, on the 18th anniversary of the Apollo–Soyuz flight, Star City and the city of Nassau Bay signed a sister-city agreement. The idea to establish such a relationship came from the Nassau Bay city manager, David Stall.

The choice of Nassau Bay was not accidental. Part of the NASA Johnson Space Center is located in Nassau Bay, and there is still a barge dock on the lake within the city limits—Clear Lake—where Apollo components used to arrive on barges. In addition, at the time Stall started promoting his idea, two Russian cosmonauts, Sergey Krikalev and Vladimir Titov, were already living in the city, preparing for their future Space Shuttle missions.* The city council of Nassau Bay recognized the significant contributions made by its residents and those of Star City to the advancement of space flight.

---

*Sergey Krikalev became the first Russian cosmonaut to fly a Space Shuttle mission in February 1994 (STS-60). Vladimir Titov was the second Russian to fly on the Shuttle (STS-63) in February 1995. During this mission, the Shuttle made a rendezvous with Mir, approaching the Russian station at 30 ft. The mission was considered a general rehearsal for the first Shuttle–Mir docking.

Therefore, it seemed appropriate to encourage cooperation and goodwill between the two communities.

Initially there were some difficulties arranging the sister-city relationship because Star City is a military base—much more like Edwards Air Force Base—rather than a regular city. However, with the intervention of two individuals, Valentina Tereshkova, the first woman in space* and past president of the Union of the Soviet Societies of the Friendship and Cultural Communication with Foreign Countries, and General-Colonel Pyotr Klimuk, chief of the Star City, the relationship was made possible. Whereas it normally takes years to establish a sister-city relationship, it took Nassau Bay and the Star City only two weeks (Ref. 67).†

U.S.–RUSSIAN OFFICIAL AGREEMENT TO COOPERATE ON ISS

In August 1993, RSA and NASA engineers developed the concept of Russian–American cooperation in the field of human space flight, ranging from the current Mir space station up to the construction of the ISS, which would combine elements of the Mir 2 and Freedom space stations. The Russian segment was supposed to include a base module, three docking modules, a science-power platform, a docking airlock unit, a service module with life support systems, and the spacecraft Progress-M and Soyuz-TM. NASA accepted the new orbital inclination of 51.6 deg. This inclination was also a compromise on the Russian side: initially Mir 2 was supposed to be launched at an orbital inclination of 65 deg for a better observation of the Russian territory (Ref. 3, pp. 498–499).

The process of the official involvement of Russia into the space station program was completed between September and December 1993. On 2 September, Prime Minister Chernomyrdin and Vice President Gore signed a "Joint Statement on Cooperation in Space." This statement delineated a concept for the future U.S.–Russian cooperation in the area of orbital space stations. It stressed, in particular, that

> [g]iven the particular importance for Russia and the U.S. of their respective efforts in developing a new generation of orbital stations for scientific and technological progress and human activities in space, the

---

*Valentina Tereshkova blasted off in Vostok-6 on 16 June 1963. She landed on 19 June 1963.

†The sister-city relationship was embraced by the residents of Nassau Bay who wanted to become involved in social activities related to residents of the Star City. They had a great interest in having community events, initially as a barbecue/picnic type of arrangement that would be very local and very personal. This activity gradually developed into a Russian festival, which grew larger and more encompassing each year. The city sponsored transportation and lodging for Russian citizens and diplomats, even if the guests had to come from Russia. The festival was warmly received by the astronaut corps and considered by NASA to be a positive example of a relationship that was above the politics of the two countries and their space agencies. Support for this yearly festival was provided by NASA and a range of local companies and organizations.

Parties regard further cooperation in this area as most important, and consistent with the interests of both Russia and the U.S., as well as the broader international community (Ref. 20, pp. 228–230).

The statement also provided an initial phased schedule for the realization of the U.S.–Russian cooperation in the space station project. The first phase encompassed "an expansion of our bilateral program involving the U.S. Space Shuttle and the MIR Space Station." Subsequent joint efforts on the second phase were to be directed "to the use of a Russian MIR module of the next generation, in conjunction with a U.S. laboratory module.... Successful implementation of this phase could constitute a key element of a truly international space station." It was also envisioned that the United States

> will provide compensation for Russia for services to be provided during phase one in the amount of $100 million dollars in FY 1994. Additional funding of $300 million dollars, for compensation of phase one and for mutually agreed upon phase two activities, will be provided through 1997. This funding and appropriate agreements will be confirmed and signed by no later than November 1, 1993. Other forms of mutual cooperation and compensation will be considered as appropriate (Ref. 20, pp. 228–230).*

The statement stressed the predominant importance of U.S. and Russian efforts in building the space station. Both sides considered Russian and U.S. elements docked together as already completing "a unified Space Station," that "can offer significant advantages to all concerned, including U.S. partners Canada, Europe, and Japan" (Ref. 20, pp. 228–230).

It is not a coincidence that the sum of the contract was determined as $400 million. According to the Chairman of the House Committee on Science, Space, and Technology George Brown,

> Russian industry was on the verge of transferring highly sensitive propulsion technologies to India in exchange for up to $400 million in hard currency. The U.S. sought to end this relationship in view of its implications for missile technology proliferation. In effect, the U.S. offered Russia the opportunity to participate in the Space Station pro-

---

*According to Congressman James Sensenbrenner, the funds were to be distributed the following way: $309 million of the $400 million committed by the United States to the RSA under the Gore–Chernomyrdin framework was supposed to go to Phase 1, up to 10 Shuttle–Mir missions planned for 1994–1997, with the remaining $91 million used in Phase 2 activities such as codevelopment of space technologies (Ref. 70, p. 27). The cost of Phase 1, which ended in July 1998, turned out to be $486 million, out of which only $295 million were paid for Shuttle–Mir operations. The rest was paid for hardware, research, and program management. Taking into consideration transportation costs, the United States spent more than $4 billion over the life of the program (see Refs. 66 and 71).

gram in exchange for a commitment to terminate sales of sensitive missile-related technology (Ref. 70, p. 2).

On the other hand, Robert Clark, then the NASA Associate Administrator for International Relations and Policy, believed that there was no link at all. He argued that nobody knew how much money the Russians actually lost because of the cancellation of their contract with the ISRO. Clark believed that the $400 million mentioned in the "Joint Statement" was one more sign of NASA's determination to bring Russia into the space station program, and that without strong technical reasons for having Russians in the space station, which NASA managed to master, the politics and the MTCR issue could not have carried the weight.[72]

ADDENDUM TO PROGRAM IMPLEMENTATION PLAN

On 1 November 1993, NASA Administrator Goldin and RSA General Director Koptev signed an Addendum to Program Implementation Plan that described "a new relationship between the U.S. and Russian space agencies which will advance their national space programs and benefit their respective national aerospace industries." The plan contained a detailed description of three phases in U.S.–Russian cooperation in the space station program. Phase One included, in particular,

> four or more U.S. astronaut flights aboard the Mir station for a total on-orbit stay time of approximately 24 months, and up to ten Shuttle flights to Mir between 1995 and 1997. The U.S. Shuttle will assist with crew exchange, resupply, and payload activities for Mir.
> Mir capabilities will be enhanced by contributions from both the U.S. and Russia. The Shuttle will bring new solar arrays to replace existing arrays on Mir. Russia will add Spektr and Priroda modules to Mir. These modules will be equipped with U.S. and Russian scientific hardware to support science and research experiments (Ref. 73).*

Phase Two was supposed to combine "U.S. with Russian hardware to create a totally new, advanced orbital research facility with early human-tended

---

*It was decided originally that the first mission to Mir would be three months long and the next three were to be six months apiece. Nevertheless, the United States was planning to send the Shuttle to Mir every three months so that there would be midterm Shuttle flights, strictly logistical in nature. But the Space Shuttle program could not support that many flights. It was decided consequently that the United States would fly only seven missions to Mir. There would be no three-month-long and six-month-long missions but rather they would be spread more or less evenly among four- to five-month-long flights. In 1996, when it became clear that the Russians would not meet the original launch dates for the ISS elements, two more long-duration missions onboard the Russian station were added, bringing the total number of Shuttle-to-Mir flights to nine.[66]

capability. This facility will significantly expand and enlarge the scientific and research activities initiated in Phase One and will form the core around which the international space station will be constructed." The final Phase Three was supposed to complete "construction of the international space station" by attaching to the station the hardware developed and built not only by Russia and the United States, but also by other space station partners— Europe, Canada, and Japan.

HARDWARE AND FINANCIAL RESPONSIBILITIES ESTABLISHED

The addendum also specified the U.S., Russian, European, Canadian, and Japanese hardware to be contributed to the station, along with the schedule for its delivery. The major Russian elements for Phase Two of the station assembly included the FGB (*Funktsionalnyi Gruzovoi Blok*) or the functional cargo module (paid for by the United States), which was to provide guidance, navigation, and control capabilities and a service module, which would become a crew compartment and provide the reboost capabilities for the station. For Phase Three, Russia was to provide one more service module, three research modules, and two ACRV/Soyuz-TM spacecraft. The major U.S. contributions for Phases Two and Three included Node 1, the Docking Node, different kinds of trusses, the habitation module, and lab outfitting. Also, a few elements of the station, such as the Airlock/STS docking adapter, Trusses 1 and 2, gyrodynes, batteries, PV arrays, and solar dynamic elements were supposed to be the product of joint U.S.–Russian efforts. These efforts might include U.S. payments for the Russian-made elements (such as FGB) or the mutual development of certain elements.[73]

The addendum also delineated the financial responsibilities of both the U.S. and Russian sides. It confirmed that NASA would award a firm fixed price contract to RSA for fiscal years 1994–1997. The funding level of the contract was to be approximately $100 million in each of the four fiscal years. According to Robert Clark, the contract covered Phase One and part of Phase Two activities. Although it was clear at that time that NASA would purchase the FGB module, no fixed price was set for this piece of hardware.[72]

FORMAL INVITATION FOR ISS PARTNERSHIP

After signing the addendum, Russia and the United States came very close to the formalization of bilateral cooperation in the ISS program. This process, however, required the consent of the other space station partners, including Europe, Canada, and Japan. European and Japanese government officials "reacted with stiff upperlips...to the announcement that the United States and Russia would join forces to build a single international space station." They privately expressed "concern that the U.S.–Russian accord would cost them millions of dollars in modifications to their own space station laboratories, and several years of delay in the launching of their hardware." The objections of the partners had a basically technical character. Some of the experiments designed by them "were chosen on the assumption that all the equipment would be launched with the laboratory module. If a changed orbit makes it impossible for the shuttle to carry a fully loaded laboratory, scientists may want to modify the kinds of experiments they place on the station."[74]

However, during a 16 October meeting in Paris, "recognizing the potential benefits to be gained, the space agencies of Canada, Europe, and Japan (CSA, ESA, NASDA), as well as NASA agreed to the need to complete an intense process at all levels that could lead to Russia becoming a full partner in the International Space Station."[75] This decision was followed by a formal invitation the United States, Europe, Canada, and Japan extended to Russia to join the program on 6 December 1993. Although the details and extent of Russian contributions still needed to be better defined, Russia had become de facto a full-fledged partner in the International Space Station Alpha.[*]

The decision to merge Russian and American efforts in the space station program was described by the Clinton administration as a major foreign policy breakthrough with Russia (Ref. 76, p. 3). President Clinton in his State of the Union address on 25 January 1994 made a direct link between space deals with Russia and the broader context of U.S.–Russian relations:

> This is a promising moment. Because of the agreements we have reached, Russia's strategic nuclear missiles soon will no longer be pointed at the United States, nor will we point ours at them. Instead of building weapons in space, Russian scientists will help us build the international space station.[77]

With reference to the space station deal with the Russians, Vice President Gore stressed that nowhere will a new partnership between the United States and Russia "be so keenly felt as in the area of high-technology cooperation."[78] The predominantly political character of U.S.–Russian cooperation in the space station program was also confirmed by John Gibbons, the White House Science Adviser. Although Gibbons acknowledged that "our work with the Russians, broadly speaking, we believe will result in lower costs for the space station program," he emphasized that Russian involvement in the program "was in no way devised as a way to make things work at $2.1 billion" (Ref. 76, pp. 42, 63).

---

[*]Russia became a full-fledged ISS partner de jure after the completion of Memorandum of Understanding/Intergovernmental Agreement process.

Cooperation with Russia in space, however, was expected to bring not only political benefits. It was thought that

> the Russian participation in the ISS Program would accelerate the assembly timetable and avoid substantial development costs in the areas of propulsion and navigation. It was also estimated that Russian contributions would nearly double the Station's on-orbit volume, allow an increase in crew size to six, and provide an earlier crew presence. After Russia agreed to participate, geographic constraints of launching elements from the Baikonur Cosmodrome in Kazakhstan necessitated a change of launch inclination from 28.5 degrees to one of 51.6 degrees. While the change negatively impacted the Shuttle's cargo carrying capacity for transport of elements and supplies to the Station by over 12,000 pounds per flight, the capability reduction was offset by the addition of 13 planned Russian assembly flights. Russian participation also allowed completion of ISS assembly to be accelerated from September 2003 to June 2002, a projected cost savings of $1.5 billion. The Russian provision of the Functional Cargo Block and an Assured Crew Return Vehicle were estimated to save another $1.0 billion. These savings were partially offset by new U.S. costs identified to integrate Russia into the Program. In total, a net estimated cost savings of $2 billion was projected from Russian involvement.

NASA stated at that time that it could develop the Space Station Alpha design within an annual fiscal constraint of $2.1 billion per year, and with Russian participation it could complete assembly of what had become known as the International Space Station for a total of $17.4 billion.[79]*

CONGRESS' CONCERNS ABOUT COOPERATING WITH RUSSIA ON THE ISS

The position of Congress concerning the space station deal with Russia, despite obvious political, economic, and technical benefits, was more complicated than that of the White House. It was expressed in the words of the Chairman of the U.S. House Subcommittee on Space, Ralph M. Hall: "The Administration's proposed joint space station program would represent a significant departure from the way the Nation's civil space program has been conducted to date, and we will need to understand the implications and costs, and also what is being proposed..." (Ref. 76, p. 1). Representatives had to weigh one set of arguments against the other.

The possibility of transferring American jobs to Russia was weighed against creating "more opportunities for U.S. businesses to work with Russian busi-

---

*Some experts believed that Russian participation in the ISS could save the program up to $3 billion (see Ref. 76, p. 173). Although these number had not been validated at that time, $2 billion was the lowest estimated saving.

ness on areas that will generate new jobs and technologies both in the United States and Russia" (Ref. 76, pp. 42, 49). The supporters of such a scenario also believed that even industry would not object to the transfer of some jobs to Russia if it could politically preserve the endangered program. "This is a radical departure," said Mark Albrecht, space policy chief under President George Bush. "But the alternative may have been to have the station canceled."[80]

The second major concern shared by lawmakers was "putting Russia on the critical path for enabling and not just enhancing capabilities, and handing over critical path functions that were never contemplated even with our long-term allies and current space station partners, the Europeans, the Canadians, and the Japanese" (Ref. 76, p. 3). Lawmakers realized that "the success of the U.S. program depends directly on financial, technical, and political factors related to Russia and other republics of the former Soviet Union..." (Ref. 70, pp. 2–3). This concern was weighed against the argument that "we have an Alpha plan independent of Russia that will enable us to move forward on the international space station that is U.S. led but has the same major participants of Japan, Canada, and the European Community" (Ref. 70, p. 36).

The third concern was about transferring funds to Russia, which along with control of the critical path, became a second major departure from NASA managerial principles.* Those who supported the idea of paying Russia for the design and construction of certain space station elements believed, like John Gibbons, that

> we are getting access to, first of all, a great reservoir of knowledge and experience that the Russians have had in operating the Mir station over a number of years, and we have an opportunity to offset at least— $100 million per year, for example, will only partially offset the costs the Russians will engage in, in having the Mir accessed by the Shuttle for docking and rendezvous experiments (Ref. 76, p. 49).

The same kind of argument was presented by Daniel Goldin, who, when questioned by Representative James Sensenbrenner about whether Russian participation would lead to an export of jobs from the United States, said that while a joint space station could save NASA more than $3 billion in operation and development costs, most NASA funds would be spent within the United States.[82]

Finally, there was concern "that Russian political problems may hurt our participation in the program." But there were two responses to this concern. First, "to begin with, political risk is a two-way street, as our European space

---

*For more information about the principles traditionally guiding the U.S. space policy, see Ref. 79.

station partners would be sure to note." Second, "more importantly, a well-constructed program, following the modular approach…, will not suffer unduly in the event of political problems. In the worst case, we would be no worse off than we are now" (Ref. 76, p. 7).

The process of evaluation by Congress of all the pros and cons of cooperation with Russia in the space station program was finished by the end of November 1993 when Clinton and Gore discussed the project with key members of Congress, including House Majority Leader Richard A. Gephardt and the chairpersons of the NASA appropriation subcommittees, Senator Barbara A. Mikulski and Representative Louis Stokes. After administration officials explained the details and Clinton expressed his full commitment, congressional leaders "unanimously agreed" to support the joint U.S.–Russian effort. "Those who met with the president," Mikulski said, "enthusiastically agree with the bold foreign policy goal he has set out to accomplish with the Russians through the space station." Representative George E. Brown, who chaired the House authorizing committee, and Senator John D. Rockefeller IV, chairman of the Senate science subcommittee, did not attend the meeting in the White House but were described as in agreement.[83]

RUSSIAN PARLIAMENT'S VIEW TOWARD COOPERATION

The attitude of the Russian parliament toward the U.S.–Russian space station deal is more difficult to assess than that of the U.S. Congress. Beginning in June 1993, parliament was engaged in an escalating confrontation with President Yeltsin, which ended with the dissolution of parliament on 21 September and its shelling on 2 October. Space issues were a low priority for Russian lawmakers, to say the least. After the shelling, the parliament did not formally exist in Russia until the new parliamentary elections took place in mid-December 1993.

However, even during such a turbulent period, state bodies did not completely neglect Russia's space activities. On 11 December 1993, the Government of the Russian Federation issued Decree No. 1282 concerning "State Support and Provision for the Space Activities of the Russian Federation" and adopted on the same day a "Federal Space Program up to the Year 2000." Along with other measures that were supposed to provide support for space activities in Russia, the decree envisaged the realization of the Federal Space Program to be supported in the same way as national defense activities.

According to the program, the broadening of international cooperation in space implied Russian participation in a number of projects, particularly in the ISS program. The international manned space program with Russian participation was to have three stages. The first stage was called Mir–Shuttle. It was to last from 1994 through 1995. This stage involved joint U.S.–Russian activities to be conducted on Mir and on the Shuttle, and also flights of Russian cosmonauts on the Shuttle and of American astronauts on Mir.

The second stage was called Mir–NASA.* It was to last from 1995 through 1997. Its main goal was to continue and deepen activities initiated during the first stage. The activities were to take place onboard the Mir station. Transport operations were to be provided by Soyuz-TM and Shuttle spacecraft.

The first and second stages were to be supported by a completed Mir space station, including the last modules to be attached to Mir—Spektr and Priroda. Overall, the Mir space station was to continue operation through 1997. The Mir budget was determined (relative to the prices in the first quarter of 1994 in millions of rubles) as 204,833.0 (1994), 259,834.1 (1995), and 1,688,178.3 (1996–2000).

The third stage was called Mir 2–Alpha. The realization of this stage envisaged the "creation of the International Space Station based on the Russian and U.S. national space programs (Mir 2 and Freedom)." The "flight tests" of the ISS were to begin in 1997, and the station was to be completed by 2005. The program assigned RKK Energia (at that time NPO Energia) as primary contractor of the Russian segment for the station.[40]

NEW ERA OF COOPERATION

AMERICAN PUBLIC'S VIEW

The attitude of the American public toward the new era of cooperation with the Russians in space, and particularly toward the U.S.–Russian space station deal, was overwhelmingly favorable. *The Orlando Sentinel* acknowledged the need to

> make the transition from competition with the former Soviet Union to cooperation. The Russian space program has unique assets, such as large rockets and an operating space station, that we could use at a fraction of the price we would have to pay to develop on our own.[84]

The Baltimore Sun described "the American-Russian venture in space" as "a visionary breakthrough that deserves congressional support—provided NASA at last puts its boondoggle days behind it."[85] *The New York Times* expressed confidence that

---

*Shuttle–Mir, NASA–Mir, Mir–Shuttle, or Mir–NASA are terms used to describe the same program involving flights of U.S. astronauts onboard the Russian space station and dockings between Shuttles and Mir. On the U.S. side, Shuttle–Mir was the term used, whereas the Russian side used NASA–Mir or Mir–NASA. In key documents, such as those issued by the Gore–Chernomyrdin commission, the program has always been referred to as Shuttle–Mir.

those who remember the terror raised in American hearts when the Soviet Union launched its first sputnik satellite in 1957, triggering the superpower race to the moon, should be pleased and relieved that the two nations are now moving toward a close in building and operating the world's next space station.

The paper noticed that although "the station will yield modest gains in understanding the health effects of living in space and the behavior of materials in the absence of gravity...given that a political decision has been made to proceed, incorporating the Russians can only enhance its value."[86] *USA Today* acknowledged that although "Russia brings a little money," it brings "a lot of knowledge of space station technology" and described the U.S. future in space with the Russians as promising.[87]

Although intensified U.S.–Russian cooperation in space was generally welcomed in the United States, some portion of the American public expressed skepticism about it. For example, *The Washington Times* saw three problems with this deal. One was from the perspective of a potentially "hostile government with very detailed insights into the state of our advanced technologies." The second problem was that, according to some space experts such as James Oberg, the unit "that Russia wants to sell us as a core module of the joint station is really an obsolete leftover from a canceled Soviet program to launch a space spy platform." The third problem was the old Soviet style in which Russian space officials conducted business. The U.S. negotiators apparently promised "to keep certain aspects of the Russian program secret, because that's the only way the Russians would release information. Since when, might one ask, is it the job of the U.S. government to keep secrets for the Russians?"[88]

RUSSIAN PUBLIC'S ATTITUDE TOWARD COOPERATION

The attitude of the Russian public toward renewed Russian–American cooperation in space, and toward the space station deal in particular, was more complicated than that of the American public. On one hand, there was an understanding that "such large-scale programs represent a source of funding for Russia's space-related enterprises, which have highly marketable intellectual potential but ran aground as soon as they tried to switch to the market." On the other hand, a losing global competition with the United States made a significant part of the Russian public again think about space exploration in terms of a space race, because space was "one of the few areas where Russia can talk to America as an equal." Many shared the view that "even a complete lack of contact [between Russia and America in space] would not have a severe impact on our manned program," whereas "the Americans are trying to save money by using Russian developments to make sure that their rather ambitious space programs don't get bogged down in Senate hearings that always accuse NASA of not treating the taxpayer with respect."[89] Some newspapers viewed Russian–American space cooperation as a direct result of concessions, such as the cancellation of the cryogenic engines deal with India, made by Russia "under pressure from Washington." The space station agreement was called by *Rossiiskaya Gazeta* "the strangest agreement that the world has seen for a long time."[90] A Russian aerospace expert, Vladimir Yaropolov, claimed in *Stolitsa* magazine that the U.S.–Russian space station deal was almost a total loss for Russia for a number of reasons. First, because of the ISS inclination, "the territory of the United States will be fully in the field of the observers. What is the practical value of such a project? The United States will have full 100 percent use of the capabilities of the orbital complex, but we'll get nothing out of it." Second, "the training of cosmonauts and their postflight examination will be fully concentrated in the hands of NASA," which would eventually result in the death of the Gagarin Cosmonaut Training Center. Third, in the author's view, "only nonprofessionals" could assert that

> this agreement is advantageous to us if for no other reason that there will be an interchange of technologies.... What can the Americans give us in the field of manned flights if our experience, our know-how, exceeds theirs by an order of magnitude!... Our Mir will be the "tent and toilet" for the Americans. The docking unit plays the role of a railroad station.

Yaropolov blamed the leadership of NPO Energia for the deal: "Speaking frankly, it was not without importance that cooperation with the Americans would bring many personal benefits for a whole group of higher directors in the aerospace branch. Even if it be in the form of constant trips to America!"[91]

However, despite a noticeable skepticism concerning space accords, many people in Russia, according to *Krasnaya Zvezda* newspaper, "believed that although these documents contain a number of discriminatory clauses, nonetheless they would immediately open access for us to the international market and give us the opportunity to earn money using the colossal potential of our space industry."[92] Another article published in the same paper, although pointing out that "for all the positive aspects of such cooperation, relations between the two sides will not be totally equal and the United States will receive more than it gives," acknowledged that "life itself, it seems, has prompted us to join forces and forge U.S.–Russian cooperation in the cause of creating a new orbital station, especially as Russia already has great work experience in this sphere."[93] Thus, although the U.S.–Russian space agreements were not welcomed by the

Russian public as enthusiastically as by the Americans, the Russian people in general did not reject the deals and accepted them in the context of the domestic and international realities of their country.

The space accords signed in the fall of 1993, and their subsequent acceptance by U.S. and Russian politicians and the public, marked an important milestone in the history of U.S.–Russian cooperation in the ISS program. Russia became a full-fledged partner in the ISS. This was clear evidence of the two sides' determination to overcome problems standing in the way of bridging the two space programs. However, the next step in the history of the Russian–American space station partnership, the "hardware building" period, subjected this determination to an equally hard test.

CONCLUSION

U.S.–Soviet/Russian space cooperation, spurred in the second half of the 1980s by *perestroika*, resulted in nearly 100 joint projects. Nonetheless, cooperation on the ISS program became "the largest and most visible sign of a new relationship in space" (Ref. 1, p. 5). This cooperation was preceded by a number of interrelated developments in the Soviet Union, in the bilateral relationship between the Soviet Union/Russia and the United States, and in the space activities of the two countries.

Mikhail Gorbachev's policy of democratization in the Soviet Union was inseparably linked to the end of Cold War confrontation. Such a transformation in Soviet foreign policy contributed to U.S.–Soviet/Russian space cooperation in two ways. On one hand, it improved bilateral relations, which eased fears in the two countries about the possible transfer of dual-use technology to a potential adversary. On the other hand, it undermined one of the major rationales for the vigorous space effort—to serve as a tool in the struggle for the global hegemony.

Although political devaluation had an impact on both space programs, the Soviet program was affected especially hard by the loss of a competitive rationale. Many Soviets saw no other reason for the space program except to beat the Americans in space. When this goal ceased to exist, a significant portion of Soviet society started perceiving the program as a burden on the country's troubled economy.

The populist leadership of Mikhail Gorbachev gave a twofold response to the people's attitude toward space exploration. The Kremlin started decreasing the Soviet space budget, but also began opening its space program to the West to become engaged in space cooperation with Western countries. Such cooperation would have enabled the Soviet leadership to achieve a number of goals. First, it would have allowed the Kremlin to share the costs of the major space projects. Second, it would have enabled the Soviet space industry to earn money through commercial contracts. Finally, like ASTP in 1975, the renewed cooperation in space would have become a symbol of a new relationship between the United States and the Soviet Union.

NASA was also interested in sharing the costs of space projects with the Soviet Union. Space cooperation with the Soviets was supported by U.S. politicians who viewed space as one of the means of promoting friendly relations with the Soviet Union. However, as opposed to the 1970s, cooperation with the Soviets did not serve as one of the rationales for the U.S. space effort. Besides, in the 1980s, NASA flew a number of classified Department of Defense missions, and some of the agency's leaders were still using "space race" arguments to convince the White House and Congress not to cut NASA's budget. This is why NASA, although generally welcoming cooperation, preferred it be limited to some particular small- or medium-size projects.

The disintegration of the Soviet Union in 1991, the emergence of Russia as the major space power among the former Soviet republics, and the rise of new space powers such as Europe and Japan introduced new elements into U.S.–Russian space cooperation. These elements contributed to the further bridging of the U.S. and Russian space programs and ultimately brought the two countries to cooperation on the ISS project. Among such elements, the most important on the Russian side were the following:

1) The deteriorating economic health of the Russian space industry, which forced the managers of the industry to pursue cooperation with the West aggressively.
2) The labor and brain drain from the Russian space industry, which encouraged the United States to create job opportunities for Russian rocket and missile specialists at home to prevent them from being hired by the countries practicing state terrorism.
3) The consideration of space by the new Russian government as one of the means to promote rapprochement between Russia and the United States.
4) The creation of a civil space managing state body—the Russian Space Agency (RSA)—whose main goal was to promote international cooperation in space.
5) Concern about the United States' competitors strengthening their competitiveness by gaining access to advanced and inexpensive Russian space technology.

An increased U.S. interest toward cooperation in space with Russia, particularly in the space station program, also resulted from developments in the U.S. space program and U.S. foreign policy:

1) The Space Station Freedom program turned out to be more expensive than estimated, and its construction encountered unexpected difficul-

ties. One of the ways to make the station less expensive and to put it into orbit earlier was to invite the Russians into the program.

2) The new NASA Administrator Daniel Goldin introduced a new policy for the agency called "Faster, Better, Cheaper," which was in tune with President Clinton's intention to stretch the federal budget. Cooperation with the Russians was supposed to help Goldin put his new policy into practice.

3) The United States was concerned about its competitors strengthening their competitiveness by gaining access to the advanced and inexpensive Russian space technology.

4) President Clinton considered cooperation with the Russians in the space station project as one of the most valuable assets of U.S.–Russian bilateral relations and also a diplomatic tool. The U.S. invitation for Russia to join the ISS program was partly a trade-off for Russia's refusal to sell its cryogenic engines technology to India.

Congress wanted the ISS to be built cheaper and faster and considered the station a means to promote friendship and partnership in relations between Russia and the United States. It supported U.S.–Russian cooperation in the ISS program.

REFERENCES

[1]U.S. Congress, Office of Technology Assessment, *U.S.–Russian Cooperation in Space*, OTA-ISS-618, April 1995, p. 5.

[2]Gartoff, R. L., *The Great Transition. American-Soviet Relations and the End of the Cold War*, Brookings Inst., Washington, DC, 1994, pp. 754, 772.

[3]Semyonov, Y., (ed.), *Raketno-Kosmicheskaya Korporatsiya "Energia" Imeny S.P. Korolyova* [Korolev Rocket-Space Corp.], NPO Energia, Moscow, 1996.

[4]*Neizvestnyi Kosmodrom* [The Unknown Cosmodrome], Orbita, Moscow, 1990.

[5]"Issledovaniye Perspectiv Rossiiskoi Kosmonavtiki s Uchyotom Geopoliticheskoi Obstanovki, Potrebnosteiy Regionov Rossii v Resultatakh Kosmicheskoi Deyatelnosti" [A Study of the Prospects of Russian Cosmonautics: The Geopolitical Situation and the Needs of the Russian Regions with Respect to the Results of Space Activity], *Kosmos e Chelovek* [Space and Man], Vypusk 2 [2d issue], Moscow Space Club, Moscow, 1995, p. 42.

[6]"'75 Space Hookup Recalled as Moscow Calls for Détente," *Washington Times*, 15 July 1985. NASA Historical Reference Collection, NASA History Office, NASA Headquarters, Washington, DC.

[7]Tucker, E., and Schrage, M., "Soviet Signs Pact with Intelsat," *Washington Post*, 28 Aug. 1985. NASA Historical Reference Collection, NASA History Office, NASA Headquarters, Washington, DC.

[8]Bohlen, C., "Soviets Open Space Center to Foreign Press," *Washington Post*, 8 April 1986. NASA Historical Reference Collection, NASA History Office, NASA Headquarters, Washington, DC.

[9]"Soviet Union Prepared to Launch Foreign Satellites," *Washington Post*, 13 June 1986. NASA Historical Reference Collection, NASA History Office, NASA Headquarters, Washington, DC.

[10]Skapars, L., "From Salyut with Love?" *Space World*, Feb. 1986. NASA Historical Reference Collection, NASA History Office, NASA Headquarters, Washington, DC.

[11]Matsunaga, S. M., *The Mars Project. Journeys Beyond the Cold War*, Hill and Wang, New York, 1986, pp. 72–73.

[12]Rogers, E., "Manned Flights Seen Crucial to Space Program's Future," *Washington Times*, 9 June 1986. NASA Historical Reference Collection, NASA History Office, NASA Headquarters, Washington, DC.

[13]Wilford, J. N., "Comet Mission: Omen for Cooperation in Science" *New York Times*, 7 March 1986. NASA Historical Reference Collection, NASA History Office, NASA Headquarters, Washington, DC.

[14]*Space. America's New Competitive Frontier*, Business–Higher Education Forum, Washington, DC, 1986.

[15]Brown, G., "Enhancing the World's Common Security," *Space Policy*, Aug. 1987, p. 172.

[16]Kohrs, R., former manager of system engineering of Space Shuttle, former director of Space Station Freedom program, director of the Center for International Aerospace Cooperation (CIAC) at ANSER Independent Public Research Inst., interview by author, Arlington, VA, 20 Feb 1997.

[17]Moser, T., former director of Space Station Freedom program, interview by author, Arlington, VA, 20 March 1997.

[18]"NASA: We're Ready to Work with Soviet Union in Space," *Orlando Sentinel*, 31 July 1985. NASA Historical Reference Collection, NASA History Office, NASA Headquarters, Washington, DC.

[19]Covault, C., "White House, Kremlin May Revive Cooperative Space Programs," *Aviation Week and Space Technology*, 6 Oct. 1986. NASA Historical Reference Collection, NASA History Office, NASA Headquarters, Washington, DC.

[20]"Agreement Between the United States of America and the Union of Soviet Socialist Republics Concerning Cooperation in the Exploration and Use of Outer Space for Peaceful Purposes," *Exploring the Unknown. Selected Documents of the U.S. Civilian Space Program*, edited by J. M. Logsdon, D. A. Day, and R. Launius, Vol. 2, External Relationships, NASA

History Office, Washington, DC, 1996.

[21]Fabian, J., President of ANSER, vice-president of International Astronautics Federation, former president of Association of Space Explorers, interview by author, tape recording, Arlington, VA, 11 March 1997.

[22]Lawler, A., "U.S., Soviets Discuss Space Crew Exchange," *Space News*, 11–17 June 1990, p. 3.

[23]Truly, R., NASA administrator (1989–1992), interview by author, Arlington, VA, 29 Sept. 1998.

[24]Shishkin, O., former minister of General Machine Building, interview by author, Moscow, 29 July 1998.

[25]"Glasnost in Space," *Science News*, 7 Sept. 1991, p. 156.

[26]Von Bencke, M. J., *The Politics of Space: A History of U.S.–Soviet/ Russian Competition and Cooperation in Space*, Westview Press, Boulder, CO, 1997.

[27]Moiseev, I., *A Report on the Economics of Space Activities in Russia*, ANSER, Moscow, 1996, p. 5.

[28]Johnson, N. L., *The Soviet Year in Space: 1990*, Teledyne Brown Engineering, Colorado Springs, CO, 1990, p. 1.

[29]Postyshev, V. M., and Moiseev, I. M., *A New Framework for the Russian Space Program: Redefinition and Reorganization*, ANSER, Arlington, VA, 1994.

[30]Congress, Senate, Committee on Appropriations. Subcommittee on VA-HUD-Independent Agencies, "Statement of Richard H. Truly, NASA Administrator, on Space Cooperation with the Former Soviet Union, 20 February 1992," draft, NASA Historical Reference Collection, NASA History Office, NASA Headquarters, Washington, DC.

[31]"From Russia with Love. The Cold War is Over!" *Space Flight 500*, A ROAD Publication, 1992.

[32]"Appointment of Y. P. Semyonov as a General Designer of Scientific-Industrial Enterprise Energia of the USSR Ministry of General Machine Building," Decree of the Council of Ministers no. 672, Moscow Region, Podlipki, Library of NPO Energia, 21 Aug. 1989.

[33]Karash, Y., "A Poletit Yaponets...," [The Japanese Will Fly, However...], *Zhurnalist* [The Journalist], Oct. 1990, p. 44.

[34]Chenard, S., "The Russian Space Industry in 1995," International Conference and Exhibition of Satellite Communications in Russia and the CIS, Moscow, 28–29 Nov. 1995.

[35]Decree of the President of Russian Federation no. 185 "Concerning the Creation of the Russian Space Agency," Duma Library, Moscow, 25 Feb. 1992.

[36]Congress, House, Sec. 314, Cooperation with the Former Soviet Union, *Congressional Record*, 5 May 1992, H 2916.

[37] "On the Priorities of Space Policy for the Russian Federation," 27 April 1993, *A Draft Statement*, Moscow, Duma Library.

[38] "On Measures to Stabilize the Situation in Space Science and Industry," 27 April 1993, *A Draft Resolution*, Moscow, Duma Library.

[39] "Law of the Russian Federation on Space Activity," Duma Library, Moscow, 6 Oct. 1993.

[40] "Federal Space Program up to the Year 2000," Duma Library, Moscow, 11 Dec. 1993.

[41] Kohrs, D., interview by author, Arlington, VA, 20 March 1997.

[42] *Report of the Advisory Committee on the Future of the U.S. Space Program*, U.S. Government Printing Office, Washington, DC, 1990, pp.1, 48.

[43] "America at the Threshold," *Report of the Synthesis Group on America's Space Exploration Initiative*, U.S. Government Printing Office, Washington, DC, 1991, p. 103.

[44] *A Post Cold War Assessment of U.S. Space Policy*, A Task Group Rept., Space Policy Inst., George Washington Univ., Washington DC, p. 34.

[45] *Budget of the United States Fiscal Year 1996*, U.S. Government Printing Office, Washington, DC, 1995.

[46] Aldrich, A., former NASA associate administrator, interview by author, tape recording, Arlington, VA, 3 April 1997.

[47] Clayton, W. E., "Official Impressed by Soyuz Proposal," *Houston Chronicle*, 26 March 1992. NASA Historical Reference Collection, NASA History Office, NASA Headquarters, Washington, DC.

[48] Warrensford, D., "Reject Soviet Technology," *Florida Today*, 24 March 1992. NASA Historical Reference Collection, NASA History Office, NASA Headquarters, Washington, DC.

[49] "White House Blocks Purchase of Soviet Technology," *Baltimore Sun*, 1 March 1992. NASA Historical Reference Collection, NASA History Office, NASA Headquarters, Washington, DC.

[50] Congress, House, Committee on Science, Space, and Technology. Subcommittee on Space, *Bilateral Space Cooperation with the Former Soviet Union: Hearing Before the Subcommittee on Space*, 102d Cong., 2d sess., 25 March 1992.

[51] "Space Station Freedom Program," *Assured Crew Return Vehicle*, Fiscal Year 1994 Estimates, ANSER/CIAC Library, Arlington, VA.

[52] Bizony, P., *Island in the Sky. Building the International Space Station*, Aurum Press Ltd, London, 1996.

[53] Beatty, R., former policy analyst at Booz, Allen and Hamilton, interview by author, Arlington, VA, 31 March 1997.

[54] "Options for Mir/SSFP Combinations," Rept. by Booz, Allen and Hamilton, 16 June 1982, CIAC Library, Arlington, VA, p. 3.

[55]Aldrich, A., former NASA associate administrator, interview by author, Arlington, VA, 4 April 1997.

[56]Contract NASW–4727, Technical Direction No: C2.1-1, 8 June 1992, ANSER/CIAC Library, Arlington, VA.

[57]"Soyuz 'Lifeboat' Passes First Test," *Space News*, 18 Dec. 1992, p. 4.

[58]Clinton, W. J., Statement on America's Space Program, issued during 1992 campaign, George Washington Univ., Space Policy Inst. Library, Washington, DC.

[59]Arbatov, A., "Russian Foreign Policy Alternatives," *International Security*, Vol. 18, No. 2, Fall 1993.

[60]Davisha, K., and Parrott, B., *Russia and the New States of Eurasia. The Politics of Upheaval*, Cambridge Univ. Press, Cambridge, England, UK, 1994, pp. 198–207.

[61]Malcolm, N., "The New Russian Foreign Policy," *World Politics*, 1994/1995, pp. 59–63.

[62]Kozyrev, A., *Preobrazhenie* [The transformation], International Relations, Moscow, 1994.

[63]Pikayev, A. A., Spector, L. S., Kirichenko, E. V., and Gibson, R., *The Soviet-Russian Sale of Cryogenic Rocket Technology to India and Russia's Adherence to the Missile Technology Control Regime*, draft of the monograph, Carnegie Endowment for International Peace, Washington, DC, 1 July 1996.

[64]Advisory Committee on the Redesign of the Space Station, *Final Report to the President*, NASA, 10 June 1993.

[65]*Perspectiva Razvitiya Kosmicheskoi Deyatelnosti v Rossiyskoi Federatsii do 2020 Goda* [Perspective of the Russian Aerospace Activity up to the Year 2000], Moscow Space Club and TsNIIMash, Moscow, 1995.

[66]Nise, J., "Space Station and the International Human Exploration and Development of Space," 1998 AAS National Conference and 45th Annual Meeting, 17 Nov. 1998.

[67]Stall, D. K., phone interview by author, 30 Nov. 1998

[68]Goldin, D., and Koptev, Y., *The Memorandum of Understanding*, 31 May 1993, ANSER/CIAC Library, Arlington, VA.

[69]Congress, Senate, Committee on Commerce, Science, and Transportation, Subcommittee on Science, Technology, and Space, *Redesigned Space Station Program: Hearing Before the Subcommittee on Science, Technology, and Space*, 103d Cong., 1st sess., 1 July 1993.

[70]Congress, House, Committee on Science, Space, and Technology, *Oversight Visit. Baikonur Cosmodrome*, Chairman's Rept., 103d Cong., 2d sess., 23 March 1994.

[71]Harwood, W., "Shuttle-Mir Outlay Touted as Bargain," *Space News*, 1–7 June 1998.

[72]Clark, R., former NASA associate administrator for International Relations and Policy, interview by author, Rosslyn, VA, 10 June 1997.

[73]*Addendum to Program Implementation Plan*, 1 Nov. 1993, NASA Headquarters, Washington, DC.

[74]DeSelding, P. B., "Europe, Japan Await Clarification on Station-Mir Effort," *Space News*, 6–12 Sept. 1993, p. 10.

[75]"Joint Statement," Space Station Heads of Agencies Meeting, Paris, 16 Oct. 1993, NASA History Office, Washington, DC.

[76]Congress, House, Committee on Science, Space, and Technology, Subcommittee on Space. *United States-Russian Cooperation in the Space Station Program: Parts I and II: Hearing Before the Subcommittee on Space*, 103d Cong., 1st sess., 6, 14 Oct. 1993.

[77]Clinton, W. J., Presidential Address Before a Joint Session of Congress on the State of the Union, 25 Jan. 1994, Office of the Press Secretary, Washington, DC.

[78]Gore, A. J., remarks by U.S. Vice President during a signing ceremony with Prime Minister Chernomyrdin of Russia, 2 Sept. 1993, NASA Historical Reference Collection, NASA History Office, Washington, DC.

[79]*Report of the Cost Assessment and Validation Task Force on the International Space Station*, Prepared under the leadership of Jay Chabrow, 21 April 1998, NASA Advisory Council, Washington, DC.

[80]Vartabedian, R., and McManus, D., "U.S. and Russia to Collaborate in Building Space Station," *Los Angeles Times*, 3 Sept. 1993. NASA Historical Reference Collection, NASA History Office, NASA Headquarters, Washington, DC.

[81]Pedersen, K. S., "Thoughts on International Space Cooperation and Interests in the Post-Cold War World," *Space Policy*, Aug. 1992, pp. 205–220.

[82]Tucci, L., "NASA Lobbies for Russian Station Work," *Space News*, 18–24 Oct. 1993, p. 3.

[83]Sawyer, K., "Clinton Wins Support for Space Station: Hill Leaders Back Venture with Russia," *Washington Post*, 1 Dec. 1993. NASA Historical Reference Collection, NASA History Office, NASA Headquarters, Washington, DC.

[84]"U.S. Needs Equal Partnerships," *Orlando Sentinel*, 10 May 1992. NASA Historical Reference Collection, NASA History Office, NASA Headquarters, Washington, DC.

[85]"With the Russians into Space," *Baltimore Sun*, 5 Sept. 1993. NASA Historical Reference Collection, NASA History Office, NASA Headquarters, Washington, DC.

[86]"Partners in Space, Not Rivals," *New York Times*, 13 Nov. 1993. NASA Historical Reference Collection, NASA History Office, NASA

Headquarters, Washington, DC.
[87] "Refuel Space Program with Foreign Partners," *USA Today*, 2 Dec. 1993. NASA Historical Reference Collection, NASA History Office, NASA Headquarters, Washington, DC.
[88] "Negotiators in Space," *Washington Times*, 20 Dec. 1993. NASA Historical Reference Collection, NASA History Office, NASA Headquarters, Washington, DC.
[89] PM 1906160992, Discussion of space-related issues, Television First Program Network in Russian, Moscow Teleradiokompaniya Ostankino, 1700 GMT, 17 June 1992.
[90] *Rossiiskaya Gazeta*, 11 Sept. 1993.
[91] Yaropolov, V., "Tombstone over Zvezdny?" *Stolitsa*, Dec. 1993, pp. 13–15.
[92] *Krasnaya Zvezda*, 27 Oct. 1993.
[93] *Krasnaya Zvezda*, 17 Dec. 1993.

Flight prototype of the Soviet Space Shuttle *Buran*. Cancellation of the *Buran* program in the late 1980s provided motivation for the Soviets to seek cooperation in space with the Americans.

The first official visit of U.S. astronauts to the Soviet Union after ASTP took place in January 1990. Pictured here in Star City are, from left, U.S. astronauts Paul Weitz and Jerry Ross, interpreter Nataly Karakulko, Soviet cosmonauts Vladimir Kovalyonok and German Titov, and U.S. astronauts Dan Brandenstein and Ron Grabe.

U.S. astronaut Jerry Ross trying on a Russian EVA space suit in Star City, January 1990.

Seventeen years later, in 1992, ASTP commanders Alexei Leonov and Thomas Stafford meet in Washington, D.C.

Goldin's first visit to Russia in 1992. The NASA administrator is inspecting the docking mechanism used by Russian spacecraft. Standing in shirt sleeves from left are: RKK Energia General Designer Yuri Semyonov, astronaut Tom Stafford, leading designer of the Russian docking units Vladimir Syromyatninok, and Daniel Goldin. (Photo courtesy of Bill Ingalls/NASA.)

Daniel Goldin, left, in the Russian Mission Control Center with Flight Director Vladimir Solovyov, far right, in 1992. (Photo courtesy of Bill Ingalls/NASA.)

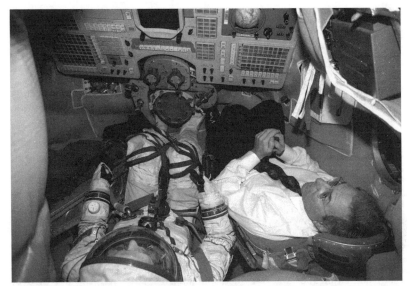

Daniel Goldin in Soyuz-TM flight simulator. The Soyuz-TM spacecraft was chosen as a lifeboat for ISS. (Photo courtesy of Bill Ingalls/NASA.)

Daniel Goldin and Yuri Koptev–RSA Director exchange pens after signing an early Memorandum of Understanding between RSA and NASA. (Photo courtesy of Bill Ingalls/NASA.)

Moscow airshow in 1994. From left: German Titov, second Soviet cosmonaut; Igor Volk, cosmonaut and test pilot; and John Fabian, President and CEO of ANSER Corporation and an important advocate for the rebirth of Soviet/Russian–American cooperation in space.

In February 1994, during his eight-day stay onboard Space Shuttle *Discovery* (STS-60), cosmonaut Sergei Krikalev performs several tasks on the aft flight deck. (Photo courtesy of NASA.)

From left: Cosmonaut Vladimir Titov, the second Russian cosmonaut to fly in the Space Shuttle (STS-63, February 1995); NASA Administrator Daniel Goldin; and Lt. Gen. Pyotr Klimuk, chief of Star City. (Photo courtesy of Bill Ingalls/NASA.)

View from onboard the Space Shuttle Atlantis as it linked up with Mir on 29 June 1995, just two and a half weeks before the 20th anniversary of the ASTP docking in Earth orbit. (Photo courtesy of NASA.)

The first handshake in space after ASTP. Mir Commander Vladimir Dezhurov greets STS-71 Commander Robert Gibson, with his back to the camera. (Photo courtesy of NASA.)

Inside the spacelab science module, the crew members of STS-71, Mir-18, and Mir-19 pose for the traditional inflight portrait. Clockwise from left: Anatoly Solovyev, Gregory Harbaugh, Robert Gibson, Charles Precourt, Nikolai Budarin, Ellen Baker, Bonnie Dunbar, Norman Thagard, Gennadiy Strekalov (at angle), and Vladimir Dezhurov. (Photo courtesy of NASA.)

STS-71 landing at NASA Kennedy Space Center in Florida. Daniel Goldin and Yuri Koptev in the foreground. (Photo courtesy of Bill Ingalls/NASA.)

Cosmonauts had no boundaries in space but they still had to deal with them on Earth. After landing in Florida in STS-71, Vladimir Dezhurov and Gennadiy Strekalov traveled to Houston in July of 1995 where they were issued visas to recognize their entry into the United States. (Photo courtesy of Bill Ingalls/NASA.)

At the celebration of the 35th anniversary of Gagarin's flight, 12 April 1996 in Star City, members of the first joint Russian–American crew since ASTP pose after receiving their awards. From left: Norman Thagard, Order of Friendship Among People; Vladimir Dezhurov, Gold Star of the Hero of the Russian Federation; and Gennady Strekalov, Order of Merit to the Motherland. Standing to the right of the crew is the author, Yuri Karash.

Signing of the sister city agreement between Nassau Bay and Star City on 15 July 1993. From left: Vadim Gorin, coordinating secretary of the Twin Cities International Association; Lt. Gen. Pyotr Klimuk, chief of Star City; Gerald Allan, Nassau Bay mayor; and David Stall, Nassau Bay city manager. (Photo courtesy of Thomas Cones, City of Nassau Bay.)

Russians and Americans strengthen their relationships by engaging in common interests separate from their official duties. Charles Precourt, U.S. astronaut and NASA representative to Star City is briefed by Russian instructor Nikolai Khvostov before flying a Russian Yak-18T at the Myachkovo airfield outside Moscow.

Michael Baker, NASA Johnson Space Center representative in Moscow, poses with Russian instructor Igor Sevbo after flying the Russian aerobatic aircraft Yak-52 at the Myachkovo airfield.

From left to right: Yuri Koptev, Russian Space Agency general director; William "Mac" Evans, Canadian Space Agency president; Dan Goldin, NASA administrator; Antonio Rodota, European Space Agency director general; Isao Uchita, National Space Development Agency of Japan president; and Yamato Inaba, parliamentary vice minister for Science and Technology after signing the IGA agreement on 29 January 1998. (Photo courtesy of Bill Ingalls/NASA.)

President Bill Clinton tours the NASA Johnson Space Center with NASA Administrator Daniel Goldin on 14 April 1998. President Clinton is wearing a flight jacket with a patch from the first U.S.–Russian joint mission (STS-60). (Photo courtesy of NASA.)

Zarya Launch 1, 20 November 1998. Baikonur, Kazahkstan (Photo courtesy of Bill Ingalls/Scott Andrews for NASA.)

FGB (Zarya) preparing for launch. (Photo courtesy of Khrunichev Space Center.)

Zarya rollout (Photo courtesy of Bill Ingalls/NASA.)

Zarya Launch 1, 20
November 1998. Baikonur,
Kazahkstan (Photo courtesy
of Bill Ingalls/Scott Andrews
for NASA.)

Two men on the U.S. side who played key roles in shaping U.S.–Russian cooperative efforts: NASA Administrator Daniel Goldin (on the stairs) and NASA Johnson Space Center Director George Abbey. (Photo courtesy of Bill Ingalls/NASA.)

The first crew of the ISS poses aboard a freighter in the Black Sea on 3 October 1997 before beginning water survival training. From left: Soyuz commander and Russian cosmonaut Yuri Gidzenko, U.S. astronaut and ISS commander William Shepherd, and flight engineer and Russian cosmonaut Sergei Krikalev. In the background is a mockup of the Soyuz spacecraft descent module used in training. (Photo courtesy of NASA and Russian Space Agency.)

Members of the second crew that will live aboard the ISS undergo winter/wilderness survival training in March 1998, near Star City. From left: Russian cosmonaut Yuri Usachev, crew commander, and U.S. astronauts Susan Helms and James Voss. (Photo courtesy of NASA.)

IBMP facility for SFINCSS-99 long-duration experiment in confinement—the institute's contribution to the ISS program.

Chapter 6

UNEASY PARTNERSHIP

While the previous period (1990–1993) in the history of Soviet/ Russian–American cooperation in space was about answering the twofold question—"to cooperate, or not to cooperate, and if yes, how?"—the next period, starting after the signing of the space station agreements between RSA and NASA at the end of 1993 and ongoing through early 1999, involved different kinds of questions, such as: "Can Russia fulfill its commitments?" and "Is it worth keeping Russia as a full-fledged space station partner, considering all the problems that Russian participation causes the ISS project?"

Such questions did not arise before late 1995. Until that time, U.S.–Russian cooperation in the space station program was still fledgling, in terms of both further formalization and the determination of each side's contribution to the ISS. In January 1994, Congressman Sensenbrenner went to Moscow where he toured aerospace facilities and design bureaus and also met with Russian space officials. After the visit, Sensenbrenner said that he now had more confidence that the Russians would maintain a strong space program in the years ahead. This was because

> (1) Russia's space program is an important source of hard currency; (2) The space program will help keep Russia's aerospace industry from falling apart due to lack of defense orders; and (3) The program gives Russia the status of a world leader and brings the Russian people self-confidence and pride.

Sensenbrenner also said that he strongly supported bringing Russia into the program.[1]

IMPLEMENTING THE ISS PARTNERSHIP WITH RUSSIA

On 18 March 1994, the United States, Canada, Japan, and the European partners met in Paris with government officials of the Russian Federation for the first time to discuss steps to implement the decision to bring Russia into the space station partnership. They stressed their interest in Russia joining the ISS program as a full partner as soon as possible.

The Russian delegation informed the participants of the key parameters of Russia's contribution to the partnership, which would enhance space station capabilities. The representatives discussed changes to the legal framework of the 1988 agreements on space station cooperation that would be needed to include Russia as a partner and modalities for negotiating those changes, including a schedule that would allow for early completion of negotiations. The first meeting of the negotiating delegations was planned for April 1994.[2]

This meeting took place on 5 April 1994, in Washington, D.C. It was the first time that the heads of the space agencies involved in the ISS met together since Russia accepted the collective invitation to join the ISS partnership in December 1993. In their joint statement, the heads of the agencies particularly "noted the remarkable progress made to accommodate Russia as a new partner and to satisfy the interests of all the partners in the new program structure. They also commended NASA for its efforts to improve program efficiency and to clarify the potential for additional partner contribution." The heads of agencies also noted the importance of concluding in a timely manner the interim agreement between NASA and RSA to facilitate Russia's early participation in program management mechanisms. They shared the determination that the ISS program would be accomplished without further delay.[3]

The "Interim Agreement for the Conduct of Activities Leading to Russian Partnership in Permanently Manned Civil Space Station" was signed on 23 June 1994. The agreement provided for initial Russian participation in the ISS program and was supposed to govern Russian participation until an Intergovernmental Agreement and a NASA–RSA Memorandum of Understanding could be concluded. NASA and the RSA also signed a separate $400 million contract for Russian space hardware, services, and data. Under this contract, NASA was supposed to purchase hardware and services from the RSA and its subcontractors for approximately $100 million per year through 1997 in support of a joint program involving the U.S. Space Shuttle and the Mir space station. Key elements of the contract included support of U.S. astronauts onboard the Mir space station for up to 21 months, as many as nine Shuttle docking missions with Mir, joint technology developments, and support for peer-reviewed science and technology research to be conducted onboard Mir. This contract also provided initial development funding ($25 million) for the FGB for use on the ISS.[4] In the spring of 1995, Lockheed and Khrunichev (the primary manufacturer of the FGB) signed a $190 million contract for building the FGB module.[5]

The process of formalizing U.S.–Russian cooperation in the ISS program was paralleled by the activity of the engineers of the two countries, who were determining the final configuration of each other's contribution to the space station. As a result of such activity, which took place in 1994–1995, it was decided that the initial ISS segment would consist of the FGB, which would be launched as the first element of the station by a Russian Proton launch vehicle, the U.S. module Node 1, and the Russian Service Module, developed by RKK Energia. Additional U.S. and Russian elements of the station, followed by European and Japanese modules, would consequently be attached to this segment (Ref. 6, pp. 499–501).

CONTINUOUS DECLINE IN THE RUSSIAN SPACE INDUSTRY

However, such managerial and engineering progress in U.S.–Russian cooperation took place against a background of continuous decline in the Russian space industry, which was seriously undermining Russia's ability to meet its space station commitments. According to RSA data, in the 1989–1995 time period, funding for the Russian space program in real terms fell more than fivefold, from the equivalent of $3.9 billion to $0.69 billion; relative to the gross national product, it declined from 0.73 to 0.29%. Precise interpretation of these figures is difficult for two reasons. First, prior to 1991 the exchange value of the ruble to the dollar was never accurately represented by the official exchange rate because of the closed nature of the Soviet economy. Second, figures before 1991 relate to the whole Soviet Union, whereas figures after 1992 describe only the Russian Federation, which has a smaller space budget as well as a smaller number of consumers of this budget, such as space industrial and research organizations. In 1996, funding was only 40% of what it had been five years earlier, using the constant value of the ruble in 1996.

However, a small budget was not the only financial problem of the Russian space program. According to statements by Yuri Koptev, the RSA received only 2.4 trillion rubles ($430 million) in 1996 from the government, one-third less than the promised 3.3 trillion rubles.[7] This funding deficit hampered operation, supply, and maintenance of the Mir space station as well as Russia's ability to meet its commitments to the ISS.

Apart from the size of the space budget, there were other indicators of the deteriorating health of the Russian space program. One of them was a decreased launch rate. From 1993 to 1994, the Russian launch rate fell to one-half the level of the mid-1980s. Another indicator was the reduced volume of production in the rocket and space industries. In 1994, it fell to 30% of the 1989 level. The average monthly salary in the space industry in 1994 was 192,000 rubles, which was equivalent to about $80. In February 1995, the average salary was raised to 296,000 rubles, but its dollar equivalent fell to $70. However, because of budgetary instability, even these salaries were often delayed for several months because of nonpayment by customers. As a result of such a drastic decline in the living standards of people employed in the Russian space industry, Russian aerospace enterprises experienced a severe loss of personnel. The ISS prime contractor RKK Energia lost about 20% of its staff.[8]

USE OF THE MIR CORE MODULE

Whereas many of the problems of the Russian aerospace sector could be attributed to the overall worsening economic situation in Russia, a great deal of them could be blamed on the Russian government's negligent attitude toward the national space industry. As stated by Nikolai Kuznetsov, one of the factory engineers in Samara, where Soyuz launch vehicles are manufactured,

> It is very difficult to witness the death of an enterprise which we built ourselves. The factory is almost on the deathbed. All these gangs of reformers—people like Gaidar [former acting prime minister of Russia], Chernomyrdin [former prime minister of Russia], Chubais [former first deputy prime minister of Russia]—were literally killing our space program, our future. Sometimes we were getting an impression that they were fulfilling somebody's evil will. I remember Chernomyrdin's visit to our factory. He was walking around and looking at everything with his "fish-like" eyes. He got excited only once, when he asked, whether it would be possible to adapt our compressors for the gas industry. Other than that, he did not show either interest, nor emotions [for the enterprise]. However, all that we needed from him at that point—just a little bit of help to finish one project. We put a lot of hope on it—both the Russians and the Westerners. This only project could have provided the whole factory with foreign investments. He [Chernomyrdin] did not give us money. Moreover, he even offended us by saying something like: "If you survive—good for you. If not—everybody would draw a sigh of relief. No factory—no problem..." (Ref. 9).*

As a result of the deteriorating economic health of the Russian space industry in December 1995, the Russian side suddenly proposed to use the existing Mir core module for the ISS instead of a service module (SM) that was to be financed by RSA and equipped by RKK Energia. The 21-ton flight vehicle is in fact a hull built in 1985, originally intended to become the core for the Mir 2 space station. When Mir 2 was canceled after the disintegration of the Soviet Union, the unit was placed in storage until resur-

---

*There are three reasons to believe that the government of prime minister Eugueni Primakov will pay more attention to Russian aerospace issues, particularly to U.S.–Russian cooperation on the ISS program. First, Primakov is more supportive of Russian heavy industry than his predecessors, including MIC. Second, Primakov, as a former foreign minister, should have a better understanding of the foreign policy value of the U.S.–Russian partnership in space. Finally, Primakov's first deputy, Yuri Maslyukov, used to supervise the Soviet space industry before the disintegration of the Soviet Union.

rected for outfitting as the SM.[10] Even in 1997, there was still SM documentation in RSA and RKK Energia that bore the words "Mir 2" on the cover. RSA made it clear that Russia could not support both Mir and ISS construction. A premature termination of Mir operations, according to the Russian cosmonaut Sergei Krikalev, was not an option for the Russians for a number of reasons; the most important reason was that besides being a considerable source of hard currency, Mir was continuously increasing the Russians' orbital long-term flight experience, thus strengthening their position in the ISS project.[11]

The proposal to use the Mir core module instead of the SM became the first indication of the serious difficulties that Russia was experiencing with its contribution to the ISS. These difficulties resulted in the Russians breaking their commitments to the ISS program approximately through the spring of 1997.

Even though the Mir space station proved its robustness and effectiveness as a science platform, U.S. politicians made it clear that using an old and rather obsolete Mir core module instead of the SM was not an option. During his 9 January 1996 meeting with the First Russian Deputy Prime Minister Oleg Soskovets, James Sensenbrenner, chairman of the Science Committee of the U.S. House of Representatives, expressed his skepticism about this idea and warned Soskovets that the further pursuit of such a plan might encourage lawmakers to kill the station. Soskovets assured Sensenbrenner that Russia would fulfill all of its commitments to the ISS project and would pay its bills (Ref. 6, p. 501).

However, taking into consideration Russia's difficulties in supporting two space stations, as well as "joint interest in the Shuttle–Mir program, enhanced cooperation, and the U.S. objective of minimizing changes to the ISS," NASA and RSA developed a "mutually beneficial" plan under which NASA would add one flight to the Shuttle manifest and remanifest two additional flights, giving a total of nine Shuttle–Mir missions. This plan, according to Daniel Goldin, would allow NASA to receive a larger return from its research investment in the Specktr and Priroda modules and help the Russians overcome their logistics shortfall in 1998.[12]

SM CONSTRUCTION FALTERING

Despite the Soskovets promise, SM construction kept faltering throughout the year. The record of the Russians' broken promises was presented by Sensenbrenner:

> In March 1996, NASA promised the Subcommittee that it would resolve this issue [the Russian funding] by mid-May 1996. It did not. In April 1996, the Vice President received more assurances that the

Russian government would pay its bills. It did not. In July of 1996, Russian Prime Minister Viktor Chernomyrdin promised Vice President Gore in writing that the Russian government would pay its bills and meet several milestones. It did not. [In February 1997,] Prime Minister Chernomyrdin promised Vice President Gore that it would give the Russian Space Agency $100 million by February 28th. It did not. On February 10th, the Russian government released a decree promising to provide a schedule of payments to the Russian Space Agency by March 10th. It did not. On March 4th, NASA promised us that it would make a decision on Russian participation by mid-March. It did not. In mid-March, NASA promised us that it would make a decision on Russian participation by mid-April....[13]

The Russian government appeared to address the problems of Russian participation in the ISS program even before the Russian participants experienced their first crisis in December 1995. On 7 August 1995, the Russian government issued Decree 791 "Concerning Realization of the Space Activity in the Interests of Economy, Science and Security of the Russian Federation." This decree committed the RSA, the Defense Industries' State Committee of the Russian Federation, the Defense Ministry of the Russian Federation, and other branches of the Russian federal government to focus their efforts within the next five years on the achievement of a number of top priority goals. One of these goals was "to fulfill the international commitments concerning the building of the International Space Station...."[14] On 26 August 1995, the Russian government issued Decree 1163 aimed at strengthening the financial basis of the Federal Space Program. This program implied active Russian involvement in a number of projects, particularly in the ISS program.[15]

However, these measures did not have any particular effect on the advancement of the SM construction. Having noted that, James Sensenbrenner and the chairman of the VA-HUD-IA House Committee, Representative Jerry Lewis, sent a letter on 8 March 1996 to Russian First Deputy Prime Minister Soskovets and requested that he take concrete steps to guarantee that SM construction would catch up with the schedule and that Russia would meet all its ISS commitments. Without such action, they warned, the U.S. Congress would have to make the decision to proceed with ISS construction without Russia (Ref. 6, p. 502). This letter was followed by a letter from Vice President Gore to Prime Minister Chernomyrdin. Gore warned Chernomyrdin that the insufficient financing of the Russian work on the ISS elements will strengthen the position of those in Congress who want to terminate U.S.–Russian cooperation in space, and he asked the prime minister to do "everything possible" to find money for the timely completion of the SM construction (Ref. 6, p. 503).

ISS AS A TOP STATE PRIORITY

The letters from the congressmen and Gore appeared to have some effect on the attitude of the Russian government toward its ISS commitments. On 12 April 1996, it issued Decree 422, "Concerning Measures Aimed at the Realization of the Russian Federal Space Program and International Agreements in the Field of Space Exploration." This decree considered "timely and adequate accomplishment of work related to international space programs, such as 'Mir–NASA,' 'Mars-96,' and the International Space Station, as the task of the top state priority." The decree transferred 1.11 trillion rubles to the civil space program from the fundamental research portion of the Russian federal budget and permitted RSA to take loans from commercial banks under state guarantees.[16] However, six days after the issuance of this decree, Yuri Semyonov had to send a letter to Prime Minister Chernomyrdin informing him that over the four previous months, RKK Energia had received only 8% of its yearly budget, and for this reason all the work at RKK Energia on Space Station Alpha had been stopped (Ref. 6, p. 515). No further funds were disbursed until the spring of 1997.

On 1 May 1996, the Russian government issued Decree 533 approving the "Concept of the National Space Policy of the Russian Federation." This concept declared "equal and mutually beneficial international cooperation in the field of space exploration" and the "broadening of international cooperation in the field of space exploration, above all with CIS countries, European States, [and] the USA…" among the main goals, principles, and priorities of the national space policy.[17] Speaking to the Russian space industry leaders on 13 May 1996, President Yeltsin said that this sector had nothing to fear. "I assure you that the state will certainly find resources to continue activities in space," he said. "Achievement in this field is one of the main factors determining Russia's technical and scientific progress and its status as a great power and a high technology country."[18]

DATES SCHEDULED FOR LAUNCH OF RUSSIAN ISS COMPONENTS

On 16 July 1996, in the absence of an intergovernmental agreement codifying the Russian contribution to the ISS and a timetable for its delivery, Vice President Gore and NASA Administrator Goldin on the one side, and Prime Minister Chernomyrdin and RSA General Director Yuri Koptev on the other, signed "The Schedule for the Development and Deployment of the Elements of the International Space Station." Even though this schedule determined dates for the launch of the FGB and the SM as November–December 1997 and April 1998, respectively, in early 1997 it became clear that Russia was definitely slipping behind schedule with its SM construction. The SM launch was postponed from April 1998 until the end of 1998.

At the end of January 1997, Congressman Sensenbrenner said that unless the Russian government started paying space station contractors by the end of February, Russia should be downgraded from a principal station partner to a subcontractor paid by the job. Sensenbrenner's remarks drew a quick response from the Kremlin, where government spokesman Alexander Voznesensky told reporters that the work on the station was included in the 1997 budget adopted on 24 January by the Duma. He vowed Russia would fulfill its obligations to the station program. The SM, Voznesensky said, would be ready for launch in November 1998.[19] However, Sensenbrenner and Science Committee ranking Democrat George E. Brown said that mere promises from the Russian government that it will uphold its end of the bargain on the SM were no longer enough. "We need actual money to go to the Russian contractors," Sensenbrenner said. "We cannot accept any further slips."[20]

On 7 February, Vice President Gore and Prime Minister Chernomyrdin issued a "Joint Statement on Human Space Flight and Science Cooperation." In this statement, they confirmed the commitment of both sides to the ISS program and encouraged continuing efforts to maintain the schedule from the beginning of on-orbit assembly through the completion of construction in 2002. At the same time, Gore and Chernomyrdin noted that "the sides are working diligently to overcome the difficulties presented by the slip of the Service Module, from April 1998 until November/early December 1998."[21]

U.S. PROPOSAL TO MINIMIZE IMPACT OF SM DELAY

On the U.S. side, the proposal aimed at minimizing the impact of the SM delay consisted of creating an interim control module (ICM).* NASA considered two options for the ICM. The first made use of U.S. hardware developed previously by the Naval Research Laboratory for a classified satellite. The second used the Russian module being developed independently by the Khrunichev factory as a backup for the FGB module—FGB 2.[22] NASA ultimately preferred the first option. NASA had also negotiated a modification to the existing NASA/RSA contract to rephase milestones for the last long-duration Phase 2 missions to the Mir station. This had provided the RSA with a total of $20 million early in 1997. This was not new money but a rephasing of committed funds from existing NASA/RSA contracts.

Other measures included the submission by NASA of a revision to the fiscal year 1997 Operating Plan to reallocate $200 million from Space Shuttle

---

*Other than providing living quarters for the crew, the SM is to provide the initial command, attitude control, and reboost functions for the First Element of the station (the FGB). The FGB has a limited lifetime in space without the SM for three basic reasons. First, it cannot be refueled for orbit maintenance if it runs low before the SM is delivered. Second, an SM launch delay of up to 8.5 months would put the FGB avionics at its certified life limit. Finally, guidance and control software on the FGB does not enable this module to dock with other elements of the station without the SM. The ICM is supposed to provide the critical SM functions to the station.

program funds to the U.S.–Russian Cooperation budget line. NASA intended to redesignate this line as "U.S.–Russian Cooperation and Program Assurance." The new budget line, according to Wilbur C. Trafton, NASA associate administrator for Space Flight, "will be the source of funds to address specific program requirements resulting from delays on the part of Russia in meeting its agreed to commitments to the ISS program and, in this context, to maintain the baseline program schedule to the fullest extent possible."[23]

RUSSIAN GOVERNMENT ATTEMPTS TO MINIMIZE IMPACT OF DELAYS

On the Russian side, on 10 February 1997, the government issued Decree 153, "Concerning Measures Aimed at the Realization of the Russian Federal Space Program and the Fulfillment of International Space Contracts." The decree approved the launch schedule of the initial elements of the ISS, including the FGB launch in November 1997 and the SM launch in 1998 (without specifying the month). The decree obliged the Russian Ministry of Finance to provide funds from the federal budget to RSA on a regular basis, particularly to enable the agency to "unquestionably fulfill commitments of the Russian Federation to the schedule of the creation and deployment of the first elements of the International Space Station." The Ministry of Economics was obliged to provide state guarantees of up to 1.5 trillion rubles for the construction of the Russian segment of the ISS.[24] Taking into consideration the limited time that the FGB can stay in orbit without the SM, RSA also proposed to delay its launch of the FGB module until June 1998.

Decree 153, however, did not have any noticeable impact on the Russian work on the ISS elements. When Sensenbrenner and Lewis went to Russia to check how the Russian government was fulfilling its ISS commitments, they did not see any progress in SM construction. In fact, they saw nothing but a copy of the aforementioned decree with a number of missing paragraphs. On 2 April 1997, the Russian government issued Decree 391, "Concerning Measures Aimed at Fulfillment of International Space Contracts," which authorized the RSA to take credits from commercial banks up to 800 billion rubles (400 billion in April and 400 billion in May).[25] Later, a NASA team, led by a former ASTP Apollo commander, General Thomas Stafford, went to Russia to gain further insight into the status of Russian funding. The Russian government promised him that it would give the RSA 400 billion rubles in April, 400 billion in May, and another trillion by the end of the year. On 9 April 1997, Sensenbrenner made an opening statement at the hearing on *FY 1998 NASA Authorization: The International Space Station*. Speaking of the situation with the SM, he said in particular that even if Russia released these funds, it would not make a difference "because so much more is necessary." Sensenbrenner estimated that the American people were misled by Russia and NASA eight times. He also made it clear that neither he nor his fellow representatives would tolerate any more misleading.[26]

YELTSIN MOVES TO FUND PROGRAM. The decisiveness demonstrated by U.S. politicians and the possibility of losing full-partner status in the ISS project forced the Kremlin to take concrete steps aimed at finding a solution to the financial problems of Russian participation in the space station program. Two days after Sensenbrenner's statement, President Yeltsin expressed his "whole-hearted support" of the Russian space program. Moreover, he ordered the establishment of direct phone communication between Yuri Koptev and himself so that the RSA general director could report immediately to the president about any kind of problem, and introduced Koptev to his aerospace adviser, Evgueni Shaposhnikov.[27] However, the most important result of the increased attention of the Kremlin to the program was the disbursement at the end of April of the first part of the loan bills guaranteed by Decree 391. The organizations involved in the ISS project started cashing the loan bills through four banks. The loans were guaranteed by the state and were to be repaid from the federal budget. On 9 June, in an interview with the Russian news service ITAR-TASS, RSA Deputy Director Boris Ostroumov said that the Russian government was about to disperse the last 700 billion rubles of the 1.5 trillion loan. The ISS subtiers were to cash the loan bills through four banks.

SM REVIEW HELD. At the end of April RKK Energia held a general designer's review with all subtier suppliers in SM construction. This event, together with the money flow to the Russian ISS contractors, enabled Russia to pass successfully the first of three major decision points set by NASA. According to Goldin,

> The first one [decision point] comes up the first or second week of May. In fact, I think on May 14th we will be holding a decision meeting with our international partners. On May 14th we will decide whether we are going to launch the Service Module, the third element, in December of '98, or we will launch the Interim Control Module, which we are having the Naval Research Lab build for us.
> There are three criteria that will determine that. First, did the Russians supply the funding they said they would supply in April? Did they hold a General Designer's review and have all 38 sub-tier suppliers come with schedules that they sign up to. And three: Has the money flowed to the sub-tier suppliers. If the answer to all three questions is yes, we will schedule the Service Module for launch. If the answer is no, we will schedule the Interim Control Module.[28]

REVISED ASSEMBLY SEQUENCE

The result of Russia's successfully passing the first decision point was a revised ISS assembly sequence. The revision was made by all of the participants in the ISS program. Key elements of this new sequence were the FGB launch, postponed from November/December 1997 to June 1998, and the SM launch, postponed from April 1998 to December 1998. The revision was made following extensive discussion, design reviews, and negotiations among RSA, NASA, and the other partners in the ISS program.

Apart from the willingness to keep Russia in the program, there were two more factors that could have contributed indirectly to NASA's and its international partners' decision to agree with Russia's postponing the delivery and launch of the SM. The first was related to the "less than stellar" performance by the space station's prime contractor, Boeing. According to Marcia Smith, "the launch delay...may have been caused by Russia's inability to deliver the Service Module, but it also reduced pressure on Boeing to complete the node on time."[28]

The second factor was related to the difficulties experienced by some of the ISS international partners in meeting their commitments to the space station program. According to Shiuchi Miura, the executive director of the Japanese National Aerospace Development Agency (NASDA), Japan had already completed 90% of the Japanese Experimental Module (JEM). If the JEM were launched according to the current schedule, the cost of its operation would drain most of the Japanese space budget, seriously hampering other Japanese space projects that were supposed to be initiated at the same time. Thus, the delay of the JEM's launch would allow NASDA at least to begin the other projects. Besides, Japan needed more time to develop a scientific program for the JEM and to gain more experience in Earth observation activities because such activities would constitute a significant portion of work onboard the module.[29]

PROBLEMS WITH MIR

Apart from the difficulties experienced with the construction of the ISS modules, Russia also encountered a number of problems related to a keystone element of U.S.–Russian cooperation in the ISS project, the Shuttle–Mir program. On 23 February 1997, a fire occurred onboard the station. In early March, cosmonauts lost control of a cargo ship as it approached Mir. They only barely avoided a collision. Two more mishaps happened in March: the failure of the prime oxygen generator and a malfunction in the station's attitude control system. In April, the failure of a cooling device caused the mechanism that removes carbon dioxide from the Mir compartment to break down, and leaks in the cooling system resulted in a rise in temperature inside the station. Most of the problems were fixed after STS-84 delivered the necessary equipment and repair materials to Mir. However, on 25 June the most serious incident happened onboard the Mir station. A cargo ship, Progress, accelerated out of control during an attempt at docking with Mir. It collided with and punctured the Mir's Spektr module, containing a considerable amount of U.S. equipment used by U.S. astronauts flying the Russian space station. The lives of the crewmembers were not endangered, but the station lost up to 30% of its power and the Spektr module was left unpressurized. Finally, on 17 July the crew accidentally disconnected a power cable, thus leaving the entire station without power. This problem, however, was relatively easily fixed. The crew reconnected the cable and repowered the gyros controlling the module's attitude.

Such mishaps triggered criticism from Congress. Congressman Sensenbrenner called on NASA to do a safety review before any more astronauts boarded the station. Or even better still, he suggested, Mir should be brought to an honorable end.[30]

However, unlike cooperation in creating hardware for the ISS, cooperation in the Shuttle–Mir program justified itself not only on the basis of potential results but also on the basis of the real and tangible outcomes. By summer 1997, the United States and Russia completed five successful dockings and one rendezvous mission. Beginning on 22 March 1996, the date when U.S. astronaut Shannon Lucid boarded the Mir station, the United States established its permanent presence in space. Through its work with the Russians, the United States gained significant knowledge in a number of areas, including how to dock the quarter-million-pound Shuttle with another large spacecraft at 17,000 mph and within a 600-lb docking force constraint. Shuttle–Mir successfully demonstrated that the five-minute launch window will not impact the assembly schedule of the ISS. The number of protein crystals that can be grown through conventional techniques has been expanded by 30 times. The program also demonstrated that joint ground and mission control operations can work effectively and collaboratively.[22]

OBSTACLES TO COOPERATION

NEGOTIATIONS

The inability of the Russian government to provide the necessary funds in time for SM construction and the problems experienced by Mir were not the only obstacles hampering U.S.–Russian cooperation in the space station program. Minor problems emerged in terms of cooperation between U.S. and Russian space officials in the summer of 1993, when joint technical work began.

Whereas the Americans viewed joint work on the ISS program (including Phase 1 of the Shuttle–Mir flights) as the second attempt at cooperation after ASTP, the Russians considered it the third attempt. Energia chief Yuri Semyonov was a program manager for the Salyut space station program in the late 1970s, when the Soviet Union and the United States were considering joining the Space Shuttle and Salyut programs. He remembered all too well how the Americans dropped their plans of cooperating with the Russians as the bilateral relationship between the two countries deteriorated, and for this reason he was extremely skeptical of working with his American colleagues again. Semyonov would remind the Americans of this fact whenever things became contentious and it looked as though his U.S. colleagues were considering a way out of the program.

Besides, the Russians considered the Americans to be "kids" in terms of building and operating a space station and in the area of long-term space flight overall. Such an attitude also created problems in terms of communication between the specialists of the two countries. Jim Nise recalled that the "first thing we [Americans] had to do was listen to the Russians, because they had stories they wanted to tell. They would not listen to anything we wanted from them until they told us what they were willing to do for us."[31]

There were always questions, such as who would be in control of the station—Kalinigrad or Houston—or what would be the composition of the crew.[32] Although the United States found the Russians insufficiently flexible with respect to these issues, the Russians apparently felt the same way about their American colleagues. According to the Russians who participated in the talks, they were "very difficult and sometimes went down dead ends." The Russians were affected not only by hot weather and the stuffy room in the basement where the negotiations took place, but also by the "intransigent and tactless attitude of the U.S. specialists to any debatable issue." These specialists usually did not bother to justify or substantiate their position, but as a rule limited themselves to the answer: "This is the decision made by the U.S. Congress" (Ref. 6, p. 499). One of the signs of the frustration experienced by some of the Russian space officials and engineers during negotiations with their U.S. colleagues was the suicide of Victor Smolin, the head of the Russian delegation at the space station negotiations, on 14 May 1994, after he had returned to Russia from Washington, D.C.[33]

CULTURAL PROBLEMS

Other obstacles had a cultural character. The origin of most of them was the residual Soviet-type mentality of the Russian space officials, who were still suspicious of the intentions of their American colleagues and oversensitive to the possible monitoring of their activity by the Russian security services. This is why they were occasionally reluctant to release previously classified data. For this reason the flow of information from the Russian side was often weak, incomplete, and late. When the Americans needed to know what was the composition of the pyrotechnic used for the explosive bolts onboard the Mir in case the Shuttle could not separate, it took them "forever" to find it. The Russians said that it was Ministry of Defense classified information. This also happened when U.S. engineers needed to know the environmental conditions inside Mir. It was very difficult for them to get this information and, even after they got it, it was quite incomplete and unclear.[32]

There was one more cultural problem that initially hindered U.S.–Russian cooperation in the ISS program.* When the Japanese, Canadians, or Europeans had problems with their contributions to the ISS, they informed the U.S. side about it and were not embarrassed to ask for help. There was never any similiar interaction with the Russians. If they had a problem, they tried to solve it themselves while hiding it from the Americans, so that the U.S. side could never be sure what the difficulty was.

Although all of these problems were relatively insignificant compared to those caused by the delays of the Russian hardware for the station, they should not be downplayed totally. Robert Clark believed that between the summer of 1993 and the summer of 1995, the ISS project could have been accelerated between six months to a year if it had not been for these minor problems.[32]

EMPLOYING MAXIMUM FLEXIBILITY WITH THE RUSSIAN GOVERNMENT AND THE RSA

Apparently, NASA was ready to deal with these kinds of problems. There was an understanding on NASA's side of the poor economic health of the Russian space program. NASA officials also realized the institutional differences between NASA and RSA; the latter was a smaller organization with a modest budget.† Besides, RSA's activity was plagued by a continuous struggle for control of the manned space program between RSA Director Yuri Koptev and RKK Energia General Designer Yuri Semyonov.

According to Nise, the Americans were caught in between two feuding organizations over who ultimately had control over the Russian manned space program. In 1992–1993, even Energia would not listen to the RSA. It

---

*The author witnessed some of the cultural problems in the summer of 1993, during his work as an intern for the Universities Space Research Association (USRA) at the NASA Johnson Space Center in Houston, Texas. He participated in the conferences on the phone between the U.S. and Russian space officials and engineers. The Russians often rescheduled talks and sometimes even failed to come to the phone, although the time of each conference had been previously agreed to with the Americans.

†In the first half of 1998, the total number of RSA employees was about 230 people, whereas the total number of NASA employees at the same time was around 19,000.

was only by the end of 1993 and early 1994 when the agreement was reached on the Russian side on how to do things that RSA became the main organization to deal with.

Because of its small size, RSA left a lot of other things out in the field for other organizations to do. This forced the Americans to have the same conversation three or four times because quite often they met another organization that was not integrated with the other space organizations in a manner expected by the Americans.[31] Additionally, many times the Americans were asked to modify their documents to a particular individual's preference on the Russian side. As a result, the U.S. documents often went out of configuration and were no longer consistent.[34]

Such problems required the U.S. government and NASA to employ maximum flexibility in its relationship with the Russian government and the RSA. One way to do this was to make a legal framework for the cooperation as flexible as possible. This is one of the reasons why neither the United States nor Russia rushed to sign an intergovernmental agreement concerning Russian participation in the ISS.

Another way to ensure flexibility was to involve officials who had enough power to bypass certain bureaucratic rules and procedures. This became necessary because of the less formal and organized Russian business style, which required that the U.S. officials accept a great deal of responsibility in dealing with the Russians.

HIGH-LEVEL SUPPORT FOR COOPERATIVE EFFORTS

Unlike ASTP, the U.S.–Russian cooperation in the Space Station Alpha program was continuously supported on the American side at the highest political level. According to James Sensenbrenner, Vice President Gore "was instrumental in bringing the Russians into the station partnership in 1993, and the negotiations relative to Russian participation have been an integral part of the summits between Gore and Russian Prime Minister Victor Chernomyrdin that have occurred twice a year since then." Gore will take the glory or the fall depending on whether the station succeeds or fails.[35] Gore's personal involvement not only facilitated a decision-making process on the U.S. side, it also provided U.S.–Russian cooperation with the necessary political support.

The high-level supervision and guidance were not the only difference between ASTP and U.S.–Russian cooperation in the ISS program. ASTP was overseen on the U.S. side by NASA Deputy Administrator George Low and ASTP Director Glynn Lunney. The U.S.–Russian cooperation in the ISS program was taken on directly by NASA Administrator Dan Goldin. There have been four reasons for this.

First, the end of the Cold War and the changed international environment improved contacts on the professional level between U.S. and Russian space officials and engineers and eliminated the need to check almost every cooperative step with a national security agency. Second, the Americans learned that personal relationships play a very important role in business relations with their Russian colleagues. Goldin made an excellent team with Koptev. American and Russian engineers and officials, after hitting a roadblock in their negotiations, would say: "Let's wait until Goldin and Koptev get together. The two of them could reach an agreement."[32] On a number of occasions, Goldin had to defend Koptev in the face of sharp U.S. political criticism of Russian performance, such as at the press conference concerning the NASA fiscal year 1998 budget. Goldin said that he had "had very long, difficult meetings with Yuri Nikolaevich over the last few days. I want to tell you there's nothing more he wants than to make this successful."[36]

Third, NASA was under much more pressure in terms of time and performance than the RSA. Congress wanted to see things done for the $2.1 billion that NASA was spending each year. However, the Russians moved at a slower pace. As a result, NASA started looking for solutions to Russian problems. As Clark put it, "We had to find a workaround. We were looking for maybe some extra money that we could give to the Russian side."[32] Such an attitude assumed an active search for different and innovative solutions to emerging problems. Examples of such solutions included NASA decisions to rephase milestones for the last long-duration Phase 2 missions to Mir to provide the RSA with a total of $20 million early in 1997, to reallocate $200 million from Space Shuttle program funds to the U.S.–Russian cooperation funding line, and to start working on the ICM as a temporary replacement for the SM.

Finally, the uncertainty about the Russians' ability to live up to their commitments placed greater responsibility and accountability on NASA in terms of handling money allocated to cooperation with Russia. When the General Accounting Office estimated in June 1994 that the United States would have to pay an extra $746 million for two additional Shuttle flights brought on by adding Russia to the ISS program, Dan Goldin disputed this by emphasizing that these two flights would be part of the eight shuttle missions scheduled for 1994. "The money for these [two] flights is in the budget. There is no additional costs due to Russian cooperation," he said.[37] Another example of NASA's financial responsibility was the $400 million contract because it was structured as payments against deliveries. It meant that NASA needed to evaluate the quality of the performance of its Russian counterparts at each stage of the realization of the $400 million deal before compensating them, rather than simply paying the entire sum of money in full, regardless of the performance of RSA under the contract.

NASA's DETERMINATION TO COOPERATE

Apart from NASA's readiness to face the difficulties, there were four more factors that strengthened the agency's determination to pursue further cooperation with the Russians in the ISS program despite all of the problems that the Russians were causing.

To begin with, NASA remained committed to the idea that the Russians could make a great contribution to the utilization of and operations planned for the ISS. According to Goldin, Russian participation in the ISS program "is in the best interest of the American people—we could gain incredible scientific capabilities; we could develop cutting-edge technology..." (Ref. 22, p. 21).

Second, Goldin's optimism concerning further cooperation with the Russians was based on the fact that the Russians had already taught NASA far more than the Russians got from NASA, and the United States probably benefited from its relationship with Russia far more than Russia had with regards to technology transfer.[35] Besides, "the Russian-produced FGB is the most mature piece of hardware we have. And it is on time and on budget. Russian industry has demonstrated that they can deliver when adequate funding is provided" (Ref. 22, p. 21).

Third, the successful completion of the ISS after its final redesign became critically contingent on the Russian contribution. This contribution was crucial, first in terms of the amount of hardware, the number of assembly launches to the station, and the spacewalk time needed to assemble the ISS, and second in terms of the "critical path" that Russian hardware, such as the SM, is supposed to constitute. Russia was not just "enhancing" but "enabling" the space station.

As of September 1994, the assembly sequence for the station showed 73 launches in a 55-month period. Of these, 27–28 were U.S. Space Shuttle launches, which constituted about 37% of the total number of launches; one was European (about 1%); and 44 were Russian using a variety of launch vehicles (about 60% of the total number of launches). The Russian segment of the station was supposed to constitute 40% of the total station mass and 50% in terms of the total number of the ISS modules. The Russians were also expected to perform 240 hours of spacewalks of the total 888 extra vehicular activity (EVA) assembly hours (Ref. 6, pp. 500–501; Ref. 22, pp. 23–28).

Finally, NASA leadership remained committed to the great foreign policy value of U.S.–Russian cooperation in the ISS program. Speaking at the press conference concerning NASA fiscal year 1998 budget, Goldin said that "we [Russians and Americans] spent too many years pointing missiles at each other, and the only way you get to know another country is to engage proactively in resolving problems."[36] Among the unquestionable benefits of

U.S.–Russian cooperation in the space station program, Goldin mentioned "the knowledge that Russia is focusing its technological expertise to benefit humanity and promote world peace" (Ref. 22, p. 21).

BRIDGING THE TWO SPACE PROGRAMS

One of NASA's main goals—"to learn as much from the Russians as possible"—resulted in the unprecedented bridging of the two space programs. Americans flying onboard the Mir space station were not considered just guests who expect to be provided with all of the necessary conditions for their scientific and research activity in space; instead they shared the bulk of the operational burden of the station with the crew. This was demonstrated by the decision to replace U.S. astronaut Wendy Lawrence, who was supposed to become the sixth American to make a long-duration flight in Mir in the fall of 1997, by her backup David Wolf. The change enabled Wolf to act as a backup crew member for spacewalks planned over several months to repair the damaged Spektr. The reasons for the replacement were that Lawrence did not fit in the Orlan suit that Russian cosmonauts use for EVAs, and she had never undergone spacewalk training, whereas Wolf met both requirements. The decision was made by both NASA and the RSA, who agreed that it would be mutually beneficial to have all three crewmembers on the Mir qualified for spacewalks in the event additional assistance was needed from the U.S. astronaut on the station.[39]

U.S. HIGH-LEVEL POLITICAL SUPPORT

One of the reasons for NASA's ability to cope with the problems and challenges presented by cooperation with the Russians was the high-level political support provided to joint activities by both the executive and legislative branches of the U.S. government. However, unlike the White House, which championed U.S.–Russian cooperation in the space station program, Congress had a more ambivalent attitude toward it. Many lawmakers supported cooperation with the Russians in the ISS program in 1993 because it promised to complete the station faster and at a lower cost and was good for foreign policy reasons. At the same time, "it was believed that Russia's addition to the Space Station program would best serve the program if Russia's role was enhancing, not enabling."[40]

CONCERNS WITH RUSSIAN DELAYS ON THE ISS

After it became obvious that Russia was experiencing problems in meeting its commitments to the ISS program, Congress started raising concerns about the impact that Russian delays could have on the space station. This had an effect on the overall attitude of the lawmakers toward cooperation with Russia in the ISS program. Although the report accompanying the NASA fiscal year 1995 appropriation bill prioritized ISS's "preeminent role in the foreign policy" over "other benefits in science, technology, and engineering,"[41] the report accompanying the Civilian Space Authorization Act for fiscal years 1998 and 1999 stressed that the ISS's goal to promote international cooperation in space

> is secondary to the Station's main focus on research. When the goal of promoting multinational ventures in space conflicts with the ability to do world-class research to benefit all of humanity, the multinational venture must give way to the needs and requirements of science.[40]

In 1995–1996, most of the lawmakers continued to support cooperation with Russia, but this support was driven mostly by the understanding expressed in the words of Senator Larry Pressler: "If, for whatever reason, the Russians were forced to withdraw from the Station, the program would collapse, and billions of dollars of taxpayers' money would be wasted."[42] Although emphasizing the benefits that cooperation with the Russians brought to the ISS program, many lawmakers, such as Senator John D. Rockefeller IV, asked the question: "Will this nation be getting enough from our partnership with Russia, in terms of cost, and time savings, hardware, knowledge, and capability for our research to make the deal worthwhile?" (Ref. 43, pp. 24–25). Senator Conrad Burns emphasized that the "recent demands by the Russians [for an additional $200 million from the U.S. (turned down) and two Shuttle resupply missions (fulfilled)] only highlight the problematic nature of the Russian participation that has troubled station supporters and opponents alike" (Ref. 43, p. 3). Later he admitted that even though the Russian involvement had introduced "troubling vulnerabilities" in the program, "our reliance on the Russians is so great that we have little choice but to accommodate their demands for changes in the baseline plan."[44]

Even in 1997, when Russia started causing considerable problems in the ISS program, including the eight-month delay of the SM and the FGB module, only those members of Congress who opposed the ISS program attempted to kill the Russian participation in it. According to one of them, Representative Tim Roemer,

> If Russia is permitted to remain our partner, the American taxpayer will be asked to bear cost overruns totaling untold billions to underwrite Russia's continued participation. This is quickly turning into a jobs program for Russians funded by Americans. Ideally, we should cancel the space station. At a minimum, we should terminate Russian participation and build the hardware components in the USA.[45]

However, the position of the majority of lawmakers toward the ISS and Russian participation in it was expressed by Representative Dave Weldon:

> I do not deny that our space program faces many difficult challenges, both technical and political—and I too grow increasingly frustrated with the Administration's inability to solve the very serious problems with Russian participation in the program—but I fear the consequences of turning our backs now on space exploration.[45]

Even the most outspoken critics of Russian participation in the program, such as Representatives James Sensenbrenner and Chairman of the Subcommittee on Space and Aeronautics Dana Rohrabacher, never proposed removing Russia from the program altogether. Moreover, they always stressed that they supported building the space station, preferably along with the Russians. Such observations could be supported by the fact that Sensenbrenner moved back the deadline for a decision on Russian participation in the ISS program several times. In early February 1997, he said that "the time has come for the Administration to draw a line in the sand and let the Russians fish or cut bait,"[46] although emphasizing that he "would prefer to have the Russians in as a full partner" (Ref. 22, p. 2). In mid-February he agreed to wait for NASA's decision about Russian participation until mid-April. However, when mid-April came and NASA was still uncertain about how to proceed with Russia, Sensenbrenner did not push the agency into making a decision. "We are now well down the road of relying on the Russians," he said. "Outright termination before we know the costs and plan to remove Russia would be premature…. There will be opportunities yet to remove the Russians from the program if the Station is not put on a better track."[48]

The Sensenbrenner and Rohrabacher proposals concerning Russian participation in the ISS program in the event Russia would not be able to live up to its commitments could be boiled down to three suggestions. First, to remove Russia from the "critical path" of the station. Second, to demote the Russians from full members to subcontractors, i.e., to pay Russia for the hardware it was supposed to build for the ISS.[49-51] In the second case, Russia would lose any decision-making power concerning ISS operation and utilization and would become a client of the ISS partners. Third, to prohibit NASA from using funds from the other NASA programs, such as the Space Shuttle, for contingency plans, in case Russia did not deliver its modules for the station on time.

The decision on the first two suggestions was not to be made until mid-August 1997—the second milestone marked by Daniel Goldin. Regarding the third suggestion, Congress showed its support-science-first attitude. While making it clear in April that "no funds or in-kind payments shall be transferred to any entity of the Russian Government or any Russian contractor to perform work on the International Space Station which the Russian Government pledged, at any time, to provide at its expense," the House on 16 July 1997 rejected an effort to delete the "Russian Program Assurance" line from the NASA budget. The effort was made by Sensenbrenner, who believed that "instead of spending money on Russian nonperformance, these additional funds should be spent pursuing the priorities that the entire House of Representatives endorsed in passing H.R. 1275 in April [1997]."[52] By turning down Sensenbrenner's proposal, the House enabled NASA to accommodate the delays in Russian hardware.[53]

OUTSIDE GROUPS FAVOR U.S.–RUSSIAN COOPERATION

Many representatives of nongovernmental research institutions and a considerable part of the U.S. public were pro-U.S.–Russian cooperation in the ISS program. According to Kenneth Galloway, Dean of the School of Engineering at Vanderbilt University,

> initial experiments in a number of disciplines have demonstrated the value of this early access to extended duration experiments that Mir affords. There has been excellent international cooperation and considerable mutual benefit at the scientific level, and this bodes well for interactions on the International Space Station.[54]

Louis Friedman, executive director of The Planetary Society, highly praised U.S.–Russian cooperation in the Shuttle–Mir program.[55] Although acknowledging that "the ISS, in purely managerial terms, would clearly be better without Russia as a partner," the Director of the Space Policy Institute, John Logsdon, wrote that "a station without Russia would be less capable and more expensive to build" and that "the positives of continuing Russian involvement outweigh the associated risks."[56] Frank Sietzen, editor of *Military Space* magazine, emphasized that "both our nations have no alternative but to work together. In our case, because we have lacked the political will to 'go it alone,' and in Russia's case because, simply put it, they are broke."[57]

Those representatives of the U.S. scientific community who criticized U.S.–Russian cooperation in the ISS project usually did so in the context of general criticism of the ISS program. Robert Park, a professor of physics at the University of Maryland, suggested using the delay in the start of construction of the ISS, caused by the failure of Russia to stand by its commitments, "to reevaluate what we are doing." The need for reevaluation, according to Park, is based on the view of the American Physical Society and "hundreds of scientists in both the physical and life sciences" that "scientific justification is lacking for a permanently manned space station in Earth orbit."[58]

Robert Zimmerman, a freelance science writer, compared Americans and Russians working in space together to two runners, who by holding hands as they sprint, "have managed only to slow each other down." According to Zimmerman, "with the passing of the Cold War, a space race need no longer be spurred by enmity or ideology. But perhaps by running in separate lanes, rather than tripping on one another's feet, we can reach the stars a little sooner."[59]

The number of mishaps that happened onboard the Mir station contributed to polarizing the U.S. public's attitudes toward cooperation in space with Russia. According to the Center for Security Policy, an independent Washington-based think tank,

> Unfortunately, it has taken the drama of a disaster that is jeopardizing the life of an American astronaut and his two Russian colleagues to precipitate widespread questions about the wisdom of a dubious U.S.–Russian space cooperation program that has been personally and insistently championed by Vice President Al Gore.

The Center came to the conclusion that

> the Russians are clearly not equipped to handle their role in preparing the International Space Station—a role that is currently part of the "critical path" to completion of this important initiative. Immediate steps must be taken to redesign the station so as to ensure that is no longer susceptible to Russian non-performance or extortion. For the time being, Russia should be scaled-back to the position of a subcontractor.[60]

The New York Times, however, supported the request from Russia that an American astronaut, Michael Foale, would help repair the crippled Mir station. Such a move, according to the *Times*, would underscore Washington's willingness to help a beleaguered space power, improve collaboration on Mir itself, and could foster greater cooperation on more Earthbound issues as well.[61]

Frank Sietzen admitted that

> Mir may be imperfect, but it remains the only long duration human space vehicle in history. It's rich with experience and data that can teach us much about how to live and work in space. So we ought to quit whining about the Russians' troubles and help salvage their remarkable laboratory of experience, with some dignity and honor, before it is lost to us—and to history.[57]

According to Elaine Camhi, editor-in-chief of *Aerospace America*,

> The accident at Mir, and the danger in which it placed its occupants, serve to remind us once again that traveling to, exploring, and living in

space is not an everyday, run-of-the-mill activity. We must bear in mind that this is, in fact exploration, not a walk in the park. The pioneers who participate understand that it is not without peril, even as they strive for safety. But they believe, as well, that the potential is also there for the flame to shed tremendous light.[62]

RUSSIA'S INADEQUATE SPACE EXPLORATION POLICY

Russia's inability to provide the necessary funds to its ISS contractors in time was caused not only by the lack of money, but also by the lack of a coherent policy on the part of the Russian government and the Russian aerospace sector regarding space exploration in general, and cooperation in the ISS program in particular. In January 1997, Russia's Ministry of Economics proposed to cut funding for Russia's manned space program. This funding was doubled compared to 1996 (although compensating for inflation would have required only a 25% increase in the funding).[63] The decision of the Ministry of Economics was heavily criticized by the Interdepartmental Expert Commission on Cosmonautics—*Mezhvedomstvennaya Ekspertnaya Kosmicheskaya Komissiya* (MVEKK). RSA Director Koptev, who was also a member of MVEKK, claimed that "if work on the manned space flight is scaled down now, in six to eight months Russian cosmonautics will cease to exist altogether." The commission proposed to safeguard future expenditures for the Russian share of the ISS under a specially protected budget line.[64]

SUPPORT FROM THE RSA FOR COOPERATION

In addition to being an advocate of space exploration, the RSA, the main Russian state space body, is a strong supporter of U.S.–Russian cooperation in the ISS program (see Figs. 1 and 2 based on an RSA analysis of the impact that the success or failure of the cooperation could have on the Russian aerospace industry and society).[65] Yuri Koptev stated that although Russia will spend only $3.3 billion as compared to the $14.4 billion spent by the United States, it will "receive the maximum possible":

> Judge for yourselves. Thirty-eight percent of the station's resources are being placed at Russia's disposal. Half the stands of scientific instruments are ours, and 30 percent of the station's energetics are also owned by Russia. Of the six crew members working permanently in orbit, we get a constant quota of three. There are a number of other privileges which make Russia's participation in this project very attractive and even profitable.[66]

One of the reasons why the RSA was interested in cooperating with NASA in the ISS project was the deep integration of those joint activities in

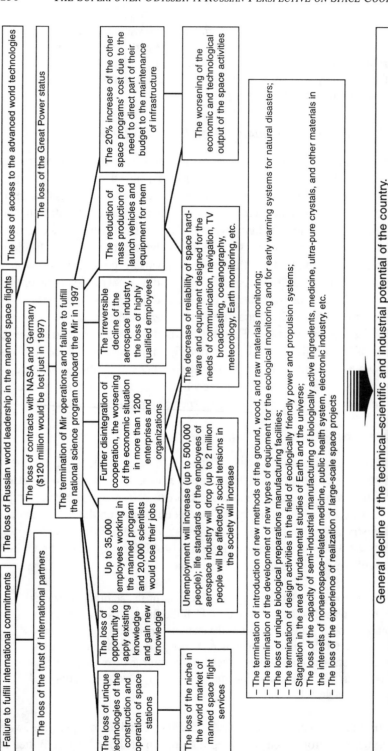

Fig. 1 Consequences of Russia's withdrawal from the ISS program.

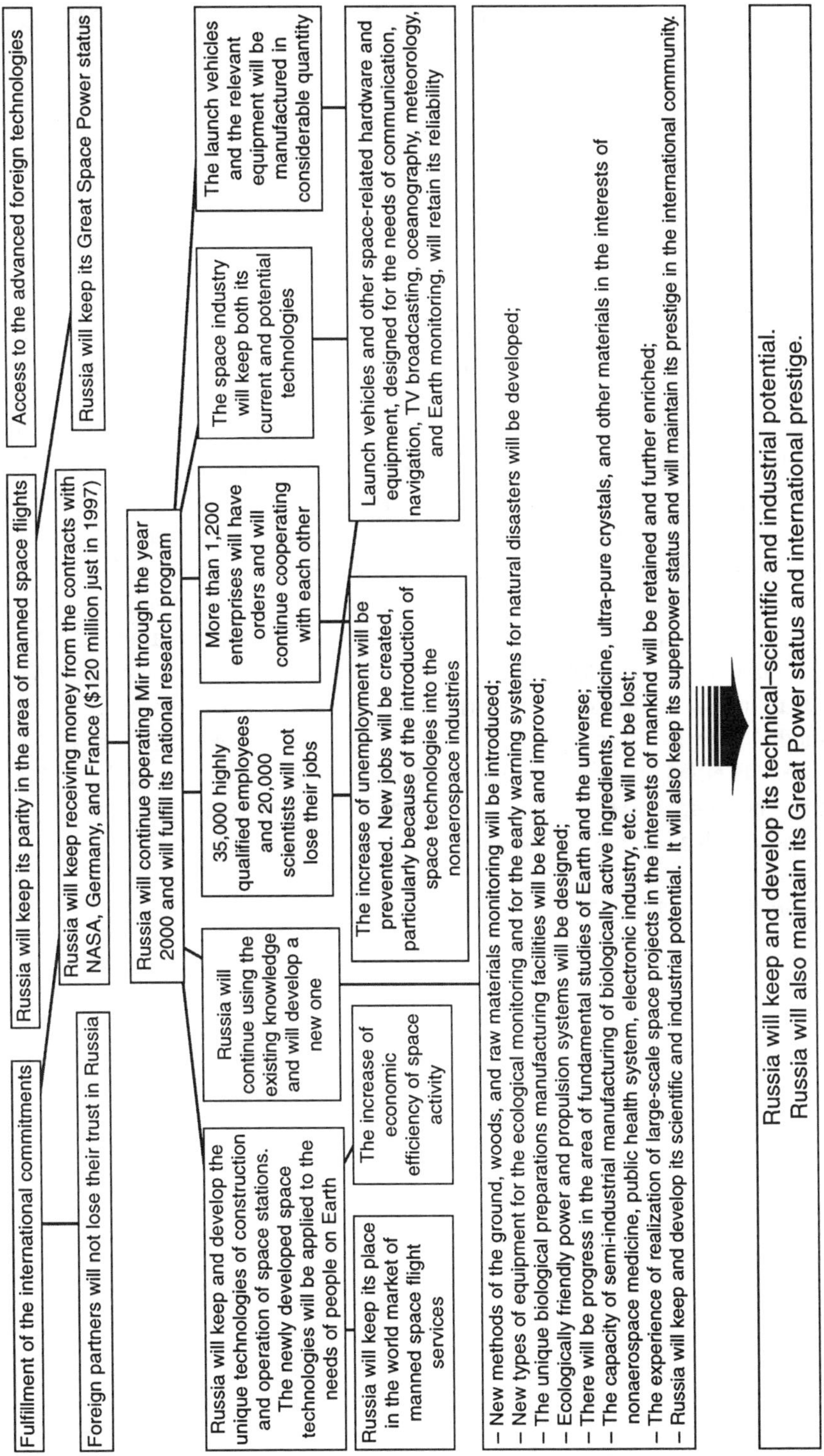

Fulfillment of the international commitments

Russia will keep its parity in the area of manned space flights

Access to the advanced foreign technologies

Foreign partners will not lose their trust in Russia

Russia will keep receiving money from the contracts with NASA, Germany, and France ($120 million just in 1997)

Russia will keep its Great Space Power status

Russia will continue operating Mir through the year 2000 and will fulfill its national research program

The launch vehicles and the relevant equipment will be manufactured in considerable quantity

The space industry will keep both its current and potential technologies

More than 1,200 enterprises will have orders and will continue cooperating with each other

Russia will continue using the existing knowledge and will develop a new one

35,000 highly qualified employees and 20,000 scientists will not lose their jobs

Russia will keep and develop the unique technologies of construction and operation of space stations. The newly developed space technologies will be applied to the needs of people on Earth

The increase of unemployment will be prevented. New jobs will be created, particularly because of the introduction of space technologies into the nonaerospace industries

Launch vehicles and other space-related hardware and equipment, designed for the needs of communication, navigation, TV broadcasting, oceanography, meteorology, and Earth monitoring, will retain its reliability

Russia will keep its place in the world market of manned space flight services

The increase of economic efficiency of space activity

– New methods of the ground, woods, and raw materials monitoring will be introduced;
– New types of equipment for the ecological monitoring and for the early warning systems for natural disasters will be developed;
– The unique biological preparations manufacturing facilities will be kept and improved;
– Ecologically friendly power and propulsion systems will be designed;
– There will be progress in the area of fundamental studies of Earth and the universe;
– The capacity of semi-industrial manufacturing of biologically active ingredients, medicine, ultra-pure crystals, and other materials in the interests of nonaerospace medicine, public health system, electronic industry, etc. will not be lost;
– The experience of realization of large-scale space projects in the interests of mankind will be retained and further enriched;
– Russia will keep and develop its scientific and industrial potential. It will also keep its superpower status and will maintain its prestige in the international community.

Russia will keep and develop its technical–scientific and industrial potential. Russia will also maintain its Great Power status and international prestige.

Fig. 2 Russia's participation in the ISS program.

the strategy and tactics of the RSA. During the first two years of Russian involvement in the ISS project, Russian space decision makers were operating very much like Soviet-style bureaucrats. They addressed dire predictions to the government and the parliament with the intention of getting more money or resolving other urgent issues. The "all-right" posture was intended to impress foreign partners to preserve their willingness to cooperate with Russia.[8] One of the best examples of such a policy was witnessed by ANSER employees during their visit to Baikonur in 1992.* An official at the Baikonur Cosmodrome told ANSER that the cosmodrome itself had cooperated in exaggerating the seriousness of the situation and collaborated with TV reporters to make a video showing the extraordinary decay of the cosmodrome to obtain more money from the Russian government.[67]

However, in early 1996, the situation changed. The RSA leadership realized that it could achieve more by publicly complaining about the problems of the Russian space program and by exposing these problems—and how they could affect Russia's ISS commitments—to the Americans.[32] Goldin would then report about the problems to Gore, and Gore would call Chernomyrdin and ask the prime minister to straighten things out.[68] Such tactics would not only ensure the Russian government's constant support of Russia's participation in the ISS project but would also bring the Kremlin's attention to the general problems of the Russian space program. In April 1996, the newspaper *Narodnaya Gazeta* quoted Koptev as saying that Russia's ISS commitments were in a critical situation. "The SM construction was virtually terminated because of the lack of funds," he said, "and the partners started panicking. If we don't resume active work in April–May, we will be shamed in the face of the world."[69]

RUSSIAN VIEW OUTSIDE THE RSA

Different representatives of the Russian aerospace community outside the RSA have had a different attitude toward Russian participation in the ISS project. Representatives of the industry, especially those whose enterprises have been involved in this project, have regarded it favorably. At the Star City celebration of the 35th anniversary of Yuri Gagarin's flight, Yuri Semyonov,

---

*ANSER (Analytic Services, Inc.) is an independent public service research institute located in Arlington, Virginia. The company's primary activity is research and analysis of aerospace-related issues. In 1992, ANSER became the first U.S. aerospace company to open an office in Moscow. ANSER also became one of the first commercial clients of the RSA. On 13 November 1992, ANSER and RSA signed a Protocol of Cooperation, according to which ANSER promised to provide the RSA with $5000 in return for a point of contact in the RSA to whom ANSER could turn to for information about ongoing space missions, Russian space organizations, and Russian space capabilities. The report about the Baikonur Cosmodrome prepared in 1993 by ANSER for the U.S. Congress helped convince U.S. politicians of the benefits of U.S.–Russian cooperation in space. In 1995, ANSER formed the Center for International Aerospace Cooperation (CIAC) to boost the company's activities aimed at promoting aerospace cooperation, particularly between Russia and the United States. With NASA's increased presence in Moscow, however, CIAC ceased operating in January 1999 and closed its office in Moscow.

chief designer and CEO of RKK Energia, the main SM contractor, said that to adapt to the new economic situation in Russia, RKK Energia had to look for foreign partners and that in 1995 Energia survived only because of increased international cooperation with Europe and the United States.[70] However, Vladimir Utkin, a TsNIIMash director and a participant in the ISS project as one of the leaders of the Utkin–Stafford safety board, perceived additional benefits in U.S.–Russian cooperation in the space station program. According to Utkin, the use of the Space Shuttle for the delivery of supplies to the Mir station decreased the number of Russian Progress cargo ships previously used for such deliveries by three to four times. This reduced Russian capacity to use the Progress ships for scientific research. Moreover, a significant portion of Mir power and research facilities, and of the crew's time onboard the station, were redirected to fulfillment of Russia's international commitments.[71] At the same time, Utkin believed that if Russia fails to fulfill its commitments to the ISS, it will be an even greater blow to the prestige of Russian cosmonautics than the Mars 96 failure. Utkin also considered Russian participation in the ISS program to be the realization of the Mir 2 program.[72]

German Titov, who was the second Russian in space after Gagarin in 1961 and is currently a Duma Deputy and member of the Committee on Conversion and Scientifically Sophisticated Technologies, has expressed that U.S.–Russian cooperation in the ISS project to be a win-win situation.[73] Those who have been actively involved in space flight operations consider the ISS to be an opportunity for Russia to continue its manned space flight program in the conditions when Russia cannot sustain the program by its efforts alone.[74-77] Yuri Glazkov, deputy chief of the Gagarin Cosmonaut Training Center (Star City), said that at least in the area of space flight training, the Russian space program has become internationalized to the extent that it is difficult to distinguish the Russian from the foreign space programs, especially from the American. In his view, Russia benefited greatly from such integration because

> our international partners have brought into the Russian program their ideas and concepts. While one can argue whether this input has boosted our activities beyond the atmosphere, it has become an injection of fresh blood into the Russian program.... While Americans living in Star City during ASTP were just guests, today they are an integral part of its life.[78]

DISSATISFACTION WITH AMERICAN LEAD ROLE. One of the manifestations of the growing professional understanding between the two space communities is an agreement on the selection of station crew commanders. The U.S. side insisted that the commander of the first ISS crew be an American even though the initial ISS segment would consist chiefly of all-Russian designed and manufactured hardware. Russian space officials agreed with the Americans, which

ignited protests among the cosmonauts. They believed that it was done mostly for political reasons, i.e., to please U.S. politicians, who would be willing
to support the ISS only if it was a visibly American-led project. One of the
Russian crewmembers of the first ISS crew, Anatoly Solovyov, left in protest.
Sergei Krikalev, a flight engineer in the crew, considered dropping out as well
because he did not want to sacrifice the success and safety of the mission to
political games. He asked how an American, who does not even speak
Russian, could learn Russian hardware, design, and operating philosophy in
the two years remaining before the first element launch.[79]

The eight-month delay of the FGB and SM launches, which gave William
Sheppard (the first station crew commander) more time for training, alleviated the problem. In May 1997, Russian and American space officials
reached a new agreement that ensured subsequent selections would be based
on experience and other similar factors. This agreement opened the door for
Russian commanders. Both sides agreed that whenever there were two
Russians and one American onboard the station, one of the Russians would
be named commander and that when there were two Americans and one
Russian, an American would command (see Ref. 80).*

RESERVATIONS ABOUT ISS COOPERATION. There are a number of representatives of
the Russian aerospace community who continue to have serious reservations
about U.S.–Russian cooperation in the ISS project. Although they cannot be
regarded as key players, they have a significant degree of influence inside
the Russian aerospace community and under certain circumstances may have
an impact on the Russian attitude toward the ISS project. Their objections
can be boiled down to three arguments.

1) Americans are buying Russian hardware and experience at the dumping price.
2) Russia, a great space power, will play the role of junior partner in the
ISS project because the Americans want to subordinate the Russians to
the U.S. space program.

---

*Such an approach, however, still needs to be implemented. The first four crews of the ISS have an
opposite composition–command relationship, i.e., when there are two Americans and one Russian
onboard, the Russian is in command, and vice versa. This is the result of two factors: First, the initial
crew for the ISS (one American, two Russians; American in command) was assigned prior to making the
preceding agreement. Second, according to cosmonaut Vladimir Dezhurov, it was decided to have a
Russian command the second crew (one Russian, two Americans) to "balance" the assignments from the
first crew—otherwise, an American would have been in command of the second crew (according to the
original agreement) and thus an American would have been in command for two consecutive crew inter-
vals. This same consideration was apparently used in the assignments for the third and fourth crews. It
is interesting to note, however, that whenever there is a Russian crewmember at the three-crewmember
phase of ISS operation, he always performs a commanding function—either expedition or Soyuz vehicle commander.

3) Cooperation in the ISS project will undermine Russian capacity for independent manned space operations to the extent that Russia will never again be able to regain it.

Academician Boris Raushenbakh, one of the leading designers of Soviet/Russian space technology, has not directly criticized U.S.–Russian cooperation in the space station program and has even accepted that currently cosmonautics is supported mostly by U.S. money. However, it is his view that the United States is doing this principally because it acknowledges Russian superiority in many fields of space science and technology and wants to take advantage of the country's desperate economic situation by buying Russia's advanced technology and experience "for pennies."[81]

According to Vitaly Sevastianov, a former cosmonaut and a Chairman of the Duma Mandate Commission,

> Russian–American cooperation in space as it now stands can seriously undermine the Russian national space program. We shall spend all our funds on designing and building of the elements for Alpha, and we will have no money for our independent space projects. As a result, we will be flying to a foreign station, and only when we are permitted to do so. Moreover, we will be spending money to do it. When Mir space station operations are terminated, our national manned space program will die.[82]

Vladimir Kovalyonok, also a former cosmonaut and current head of the Zhukovski Air Force Engineering Academy, said that

> the Americans are trying at all costs to oust us from the leading position which Russia occupies today in the sphere of world cosmonautics. We have opened our doors wide to the United States: you want to know about the methods of organizing long flights?—Fine, help yourselves; you want to work on Mir?—go ahead! They are using our station for around $300–400 million, which means practically for free. It took us more than a decade to construct it, we spent hundreds of millions of rubles, and we have sacrificed our health for it: it is no secret that many cosmonauts have contracted occupational illness....
>
> The Americans are insisting that it is necessary to change the station's ideology, architecture, and design, as a result of which we will have to create some different kind of module apart from the one that is already functioning. I view this demand as an attempt to dissipate our production and scientific forces and thereby bring the Russian "train" to a standstill. What does this mean? When we get embroiled in financial and other problems they will tell us: sorry, friends, you have fallen behind us in fulfilling the program and therefore we will fill the station with our own equipment.[33]

Viktor Savynykh, a former cosmonaut and current president of the Moscow University of Geodesy, Aerial Photography, and Cartography, believes that as soon as Mir drops out of service, Russian cosmonautics will cease to exist. In his view, Russia is working now not for its own space program, but for the U.S. program, and the United States has to cooperate with Russia because the Americans would not be able to build the station without Russian help.[83]

A number of arguments, such as the unavoidable dependence of the United States on Russia in building the ISS and the assertion that the United States is cooperating with Russia for the purpose of making a serious competitor serve the goals and interests of the U.S. space program, are shared not only by the critics, but also by supporters of U.S.–Russian cooperation in the space station project.

Anatoly Kiselyov, a director of the Khrunichev State Research and Production Center, which receives more than 90% of its revenue from commercial contracts, including the manufacture of the FGB module,[84] said that Americans would not be able to build the ISS without Russia's involvement.[85] The same view is shared by Yuri Semyonov, chief designer of RKK Energia and by many Russian cosmonauts.[86] RSA General Director Koptev stated that despite all of the benefits of cooperation with Russia in space, "the United States and Europe would draw a sigh of relief if tomorrow Russia ceases to be a great space power."[87]

ISS DEPENDENCE ON RUSSIA

Although representatives of the Russian government have never mentioned that the dependence of the ISS project on Russia or the suspicion that the United States chose to cooperate with Russia for the sake of neutralizing a strong competitor in space activities were behind the delays in Russian hardware, such arguments might have convinced Russian decision makers that the United States and other ISS partners would have no choice but to accept Russia's inefficiency in meeting its commitments to the space station program.*

The lack of funding and the understanding that the Russian role was "indispensable" to the ISS project may not be the only explanations for the inadequate performance of the Russian enterprises involved in the space sta-

---

*RSA Deputy General Director Alexander Medvedchikov, meeting with ANSER representatives in Moscow said, "The International Space Station Program shows the importance of Russia—before Russia became involved, nothing had happened after years of planning the station."[88]

An indirect confirmation of the Russians' belief that the United States and other international partners are totally dependent on Russia in building the ISS was given by Daniel Goldin on 6 May 1998 during his testimony before the House Committee of Science. He said, "I don't believe the Russians thought we were serious about building the Interim Control Module. I invited the head of the Russian Space Agency along with the heads of the Russian corporations building their portion of the space station to [see] the Interim Control Module. They were surprised by what they saw."[89]

tion program. Todd Breed, the NASA representative in Moscow, believed that the reason why RKK Energia—the enterprise programmatically responsible for building, testing, and operating the SM—was in a "state of near paralysis" and unable to decide for sure what to do about the SM was because it consisted of

> crusty old veterans of the Cold War who spend a great deal of time reliving the glory days of Sputnik, Gagarin, and Mir…. It is plainly obvious that [RKK Energia] has no real desire to work with Americans (with the obvious individual exceptions down to the one-on-one level) and really wishes we would go away.
>
> So, the lack of progress on the SM is explainable by the fact that the Russian government and [RKK Energia] charged with its production deep at heart simply does not want to do this. There is a sense of almost desperation in this country concerning the loss of autonomy in manned space operations which being a partner in the ISS program represents.[90]

Despite the number of questions that U.S.–Russian cooperation in the ISS project raises, it is supported overall not only by the ruling political elite, but also by all of the major opposition political forces in Russia, although with some reservations. Gennady Zyuganov, the leader of the Russian Communists, head of the largest faction in the Duma, and President Yeltsin's main rival in the presidential elections in Russia June–July 1996, has considered Russian–American cooperation in space to be "one of the greatest achievements in Russian–American relations in the past few years":

> From a political standpoint, it proves that Russia and the USA can cooperate in a stable and productive manner in the field of high technology, despite all the fluctuations in bilateral relations. From an economic standpoint, it helped the main part of our space industry survive in the hostile environment of the newly-born market.[91]

At the same time, Zyuganov shares widespread concern about Russia possibly losing its independence in space and believes that in the future cooperation must also be justified on the grounds of strengthening Russian space autonomy:

> Despite the unquestionably positive moments of Russian-American cooperation in space, there have been those that have aroused my concern. I fear that Russian cosmonautics, as a result of the difficult situation it faces, has found itself a junior partner of NASA and will remain so, even after the situation improves. It could happen if Russian cosmonautics loses its direction and blindly follows the U.S. or another foreign space program. If Russia loses its independence in space, it will not only

slow down the development of its space technology but also seriously undermine its national prestige. It must never be allowed to happen. At the moment, Russia must fulfill all obligations relating to the Alpha space station and honor all standing international contracts. In the future, any international contract must be signed only if it strengthens the independence of Russian cosmonautics and consolidates its standing as the premier space nation today.[91]

Although Zyuganov's opinion is dominant in the Communist Party, especially taking into consideration its rigid hierarchical structure, there are members of the party who have a different view of U.S.–Russian cooperation in space. The aforementioned Vitaly Sevastyanov and German Titov are both members of the Communist Party; however, it has been Sevastyanov's opinion that U.S.–Russian cooperation is unfair—or as he put it, "Rich guys order music and make us dance what they want us to dance. Even now we have an economic and technological potential to execute a more independent policy. We have sold ourselves to Americans."[82] Titov has had a different view of this cooperation. In his opinion, "the most important thing is to work together":

> Of course, some particular tasks can be realized by national efforts, but a broad scale venturing into space can be achieved only by global efforts. I believe we wasted many years by duplicating each others systems instead of dividing work on designing and building space systems among spacefaring countries. Today they talk about losers and winners in international cooperation in space.... But we are arguing about small differences in the size of pieces of a pie which has not yet been baked. I believe we should bake this pie first and then we shall discuss sizes of pieces.[73]

Alexander Lebed, a representative of noncommunist nationalistic forces, who finished third after Zyuganov and Yeltsin in the first round of presidential elections in Russia in June 1996 and was appointed Secretary of the Security Council of the Russian Federation, also has supported U.S.–Russian space cooperation.* He also believes that this cooperation

> has not realized 10 percent of its potential yet. To make joint flights to the Mir space station or to give rides to Russian cosmonauts in the Space Shuttle is fun, but it does not bring any profit. It is much better to supply Russian rocket engines for the new American launch vehicle, or to build a core module for the ISS in Russia. But the most important

---

*Alexander Lebed was fired by President Yeltsin from his position as Secretary of the Security Council in November 1996. In May 1998, he was elected as Governor of Krasnoyarsk region of the Russian Federation.

thing for Russia is to learn how to make full use of its competitiveness in the fields where it already plays a leadership role, and to strengthen this leadership by equal cooperation with the United States in the fields where it lags behind.[92]

The support for U.S.–Russian cooperation in space was also expressed by a representative of a supreme judiciary power in Russia. Valery Zorkin, Justice of the Constitutional Court (the equivalent of the U.S. Supreme Court) of the Russian Federation, has viewed two kinds of benefits of Russian–American cooperation in space: "First, this cooperation will promote trust and understanding between the two countries. Second, it will enable Russia and the United States to save on space exploration by pooling their resources together." In Zorkin's view, this "cooperation may bring negative results in two cases. First, if Russia and the United States use each other's technological achievements and experience to create weapons to use against each other. Second, if the U.S.-Russian space partnership falls apart, it may have a negative impact on the overall relationship between the two countries."[93]

PUBLIC'S VIEW OF SPACE PRIORITIES AND ISS COOPERATION

Generally, the Russian public has had a rather complex attitude toward U.S.–Russian cooperation in the space station program. A significant part of Russian society is pro-cooperation and has considered it beneficial to the Russian space program. An article in the newspaper *Rossiiskaya Gazeta* stated that Russia is not able to support its own space activities to the extent it did in the past, which means that Russia should consider its space priorities. The country gained enormous experience in the area of manned space flight and clearly holds the leading position in this field. Thus, it would be wise to continue developing Russian cosmonautics in this direction. Russian experience and the hardware that the country is building for the ISS would give Russia a "key" to the station, which means that Russian participation in the ISS project should be welcomed.[94] However, an article in the newspaper *Segodnya* expressed concern about Russia being degraded to the subcontractor level because such a move would result in the loss of Russia's right to send cosmonauts to the station and to conduct its own experiments onboard the ISS at its discretion.[95]

Yaroslav Golovanov, the leading Russian space writer and journalist, strongly criticized those who advocate Russia's withdrawal from the ISS project and the "reanimation" of the Mir 2 space station project. In his view, such plans if realized would result in the isolation of Russia from world space activities, and ultimately in the technological stagnation of the Russian space program.[96] An article in the newspaper *Rossiiskiye Vesti* called the possible Russian withdrawal from the project "Russia's global shame,"[93] and another article in the *Moskovskiy Komsomolets* newspaper referred to the Americans as Russia's "best partners" in space.[98]

Some Russian people still consider U.S.–Russian cooperation in the ISS project a continuation of the space race. Yana Yurova, in an article published in the *Moskovskiy Komsomolets* newspaper, wrote that Russia decided to build more modules for the space station than the United States because it wanted to "beat the Americans" in the ISS project.[99]

There are a number of people who, although not objecting to U.S.–Russian cooperation in the ISS project, believe that Russia should not push itself too hard to fulfill its space station commitments on time. Such people usually support their positions with two arguments. First, the Americans just "blow hot air" over Russia's so-called inefficiencies for political reasons—to show that NASA cares about the taxpayer's money and to press Congress to increase NASA's budget to enable the agency to build replacements for the missing pieces of Russian hardware.[100,101] Second, the Americans (not to mention the other ISS partners) are "children" in terms of manned space flight experience compared to the "adult" Russians. They would not be able to build the station without Russia anyway, or even if they managed to do so, the station would have less capacity, would be more expensive, and would be built at least two years behind schedule. Advocates of such a position also believe that Russia should occupy a privileged place within the ISS project.[102,103]

The most ardent criticism of U.S.–Russian cooperation in the ISS project came from hardline communists. An article in their newspaper *Sovetskaya Rossiya* argued that the United States just pretended to help Russia during the so-called transitional period of the Russian economy. In reality, the United States was "taking the cream" of Russian science and technology, particularly in the field of space exploration. Having done that, the United States could now easily blackmail Russia because it no longer needed Russia to achieve its goals in space.[104]

U.S.–RUSSIAN COOPERATION: PASSING THE POINT OF NO RETURN?

In August 1997, Russia was to meet the second of the three decision-making points set by NASA Administrator Daniel Goldin:

> Assuming success in May, we then have another decision point coming just before fall. We will then have to recertify that the Russians do what they said they are going to do. If they have, we proceed forward. If they have not, we then have to take the next step, which could be a relatively costly one, and that is commit to a permanent propulsion module, and attitude control system that the U.S. would build; and secondly

accelerate the life support systems that we have in the U.S. habitation module, and put them into the U.S. laboratory....[28]

Based on the criterion for the second milestone, it was possible to conclude that meeting it would mark a certain "point of no return" in U.S.–Russian cooperation in the ISS program, at least as an interaction between two equal partners. After this, the United States and other ISS partners would become dependent on Russia for a critical capacity of the station that Russia was supposed to provide with its own money.

According to Lynn Cline, NASA deputy associate administrator for External Relations, Russia was meeting the second decision-making point overall by August 1997, although with some reservations. These reservations were caused by the extremely complex and contradictory Russian economic and legal environment that affects Russia's participation in the ISS program.[105]

ECONOMIC PROBLEMS CONTINUE

Money kept flowing to the SM constructors, although it flowed irregularly. However, President Yeltsin on 12 May 1997 signed a decree, "On the Termination of Guarantees and Securities at the Expense of the Federal Budget." This decree terminated "the provision of guarantees and securities on the credits of commercial banks, used by the subjects of the Russian Federation as well as by other recipients of federal budget resources to cover expenses stipulated by the federal budget."[106] Two government-guaranteed commercial loans to the RSA of 400 billion rubles each had already been issued and would not be affected by the decree. However, there was concern as to whether other commercial loans, covering the remaining 700 billion rubles of the 1.5 trillion earmarked for the ISS, would be issued as planned at the end of summer 1997. President Yeltsin's decree of 8 August 1997 allowed the Russian Finance Ministry to appropriate $99.5 million for the RSA through credits with foreign banks "in order to ensure continuous financing of the development and manufacture of space equipment for the Russian segment of the International Space Station." Moreover, Yeltsin required the government of the Russian Federation to "provide for payments on credits [issued to the manufacturers of ISS hardware] in the draft 1998 Federal Budget from resources used to pay for state debts."[107] In an attempt to show his firm support for Russia's participation in the ISS project, Yeltsin also visited the Khrunichev space center and inspected the SM and the FGB. He also advised Khrunichev Director Anatoly Kiselyov to raise the salaries of the workers and signed a decree exempting 300 young employees from mandatory service in the armed forces to prevent a labor drain from the enterprise.

Money flow was an important indicator of how Russia was meeting the second milestone, but it was not the only one. Two other indicators were the critical design review of the SM and its shipment from the Khrunichev space center to RKK Energia for final completion. Russia was supposed to take these steps in the fall of 1997. However, in late summer 1997, it was already clear that Russia would not be able to take these actions on time.

NASA's POSSIBLE REPLACEMENT OF THE SM

During 1997, NASA continued working on the possible replacement for the SM, but it was supposed to focus on ICM modification and the extension of life support systems into the FGB rather than construction of a new permanent propulsion module. The agency still believed that Russia would be able to supply the SM on its own within the acceptable time frame. According to Cline, NASA leadership had no doubt that the benefits of Russian participation in the ISS program outweighed the problems that Russia was causing. In the unlikely event that Russia would not be able to deliver the SM at all, other parts of Russia's contribution to the program, including the Soyuz-TM lifeboats, cargo ships necessary for the resupply of the station, the ISS assembly spacewalks scheduled to be performed by the Russians, and experience gained in the operation of orbital outposts, would still provide NASA with a sound rationale for keeping Russia in the program. NASA leadership realized that the current economic situation in Russia would probably never allow the country to demonstrate 100% performance in the ISS program, but the agency could accommodate inefficiency to a certain degree on the part of Russia as an ISS participant as long as Russia would be able and willing to provide the ISS with a number of critical functions. This was the NASA strategy in terms of cooperation with Russia on the ISS project, which was also fully supported by the White House.[105]

According to Daniel Goldin's April 1997 statement, NASA would "make a fundamental decision to completely change [its] relationship with the Russians" only if, in addition to the SM, they were also unable to supply all of the support services, such as space tankers, cargo ships, and crew rescue vehicles (the third milestone).[28] The general economic situation in Russia in 1998, which seriously affected Russian participation in the ISS program, brought NASA very close to making such a decision.

RUSSIA'S FULL-MEMBER STATUS IN THE ISS

Early 1998 was marked by a strengthening of a legal framework of U.S.–Russian cooperation in the ISS program. On 29 January, Russia, the United States, and other international partners signed an *Agreement Among the Government of Canada, Governments of Member States of the European*

Space Agency, the Government of Japan, the Government of the Russian Federation, and the Government of the United States of America Concerning Cooperation on the Civil Space Station.* NASA and the RSA also signed a bilateral *Memorandum of Understanding Between the National Aeronautics and Space Administration of the United States of America and the Russian Space Agency Concerning Cooperation on the Civil International Space Station.* These documents finally codified Russia's full-member status in the ISS program. The objective of the agreement was

> to establish a long-term international cooperative framework among the Partners, on the basis of genuine partnership, for the detailed design, development, operation, and utilization of a permanently inhabited civil international Space Station for peaceful purposes, in accordance with international law.[109]

The agreement and the memorandum established the critical role of the United States and Russia in building the space station while emphasizing "the lead role of the United States for overall management and coordination" of such activities.[109] They stated that the United States and Russia and their respective agencies, "drawing on their extensive experience in human space flight, will produce elements which will serve as the foundation for the international Space Station."[110] The memorandum and the agreement also codified each partner's contribution to the station. The U.S. contribution included:

> [O]ne permanently attached Habitation Module with complete basic functional outfitting to support habitation for four crew members, including primary storage of crew provisions and the health maintenance system;
>
> [O]ne permanently attached multipurpose Laboratory Module, located so as to contain the optimum microgravity environment of the Space Station payload accommodations, with complete basic functional outfitting, including accommodations for International Standard Payload Racks and provisions for storage of NASA spares, and secondary storage of crew provisions;
>
> [O]ne permanently attached Centrifuge Accommodation Module with complete basic functional outfitting, a centrifuge rotor, and accommodations for International Standard Payload Racks which will contain a glovebox and specimen habitats;

---

*According to John Schumacher, NASA associate administrator for External Relations, the agreement facilitated cooperation in the ISS program, particularly between Russia and the United States, by regulating the tremendously different operating environment of the space station. However, one should not forget that U.S.–Russian cooperation in space was really pushed along by space agreements that were signed between Russia and the United States in the 1992–1993 time frame.

[T]hree Nodes which will provide pressurized volume for crew and equipment and connections between Space Station pressurized elements;

[T]russ Assembly which provides Space Station structure for attaching elements and systems;

[F]our accommodation sites for external payloads attached to the Space Station Truss Assembly;

[S]olar Photovoltaic Power Modules and associated power distribution and conditioning equipment which serve as the primary Space Station electrical power source, providing an average of 75 kW;

[O]ne FGB Energy Block, a self-sufficient orbital transfer vehicle which contains propulsion, guidance, navigation and control, communications, electrical power, thermal control systems, and stowage capacity;

[O]ne airlock for purposes of crew and equipment transfer with the capability to accomodate U.S. and Russian space suits;

[C]rew rescue vehicle with capabilities to support the rescue and return of a minimum of four crew;

[L]ogistics carriers which provide the delivery of water, atmospheric gases and crew supplies and delivery and return of dry cargo, including crew supplies, logistics and scientific equipment; and

[O]ne Mobile Transporter which will serve to provide translation capability for the Mobile Servicing Center.[110]

The Russian contribution included:

Service Module providing a capability for attitude control and reboost with complete basic functional outfitting to support habitation of three crewmembers;

[T]wo Life Support Modules to accommodate additional equipment to support Space Station crew and supplement the life support functions present in the Service Module;

[T]wo Docking Compartments to support EVA for assembly and operations;

Universal Docking Module which includes gyrodynes to provide docking and pressurized access to the Russian elements and a capability to support research activities;

Science Power Platform which will provide an average of 19 kW and which includes Autonomous Thrusting Facilities, power distribution and conditioning equipment, accommodation sites for external payloads, and a remote manipulator system;

[T]wo Research Modules with a complete set of equipment to support research activities;

Soyuz TM vehicle to provide on-orbit shelter, crew rescue and emergency crew return functions in accordance with technical capabilities of one permanently docked Soyuz TM vehicle;

Progress vehicle to provide Space Station reboost capabilities and delivery of infrastructure elements, propellant, water, atmospheric gases and delivery and return of dry cargo, including crew supplies, logistics and scientific equipment; and

Docking and Stowage Module to accommodate additional stowage and support Soyuz docking.[110]

In addition to the flight elements, both Russia and the United States were supposed to provide space station-unique ground equipment (Ref. 109, Annex; Ref. 110, Articles 3.2, 3.3, 3.5).

RUSSIA'S COMMITMENTS THREATENED BY UNDERFUNDING

The realization of the agreement and the memorandum, however, turned out to be seriously threatened just a few months after they were signed. The threat came from Russia's inability to meet its commitments because of the gross underfunding of the Russian space industry.

Only $180 million of the $250 million U.S. dollars promised by President Yeltsin in 1997 in government-guaranteed commercial loans for ISS had been released by January 1998.[111] In April 1998, President Yeltsin called on the new Acting Prime Minister Sergey Kirienko to "strictly" adhere to plans for funding space activities and promised increased funding for Russia's space program.[112] However, this promise like many others made by the Russian president, particularly regarding the Russian space program, turned out to be an empty pledge. On 28 April, Yuri Semyonov, Energia CEO, said that the Russian government had transferred only $20 million of the $100 million needed to complete the work on the SM.[113] The RSA received only one-third of the $99.5 million that was budgeted for the work on SM in 1997. To fulfill its obligations, the RSA needed at least 1.5 billion rubles ($250 million) in 1998. The sum was not included in the 1998 budget and the RSA tried—without any noticeable success—to obtain it as commercial loans guaranteed by the state.[114] Moreover, Minister of Finance Mikhail Zadornov accused the RSA of mismanagement of finances because the agency overspent 130 billion rubles in 1997. "Your appetite will never be satisfied as long as you have such approach to the management of your funds," said Zadornov to Koptev.[115] In June 1998, Khrunichev Director Anatoly Kiselyov said that the government had not paid 1.3 billion rubles (an equivalent of $211 million) to make SM ready for launching.[116]

EFFORTS TO FIND SOURCES OF FINANCING

The Russian government tried to take certain actions aimed at improving the economic health of the Russian space program, particularly the part involved in building the ISS. On 16 and 22 July, meetings took place in the

Office of the Deputy Chairman of the Government of the Russian Federation Boris Nemtsov regarding the use of additional sources of financing for programs within the framework of the Federal Space Program. One of the goals of those meetings was to mobilize a broad spectrum of the Russian governmental bodies for the solution of economic problems of the Russian space program. The meetings were attended by RSA General Director Yuri Koptev, Deputy Minister of Finances Andrei Astakhov, Deputy Minister of State Property Victor Pylnev, First Deputy Minister of Defense Nikolai Mikhailov, Vice President of the Russian Academy of Sciences Nikolai Laverov, First Deputy Chairman of State Communications Committee Naum Marder, Deputy Director of the Security and Disarmament Department of the Ministry of Foreign Affairs Vladimir Pavlinov, and other high government officials. RKK Energia General Designer Semyonov, Khrunichev General Director Kiselyov, and TsNIIMash Director Utkin were also among the attendees. The most important steps toward solving Russia's ISS funding shortfall included:

> Sales of stocks of Russia's national telecommunications stockholding society Svyazinvest and of cellular phone services. The estimated amount of money which was supposed to be allocated to the Federal Space program was $170 million;*
> Russian government agreed to sell 13% of its 38% share of RKK Energia's stocks. This sale was supposed to generate at least $100-120 million. Of this amount of money, 90% was supposed to be spent on Mir and ISS;
> Contracts signed between RSA and its contractors were supposed to envisage a possibility of getting up to 3 months credits from commercial banks for Mir and ISS programs. The interests on these credits were supposed to be included in the cost of the Mir and ISS related works.[118]

These actions, however, did not bring any tangible results either. It became evident when the Soyuz-TM-28 mission with cosmonauts Gennady Padalka, Sergey Avdeev, and Yuri Baturin was rescheduled from 3 August to 13 August. The flight was postponed because electricity to the Baikonur Cosmodrome had been cut off because of an outstanding bill. RKK Energia had taken out $33 million in commercial credit to finance the Soyuz-TM-28 launch.[119] One of the sponsors of the flight was the Russian MENATEP bank. On 5 August, the debt

---

*Realization of this part of the plan was seriously impeded by the economic crisis that hit Russia in the fall of 1998. At the end of October, the Russian Federal Property Fund (the vendor of state property) officially cancelled a commercial tender for a stake of 25% minus two shares in Svyazinvest. The tender was cancelled because the starting price for the stake, which had been set at 6.49 billion rubles, fell sharply because of the ruble's devaluation. Such development became one more example of how the general economic situation in Russia negatively impacts Russia's ability to meet its ISS commitments. See Ref. 117.

of the Russian government to RKK Energia increased to $120 million with virtually no payments having been made since the beginning of 1998. Of the $340 million needed for Russian ISS contributions in calendar year 1998, only $160 million had been budgeted, which would provide for only the most essential items. No funding had been provided to the RSA since early in 1998, and the total 1998 funding allocated to the RSA by August 1998 was only $20 million. Production of Soyuz and Progress vehicles had virtually ceased by the fall of 1998 because of nondelivery of components.[120]

ECONOMIC PROBLEMS EXACERBATED BY NEW ECONOMIC CRISIS

The economic problems of the Russian space program were greatly exacerbated by the worst economic crisis to hit Russia during the past two years. 17 August 1998 was described by Russians as "Black Monday." The price of currency jumped from 6.3 up to 10 rubles per dollar. The long-discussed and expected devaluation of Russian national currency became a reality. The Central Bank lifted a limit at which the ruble could fluctuate against the U.S. dollar, allowing it to float, through 31 December 1998, in a corridor between 6.0 and 9.5 rubles to the dollar.[121,122] The devaluation of the ruble resulted in a further decrease of the scarce budgetary allocations to space activities in Russia, particularly to its ISS obligations. The original RSA budget for 1998 amounted to 3.670 billion rubles ($614 million U.S. dollars at the exchange rate of 5.98 effective early in 1998). Later RSA General Director Yuri Koptev had outlined a revised RSA budget for 1998 of 2.7 billion rubles (approximately $450 million).[123] As a result of the financial crisis in Russia and the depreciation of the ruble to approximately 16 rubles per dollar, the actual 1998 RSA budget had, in effect, been decreased to $230 million. Thus, proceeding from the latter dollar figure of $230 million, the promised amount became 37% of the total budgetary allocations for 1998.[124] One of the signs of the critical economic condition of the Russian space program was a delay of the Progress cargo ship launch from mid-October to 25 October—the last cargo ship to go to Mir in 1998.

SOME PROGRESS MADE IN MEETING ISS COMMITMENTS

Despite all of the hardships it faced in 1998, Russia made some progress in meeting its commitments to the ISS program. This progress had a two-dimensional character—on one hand, Russia continued building the SM and preparing FGB for launch and, on the other, started looking for ways to restructure its contribution to the ISS as well as for outbudget sources of financing this contribution. In early June the agency had shipped the SM from the Khrunichev factory to the Energia facility for checkout. The RSA had also reported that it had made near-term core ISS obligations its top priority, including the launch of the FGB in November 1998. The agency had outlined a plan to deorbit the Mir space station earlier than originally planned so that it could focus its resources more fully on its commitments to the ISS program.* The RSA invited the international partners to buy time onboard the SM with the Russian-allocated astronaut time and operating time to pay for the completion of this module. The Russians also stated that they were planning on selling some of their cosmonaut time to the Ukranians in exchange for having the Ukranians build one or more modules.[89,127] As a possible way to meet the SM launch deadline, the RSA even considered completing its outfitting after putting the module into orbit.[128]

RESTRUCTURING TO REDUCE NUMBER OF LAUNCHES

To decrease the number of launches necessary to put the Russian ISS segment into orbit, Russia also restructured its segment. In 1993, Russia decided to include in it a life support module (LSM; Russian acronym MZhO). LSM was supposed to become part of the Mir 2 space station that was never built. In 1994, the Russians also decided to incorporate a docking and stowage module (DSM; Russian acronym MSS) into their contribution to the ISS. Originally, LSM and DSM were supposed to be launched by Ukranian Zenit boosters (each module by one launch vehicle). In 1995, however, because of doubts about Zenit reliability and uncertainty in Russian–Ukranian relations, the RSA made the decision to replace the Ukranian booster with Russian-made Soyuz launch vehicles.† Because Soyuz is smaller than Zenit, each LSM and DSM was "split" into two smaller modules, overall requiring four boosters to be put into orbit.

By January 1998, because of the deteriorating economic health of the Russian space program, it was decided to save on the number of launch vehicles necessary to put the DSMs into orbit. For this reason, DSM 1 and 2 were merged together again in one big FGB-type module (FGB 2).‡ This big DSM will be put into orbit by a Proton launch vehicle. In April 1998, to decrease the number of launches even further, functions of LSM 1 and 2 were distributed between a big DSM and a universal docking module (UDM; Russian acronym USM). UDM will also be launched by Proton, and now

---

*During his 12 December 1997 meeting with the staff of *Florida Today*, Daniel Goldin mentioned the Mir's deorbiting together with the Russia's adherence to MTCR and living up to its commitments to the ISS as the major conditions for successful U.S.–Russian cooperation in the space station program. Goldin did not completely disregard the possibility of the integration of some of Mir's newest modules, such as Priroda, into the ISS. Such integration could take place by redocking these modules from Mir to the ISS. In July 1998, House Science Committee Chairman James Sensenbrenner sent a letter to Russian Prime Minister Sergey Kiriyenko encouraging him to ensure Mir's safe deorbit and to concentrate the efforts of the Russian space industry on the ISS program. Sensenbrenner raised the same issue in his letter to Vice President Al Gore asking him to obtain commitments from Russia to safely deorbit Mir and fully honor all of its obligations to the ISS. The letters were sent shortly before the Gore–Kiriyenko meeting.[125, 126]

†Current RSA plans do not envisage using Zenit before 2002.

‡The original plans included using FGB 2 as a supply vehicle like a Progress, although much bigger. After fulfilling its delivery and refueling functions, FGB 2 was supposed to be disposed. Now, FGB 2 as a DSM will stay with the station for the duration of the ISS lifetime.

there will be no separate LSM launches. DSM was tentatively scheduled to be put into orbit between June and September 2001, which was earlier than the previously scheduled DSM 1 and 2 launches (February and May 2002, respectively). LSM 1 and 2 were baselined for launch in January and March 2003, respectively.

As a result of all these changes, the Russian segment became more compact. The three available docking units—two side units on DSM and one side unit on UDM—could be used for the docking of Ukranian and Chinese modules.* A possibility of attaching these modules to the Russian segment was discussed by the managers of the Russian space program. Figure 3 shows the Russian ISS segment.

STRUGGLE FOR CONTROL OF THE RUSSIAN SEGMENT

The changes in configuration of the Russian segment reflect to some extent a struggle for the control of the segment between two leading Russian space hardware design and manufacturing enterprises—RKK Energia and Khrunichev enterprise. Energia traditionally wants to have as much control as possible over the Russian manned space program. For this reason Energia is very reluctant to pass to Khrunichev the design and manufacturing of the elements that it was supposed to provide.[†] LSM 1 and 2 and DSM 1 and 2 were originally based on Energia-designed and manufactured Progress-M cargo spacecraft. Energia also has general supervision of the launch of its hardware on Soyuz boosters. The LSMs were supposed to be launched by modifications of Soyuz launch vehicles—Soyuz-U-FG[‡] and Soyuz 2 "Rus." Both boosters have not been flown and tested yet.

Energia's resistance to reassigning to Khrunichev the delivery of modules for the Russian segment is hardening in view of Krunichev's increasing role as a primary manufacturer of hardware for the Russian segment of ISS. Khrunichev has already made the ISS FGB and the bulk of the SM (Energia is responsible for its final outfitting) and is currently manufacturing UDM and

---

*There is an agreement between RSA and the Ukranian National Space Agency regarding development of the Ukranian RM. Khrunichev and Energia are participating in the contest for the development of the module. Ukraine, however, might prefer Energia's project, because it will be designed to be launched by the Ukranian Zenit 2 booster. According to RSA Deputy General Director Medvedchikov, personally, he would not be against the Ukrainian module replacing the Russian module. The final decision will depend to a great extend on the research capabilities of the proposed Ukranian module.

[†]Energia's resistance to the delegation of some of its managerial authority to Khrunichev might partially be explained by Energia's traditional perception that Khrunichev is a minor partner. In 1978–1986, Khrunichev was a branch of Energia.

[‡]FG in the name of the new type of Soyuz booster stands for *forsunochnaya golovka* (fuel injector's head). Such heads will be installed in the engines of first and second stages of Soyuz-U-FG.

It is important to keep in mind that most of the decisions regarding the restructuring of the Russian contribution to the ISS are not final and might be reconsidered in the future.

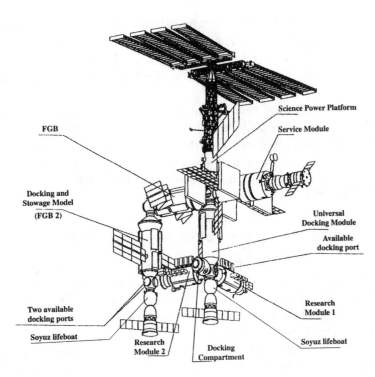

Fig. 3 Russian segment of the International Space Station as of the end of 1998.

DSM. Energia is still in charge of providing the science power platform, two docking compartments, and the Soyuz and Progress vehicles that will be rotated after performing their functions and disposed. Moreover, at some point, Soyuz will be replaced by a U.S. rescue vehicle and Progress' functions will be partially performed by European ATVs (automated transfer vehicles). Anatoly Kiselyov also proposed developing a heavy FGB-type cargo spacecraft that could do the work of two and a half to three Progress vehicles. If accepted, such a plan would save up to three Soyuz boosters a year because a Khrunichev-proposed spacecraft would need only one Proton launch vehicle to be put into orbit. Although Energia's nondisposable contribution to the ISS is much smaller than Khrunichev's, Energia formally remains a Russian ISS prime contractor.*

Energia's resistance to the strengthening of Khrunichev's role in the ISS also resulted in uncertainty regarding the configuration of two Russian

---

*Although Khrunichev manufactures significantly more hardware for ISS than Energia, a number of space professionals question Khrunichev's ambitions to become a Russian prime contractor (or Boeing prime subcontractor) instead of Energia. Unlike Energia, Khrunichev does not have enough experience in the design of huge space complexes.

research modules (RM 1 and 2; Russian acronym EM). Energia insisted on developing RMs based on Progress-M spacecraft and launching them by Soyuz 2 boosters. However, Khrunichev proposed building one big RM instead of two relatively small RMs. The big RM, like DSM and UDM, would be based on an FGB-type module and launched by a Proton booster, saving two more Soyuz 2 launch vehicles in addition to four Soyuz boosters that have already been saved by merging together two small DSMs and distributing the functions of LSMs between a big DSM and UDM. Thus, the restructuring of the Russian segment might save six Soyuz boosters overall. If Kiselyov's proposal regarding replacement of the Progress vehicles by a bigger FGB-type vehicle is accepted, the number of saved Soyuz boosters will increase even further. Unfortunately, according to Khrunichev's Senior Expert of Public Relations Konstantin Lantratov, the struggle between Khrunichev and Energia for the right to manufacture RMs has nothing to do with science, which these modules are supposedly being developed for.[129]

GROWING COST OF THE ISS

Although Russia was trying to improve its performance, it was still far from what the United States and other international partners were expecting in 1998. NASA was hit especially hard by the continuous delays of Russian hardware because of its managerial role in the program, because it was bearing the major financial burden in building the station, and because it was already coping with the growing cost of the ISS. According to the *Report of the Cost Assessment and Validation Task Force* (CAV) prepared under the leadership of Jay Chabrow,

> the Program size, complexity, and ambitious schedule goals were beyond that which could be reasonably achieved within the $2.1 billion annual cap or $17.4 billion total cap…. The Program should plan for the development schedule to extend an additional two years [beyond 2003–early 2004] with additional funding requirements of between $130 million and $250 million annually, including the period beyond Assembly Complete…. This level of funding and schedule extension results in a total assessed cost of approximately $24.7 billion from the 1994 ISS redesign through ISS Assembly Complete.[130]

Allen Li, associate director, Defense Acquisitions Issues, National Security and International Affairs Division at the U.S. General Accounting Office, stated that

> Since June 1995 through June 1996, total Space Station estimated costs have increased from $93.9 billion to $95.6 billion. In particular, the development cost estimate has increased by more than 20 percent, in-

house personnel costs have more than doubled, and eight shuttle flights
have been added to the development program (Ref. 127).*

Not all of the additional costs of the program could be attributed to
Russian delays. Among problems on the U.S. side that contributed to cost
growth in 1998, Chabrow and Li mentioned the U.S. laboratory and other
elements that continued to incur schedule erosion, the destaffing plan that
had not been met, the growing cost of the prime contract (Boeing), multi-ele-
ment testing that had been extended by software and hardware problems, and
the development of the crew return vehicle (CRV).[127,131] The April 1998
Performance Report disclosed that there was an additional prime contractor
cost growth of $80 million over and above fall 1997's $600 million overtar-
get estimate at completion.[127] However, Science Committee Chairman James
Sensenbrenner estimated that the Russian share in the cost increase of the
station constituted $1.2 billion because of Russia's failure to honor its com-
mitments to the ISS and to meet its schedule.[131]

Russian continuous poor performance, which threatened the cost and
schedule of the ISS and created difficulties for NASA in its relationship
with the Congress and its international partners, caused a significant degree
of frustration and disappointment for NASA. During the May 1998 *Hearing
on ISS Problems and Options*, Daniel Goldin openly said that he was "very
frustrated and angry at the leadership in Russia who doesn't do what they
say they are going to do, and they are moving to hurt the tremendous pride
of the Russian people." Goldin also indirectly admitted that it was a mistake
to put Russians in the critical path by saying "in retrospect, I wish we built
a propulsion module. We did not, and I accept the responsibility and
accountability for not doing this."[85] *Florida Today* characterized Goldin's
words as

> a turning point in the stormy relations between Congress, NASA and
> the White House over Russia's involvement in the Space Station. No
> longer was [Goldin] asking for "one more chance" as he and his top
> managers have done so many times in the past.[132]

Goldin's statement sparked a negative reaction from some of the RSA
decision-making officials who considered it a revision of the basic principles
of friendship, cooperation, and mutual assistance that were underlying the
NASA–RSA relationship for almost six years. RSA officials believed that

---

*Li provided a smaller figure for the ISS estimated development cost than Chabrow—$21.9 billion.
The difference is because of the different estimated time of the end of the station's assembly. Whereas
Chabrow believed that it would be sometime around 2005, Li was still basing his assessment on the offi-
cial time of ISS completion—2003. Sensenbrenner, however, believed that it might not be ready until
February 2007, more than 4.5 years behind schedule. See Ref. 89.

NASA was just trying to blame all of the ISS program mishaps on Russia to hide the problems that were caused by U.S. contractors.

The conclusion that NASA was revising the basics of its relationship with the RSA turned out, however, to be premature.* At the same hearing Goldin reiterated that "from a technical standpoint, the station will suffer significant loss from the absence of the Russians. And...from other broader considerations [the United States] would suffer loss."[89]

To prevent such losses from happening and to make it easier for Russia to meet its ISS commitments, NASA and its international partners rescheduled the space station assembly plan one more time. In meetings of the Space Station Control Board and the Heads-of-Agency on 30–31 May at NASA Kennedy Space Center, all station partners agreed to target launch dates of 20 November 1998 for the FGB, called *Zarya* (the Russian word for sunrise) and April 1999 for the SM (previously scheduled to be launched in June and December 1998, respectively).[134]

The seriousness of NASA's intentions to continue down the path of cooperation with Russia in the ISS program was confirmed in early June 1998, when the U.S. space agency formed the Office of Human Space Flight Programs in Moscow. The main task of the office was to oversee the transition from Phase 1 of the Shuttle–Mir program to the assembly and operation of the ISS. Astronaut Michael Baker became the first chief of the new office. Baker's main role was to serve as NASA's lead representative to RSA and its contractors on operational issues as part of NASA's Human Exploration and Development of Space initiative (HEDS).

This development allowed one office to serve as liaison with the Russians for all human space flight operations and initiatives and also consolidated preparations for the ISS assembly, including mission operations, crew training, logistics, and technical liaison activities with Russian space organizations. Other activities in Russia that fell under Baker's leadership included the oversight of astronaut training at Star City and all NASA mission operations at the Mission Control Center in Korolev City.[135]

SUCCESS IN PHASE I

Although U.S.–Russian cooperation in Phase II of the ISS program experienced difficulties, cooperation in Phase I ended in success. On 12 June, STS-91 landed at NASA Kennedy Space Center (KSC) bringing back Andy Thomas— the last U.S. astronaut to fly in Mir. The Shuttle landing culminated 977 total days (812 days in a row) spent in orbit by the seven U.S. astronauts who stayed

---

*In October 1998, RSA General Director Koptev said, "Goldin is a good partner interested in cooperation with Russia. We have witnessed this on numerous occasions. In particular, when problems multiplied on Mir last September, October and early November, numerous congressmen demanded that the Mir-NASA program be discontinued. At that time Goldin convinced Congress that the program had to continue."[133]

aboard Mir since the beginning of the Shuttle–Mir program;* of those, 907 days were spent as actual crewmembers. Frank Culbertson, U.S. director of the Shuttle–Mir program said, "When we began the program...we all had very high expectations of what we would learn and what we would experience in Phase I, particularly of how that would relate to what we would do in future cooperative ventures. I believe that we far exceeded these expectations." Culbertson estimated that those expectations were exceeded tenfold in the quantity and quality of lessons learned. John Uri, a Shuttle–Mir mission scientist said, "we've conducted over 100 unique investigations in seven major disciplines, ranging from space science to biomedical, to risk mitigation, material processing." They involved 150 principal investigators and numerous co-investigators from government, universities, and the private sector.[136]

The Shuttle–Mir program became the first serious test of the two countries' commitment to cooperation with each other in space. Frank Culbertson stated that one of the reasons why the Russians did not release money for their contribution to ISS until summer 1997 was their insecurity about the U.S. intentions regarding the ISS. According to Culbertson, the Russians might have been afraid of spending millions of dollars on the Russian segment of the station and then having it cancelled. However, when they saw that their American partners were willing to remain with them through all of the mishaps onboard Mir, they became convinced that the United States was a reliable partner who could be trusted from a long-term perspective and, as a result, released more money (albeit less than what was planned) for the SM.[136]

Major ISS-relevant lessons from Phase I included: 1) dealing with emergencies, 2) experience with special design aspects, 3) experience with station-peculiar operational procedure, and 4) joint international ground operations.[137] Goldin confirmed that it would not be possible to build the ISS without the Shuttle–Mir missions.[89] Randy Brinkley, ISS program manager, said that he believed remaining ISS challenges would be no more difficult than those that had been overcome during Phase I.[138] George Brown, ranking member of the House Science Committee, having admitted that he "did not believe at all times that [Shuttle–Mir] was going to be successfully completed," called it a "significant accomplishment."[127]

---

*The reason for the difference between a cumulative time in orbit and days spent in a row is because of a lapse between the end of Norm Thagard's flight (7 July 1995) and the begining of Shannon Lucid's flight, the second U.S. astronaut to fly on Mir (22 March 1996). Thagard was launched from Baikonur Cosmodrome in a Soyuz-TM-21 spacecraft on 14 March 1995, inaugurating the flight part of the Shuttle–Mir program. He logged 115 days during his long-duration journey in orbit. Thagard returned to Earth together with his Russian crewmembers Vladimir Dezhurov and Gennady Strekalov in STS-71 (the first Shuttle–Mir docking), which landed at KSC. Lucid currently holds a long-duration flight record among women—188 days.

The Russian side also benefited from the Shuttle–Mir program. However, because Russia was no longer considering building and operating a shuttle-type spacecraft—the major U.S. contribution to the program in terms of knowledge and experience—the main interest of the Russians in Shuttle–Mir was the need to save the Russian space program from total disintegration because of a lack of money. According to cosmonaut Valery Ryumin, RKK Energia deputy general designer and leader of the Russian section of the Shuttle–Mir program,

> [This cooperation] was useful both for us and for them. It was useful for [the Americans] because [flying on Mir] was the only opportunity for them to prepare for the ISS long-duration flights. They gained a considerable experience of long-duration flights. Their flights were not very long-lasting from our point of view, however, 4–5 months is not bad either.
>
> We [the Russians] also gained experience from this cooperation. The Soyuz–Apollo experience of 1975 became very obsolete. A lot of things changed since then. The hardware became a little bit different. Thanks to this [Shuttle–Mir] program, we got some money which our state never has in sufficient amount. And this money helped us to survive. Sure, it was not a 100% rescue; however, we could take money credits to fulfill the contract that supported the existence [of the Russian space program].[139]

NASA CONTINGENCY PLAN

One of the ISS "remaining challenges" mentioned by Brinkley was certainly a delay of critical components of Russian hardware. To solve this problem NASA accelerated work on a contingency plan aimed at the replacement of Russian critical elements of the ISS. On 30 July, NASA gave a briefing to the Office of the Vice President, the Office of Management and Budget, and the Office of Science and Technology Policy. The agency presented three options regarding further structuring of cooperation with Russia (see Fig. 4).

Option 2, which integrated a comprehensive contingency plan into continuing U.S.–Russian joint work on the space station, was characterized as the "best chance for long term program stability." Its overall cost was $510 million ($325 million for the propulsion module, plus $95 million to launch it to the station). The cost of Shuttle modifications was estimated as $90 million. The Clinton administration, however, approved only the modification of the Space Shuttle fleet in early August.[140]

The administration's decision did not mean, however, that the White House's support for the ISS or U.S.–Russian cooperation in the ISS program began to erode. Members of the Clinton administration confirmed that the White House's approach to these issues remained unchanged. Associate Director for Technology in the Office of Science and Technology Duncan Moore stated that a number of benefits "are already flowing to the ISS pro-

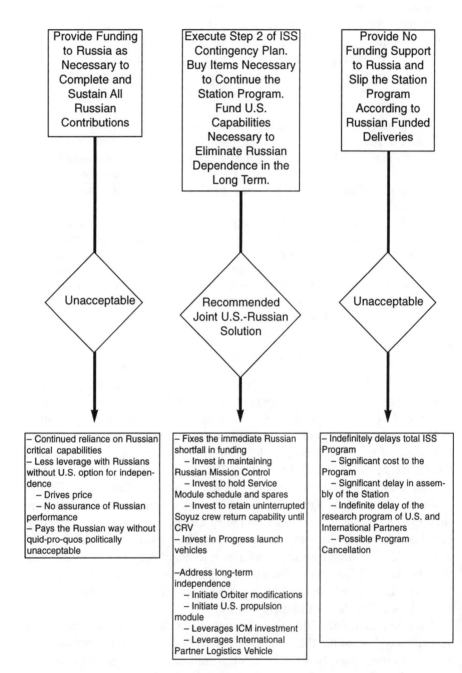

Fig. 4 NASA's three options for resolving the Russian issues. (Ref. 164).

gram and with continued Russian participation, will benefit partnership."[89] Director of Office of Management and Budget Jacob Lew acknowledged that his office was "concerned about the funding shortfalls the Russians are experiencing, but we are committed to seeing the partnership through with them. We believe it is the right thing to do." The reason why the White House did not rush to approve NASA's contingency plan was explained by Lew. He cited NASA's search for a backup for the Russian Service Module as an example. In 1995, NASA came up with a number of options ranging in cost from $500 to $750 million. NASA did not pursue any of those options, but in 1997 was able to proceed with the development of ICM, which would cost much less than the original options. "Time and patience," concluded Lew, "allowed us to pursue a better choice. We believe the same is true here."[120]

ACCELERATING THE CONTINGENCY PLAN

The economic crisis in Russia in August–September 1998 forced NASA to accelerate work on the development and realization of contingency plans. The RSA and its enterprises acknowledged that funding delays had pushed the launch of the SM from April 1999 to no earlier than the summer of 1999.* Another delay of Russian hardware resulted in another assembly sequence update, which postponed the launch of the SM to July 1999 with the first two elements (FGB and Node 1) of the station to remain on schedule. However, just revising the schedule did not give a sufficient guarantee of an SM launch in the summer of 1999. NASA concluded that "without immediate financial assistance to Russia, timely delivery of critical Russian components for ISS, including the Service Module, is not possible, placing the first launches of ISS hardware at further risk." NASA therefore recommended that "funding be provided immediately to RSA [to] help enable completion of the Service Module by Summer 1999, and to facilitate the expeditious delivery of other important deliverables, such as near-term Progress and Soyuz launches." The RSA, and all Russian enterprises involved in the SM, Soyuz, and Progress supply for the ISS

agreed at a Russian General Designers Review, held on September 28, 1998, in Moscow, that if the modification of the NASA-RSA contract for $60 million is executed immediately, they could keep working toward: a First Element Launch of the U.S.-provided, Russian-built Control Module

---

*At the General Designers Review on 28 September 1998, Yuri Semyonov, RKK Energia general designer, said that "Over the past 5 months, much work has been done although problems do exist.... We've had three governments over the last 5 months. Each government has promised help which has not materialized.... Main problem facing the community is financing and is the main driver in meeting or not meeting schedule. We must continue to work 24 hours a day, 7 days a week, no holidays especially on software to deliver the flight version [FGB] by November. The April 1999, launch date [for SM] cannot be made, but we must try for May, June or July."[141]

(FGB) on November 20, 1998; a summer 1999 launch of the Service Module; and availability of requisite Soyuz and Progress vehicles.

Based on the conclusion of the General Designers Review, NASA proposed to revise the fiscal year 1998 Operating Plan to reallocate $60 million in fiscal year 1998 ISS resources to the Russian Program Assurance line to enable Russia to fulfill its ISS obligations.[142]

Sixty million dollars did not constitute "humanitarian aid" to the Russian space program. In return for this funding, NASA secured goods and services of equivalent value, "not otherwise covered under the existing contract with RSA." NASA's plans include getting up to 100% of the research time previously allocated to Russia during the ISS assembly, in addition to that already allocated to the United States.[136] According to NASA Associate Administrator for External Relations John Schumacher, NASA will basically double its research time and storage space through the assembly phase of the ISS.[143] To be sure that the $60 million would be spent on the Russian contribution to the ISS, NASA tied this money to "confirmation of RSA's completion of milestones necessary to ensure the completion of critical early assembly activities related to the final integration and launch of the Service Module and initial Soyuz and Progress spacecraft."[131] NASA also considered spending another $600 million over the next four fiscal years to buy goods and services from Russia to ensure timely delivery of the critical Russian components for the space station. Russia's 1998 estimate of the cost of its ISS activities was $300 million per year. Although NASA hopes that Russia will be able to provide all of this money, the U.S. space agency has made plans to cover half of Russia's yearly expenses ($150 million) by buying goods and services from it.[143]

One such good could be a Soyuz spacecraft. The RSA committed to providing the ISS with the continuous presence of one vehicle as a lifeboat (assuming that there will be a rotation of spacecraft every time a new crew from Baikonur arrives at the station). Soyuz has a three-person capacity. However, in the future, the space station will accommodate up to seven people. For this reason NASA's plans include developing a reusable CRV that should accommodate this number of crewmembers. It is currently based on the X-38 technology demonstrator design. The new CRV, however, might not be ready until 2003. Thus, NASA may buy more Soyuz spacecraft to have two of them continuously docked to the station to allow early six-person operation of the ISS. Current NASA plans include purchasing two to four Soyuz spacecraft until NASA's CRV is ready.*

---

*According to the RSA spokesman Sergei Gorbunov, NASA approached the RSA with a proposal to buy two escape capsules for $100 million. The proposal was turned down by the Russians because the offer was too low.[144]

Congressional Approval for the Contingency Plan

Although NASA legally was not required to get congressional approval to transfer $60 million from the space station budget to the RSA, politically NASA would have made a big mistake if it had gone against the will of Congress, particularly of the House Science Committee.* A long-standing critic of U.S.–Russian cooperation in space, the Committee's Chairman James Sensenbrenner had a rather complex attitude toward NASA's contingency plan.† He believed that such a "plan should not go kicking the Russians out of the program...because that would become prohibitively expensive,"[89] but he had very serious reservations about giving the RSA any further financial help from NASA. During the 7 October 1998 committee hearing on the ISS, he stated that he could not go along with NASA's request to bail out the Russian space program.

> I've seen nothing since passage of the Sensenbrenner–Brown amendment‡ that would lead me to believe that NASA, the White House, or the Russians would make good use of the money....
> The plain truth is that the White House is addicted to the Russians. I'm beginning to think it doesn't care whether the Space Station gets built, so long as the Russians are happy. The problem is that our relationship with the Russian space program is fundamentally flawed and is hurting our national interest. What makes me particularly angry is that all of the talent, the creativity, the energy, and the passion that exist for space within NASA is being wasted in frantic efforts to create ad-hoc, short-term bandaids that enable the White House to indulge its addiction to Russia instead of being channeled into actually building our Space Station and opening the space frontier.

Sensenbrenner estimated that the $660 million ($60 million in the near term plus $600 million over the next four years) that NASA wanted to

---

*In 1998, the NASA–Congress relationship regarding the ISS program maintained basically the same structure as during the previous years. NASA was criticized, albeit with increasing strength, for the growing cost of the ISS, the delays of its assembly on orbit, and the inability to assure provision of Russian elements on schedule. However, Congress again demonstrated its willingness to overlook Russian delays and contractor's overruns on 29 July when House members defeated Tim Roemer's annual attempt to kill the station by a vote of 109–323. The reasons for keeping Russia in the program remained generally the same: concern about the negative impact of the disintegration of U.S.–Russian cooperation in space on the overall context of the bilateral relationship, and concern about further cost growth of the ISS without Russian participation.

†On 1 October 1998, Sensenbrenner was presented with the 1998 "Excellence in Programmatic Oversight" award, which recognizes members who hold federal agencies and programs accountable to U.S. taxpayers. As the only standing Committee Chairman to receive the award, he was honored for, among other things, exposing the administration's failures in handling Russian participation in the ISS program. Through nine hearings on the subject, "the Chairman worked tirelessly on a bipartisan basis to require the Administration and NASA to develop clear-cut plans in dealing with Russian delays and waste."[145]

‡The Sensenbrenner–Brown amendment to the House bill authorizing NASA's budget for FYs 1998 and 1999 prohibited the payment of any funds to the Russian government for the purpose of enabling the RSA to build ISS components that the Russian government pledged to provide at its own expense.

transfer to RSA would go on top of the $472 million that the United States paid Russia for access to Mir, on top of the $210 million that the United States paid Russia for FGB, on top of the tens of millions the United States paid Russia for administrative support, and on top of the $1.2 billion that the United States lost because Russia did not meet its commitments in time.* Sensenbrenner repeated his requirement to NASA "to take the Russian government out of the critical path—now."[131]

Whereas in the past Sensenbrenner tried to convince NASA and the White House to follow his recommendation, in October 1998 he made approval of NASA's request regarding allocating $60 million to Russia contingent on significantly restructuring NASA's relationship with the RSA. He promised to "view near-term payments to Russia more favorably" should "the Administration propose payments to Russia as part of a comprehensive plan ending our dependence on Russia."[146]

Sensenbrenner's position was supported by the Senate. Chairman of the Senate Commerce, Science, and Transportation Committee John McCain and Chairman of the Subcommittee on Science, Technology, and Space Bill Frist sent a letter to Daniel Goldin on 13 October, stating that they could not support his "efforts to spend $60 million at this time without a comprehensive plan from NASA on how the agency intends to handle this complex situation both in the near and long-term."[147] Chairmen and ranking members of the House and Senate appropriations subcommittees also shared the House Science Committee Chairman's approach. Senate VA-HUD-IA Subcommittee Chairman Christopher Bond and Ranking Member Barbara Mikulski, and House VA-HUD-IA Subcommittee Chairman Jerry Lewis and Ranking Member Louis Stokes basically approved the use of $60 million from fiscal year 1998 to help Russia solve its financial problems, but made it clear that "none of the funds may be expended until the Congress receives a plan which eliminates United States reliance on Russia at the earliest possible date."[148] On 14 October Sensenbrenner introduced H.R. 4820, *Save the Space Station Act of 1998*. The specific U.S.–Russian cooperation-related provisions of the bill included:

> 1)Prohibiting additional payments to the Russian Space Agency to meet its existing obligations unless Congress concurs that additional payments serve the taxpayers interests;
> 2)Requiring the Administration to develop a contingency plan and report that plan to Congress for removing each element of the Russian

---

*It is important to know that, "since the redesign of the ISS in 1993, with the exception of $100 million provided by Congress in the fiscal year 1998 VA-HUD-Independent Agencies Appropriations Act (P.L. 105-65) for NASA's contingency effort against Russian shortfalls, all additional resources required for the ISS have been identified within NASA's budget."[119]

contribution from the critical path for assembling the International Space Station.[149]

To meet Congress' requirements, NASA came up with a large-scale contingency plan that was also called the "incrementally buying down risk" approach. This plan addressed near-term reliance on Russian capabilities while accelerating U.S. contingencies for long-term independence. The three-step approach provided the most flexible Russian contingency planning while allowing sequential implementation of these plans as warranted:

Step 1. Protect Against Further Service Module Delays
(estimated cost $250 M)
Modify FGB tug for extended life and Progress refueling of FGB ($40 M)
Build ICM and provide temporary reboost and attitude control capability ($120 M)
Contingency reserves ($30 M)
Other ICM related costs ($40 M)
Airlock 02 mods/Mission Control Center mods ($20 M)

Step 2. Protect for Further Delays of Service Module
and Fewer Progress Vehicles
(estimated cost $750 M)
Modify Shuttle for reboost ($90 M)*
Build U.S. Propulsion capability to provide permanent independent reboost and attitude control ($350 M)†
Additional Shuttle logistics flights ($100 M)
Russian near term contingency
    1998—$60 M for research crew time and stowage (Requesting concurrence)
    1999—Up to $150 M for goods/services
CRV Development

Step 3. Protect Against No Delivery of Service
Module/No Progress Vehicles
Purchase required services (such as Progress/Soyuz) through 2002 as necessary and/or assume additional Space Shuttle missions

---

*In early 1997, NASA and the RSA were looking at using Russian-provided fuel tanks carried by the Shuttle to provide a refueling capability for the ISS. In 1998, it was decided that Shuttle missions could carry more propellants in their orbital maneuvering system (OMS) tanks to provide ISS reboost. According to Goldin, the Shuttle would be able to handle about 50% of the reboost needs of the Station during assembly, with Progress spacecraft handling the remainder. Each Shuttle "could come up to the station and raise it 10 to 13 miles at a shot."[127]

†The U.S. propulsion module is one of the most important replacements for the Russian critical elements of the station. It is supposed to provide permanent reboost and attitude control. According to Schumacher, the propulsion module is not to replace a Russian capability; it is to have a U.S. backup capability so that if there is a continuous problem on the Russian side, the Americans could refuel the station.[108]

Accelerate life support development for outfitting in U.S. lab

Increase use of ATV/HTV*

Increase aggressiveness of CRV development[150]

NASA also outlined in its plan eight Russian critical path functions that would be augmented or replaced as necessary by similar U.S. functions.[120,150]

Although Sensenbrenner did not consider the NASA contingency plan as a "comprehensive strategy" to end U.S. dependence on Russia, he was "particularly relieved" that the White House and NASA "[have] had a change of heart" regarding continuous reliance on Russia in critical elements of the station. As a result of the U.S. space agency and the Clinton administration's willingness to follow congressional recommendations, the heads of the House and Senate appropriation committees verbally approved the transfer of money to the RSA. The initial transfer constituted $16 million, or 27% of the $60 million, with the rest of the money scheduled to be transferred before the end of 1998.[151]

In January 1999, Goldin said, "The $60 million has made a huge difference. I was over there [in Russia] in the end of November and we had serial numbers on a whole bunch of Soyuz and Progress hardware; I saw it in the factory. The testing was just going great guns on the Service Module. Hardware on the Service Module that was missing was being worked out." However, In January 1999 it also became clear that the SM would face another delay, possibly until September.[182]

STATUS OF RUSSIA IF ISS COMMITMENTS NOT MET

In June 1998, Daniel Goldin outlined a possible evolution of the Russian status in the ISS program if Russia would fail to meet its ISS commitments:

> The Russians have signed up to perform a very specific set of tasks. In return for performing that very specific set of tasks, they have a number of opportunities for utilizing that Space Station. As they diminish the number of tasks they perform, their ability to utilize the Space Station will diminish proportionally. At some point in time, it comes to a point where they are not a partner, if they give too little. And they could run the risk of becoming a subservient partner or no partner at all.[127]

---

*The ATV was developed by Europe; the HTV (hope transfer vehicle) was developed by Japan. The Europeans plan to deliver six ATVs that could deliver propellant equivalent to about 16 Progress flights. The HTV will provide dry goods and complement the capability of the ATV. Both are in the early requirements definition phase and are not planned to be available earlier than 2003.

Did NASA's decision to buy storage space and research time inside the Russian segment during the assembly phase of ISS program constitute a step toward making Russia a subservient partner/contractor? Schumacher was certain that

> Whatever NASA is doing now, it does not demote Russia to a contractor's status. It is quite the opposite. It doesn't affect Russian contribution to the Station. We have yet to cross the line into buying things or paying for things that the Russians were supposed to contribute as a part of ISS. We have not modified Russian partnership at all and we are not getting into Russian contribution to the Station. If we have to continue down that path buying more goods and services and that type of thing, we eventually might get there. But we believe that Russia has a long way to go from a partner to a contractor.[143]

Overall the NASA–RSA agreement regarding the leasing of space and time inside the Russian segment could be viewed, according to Lynn Cline, as an "implementing arrangement" under the terms of the Space Station agreement. The IGA and MOUs clearly state that among themselves the partners have the right to barter or sell their research time and resources under terms and conditions mutually agreed upon (Ref. 152).*

One of the signs that NASA was treating Russia as a full partner and not a subservient or a hardware manufacturing contractor is the nature of the purchase of the Russian goods and services. Russia did not transfer ownership rights on its storage space and research time to NASA, but simply leased them to the U.S. space agency. Moreover, this lease could be terminated at any time "should the situation improve in Russia and RSA would be able to buy back some of that time and space."[143]

Moreover, NASA gave RSA a lot of flexibility in terms of fulfilling its responsibilities to the U.S. space agency as outlined by the Protocol commitments. First, the two space agencies agreed that "it would be desirable for RSA to participate in the Baseline Data Collection process. Therefore, if NASA and RSA agree, hours spent by RSA cosmonauts on these activities may be accounted for as part of the hours on that increment...."
Second,

> With regard to research crew time, after provision of the first 2400 hours, or at assembly complete, whichever occurs first, NASA and RSA will review the status of the crew time transfer. Depending on the results of this review, they may agree to continue the transfer of crew time in

---

*For more information about the rights of the ISS partners to barter or sell any portion of their respective allocations, see IGA, Article 9, and MOU, Article 8.3.

post assembly timeframe until the entire 4000 hours are transferred. Absent agreement on this point, the procedure set forth in paragraph below, will apply.

With regard to stowage and other deliverables under this modification, NASA and RSA will first look for additional goods and services for RSA to provide instead. In the event they are unable to reach timely agreement on alternative goods and services, NASA and RSA will resolve the issue in a timely manner in the context of the NASA/RSA Balance of Contributions.

NASA also decided to spend some of the crew research time allocated from RSA to deepen Russia's integration into the ISS cooperative research activities:

> RSA will participate in the multilateral strategic research planning working groups which select cooperative research. If a NASA/RSA cooperative payload is selected, NASA will endeavor to have a Russian crew member participate in the performance of this research. NASA will contribute the research crew time for such cooperative research out of the additional allocation of research crew time NASA is acquiring under [this Protocol].

Finally, "if NASA and RSA agree, through the completion of assembly, RSA may buy research crew time from NASA at the same price."[154]

Schumacher believed that NASA had three basic reasons for maintaining Russia's full partner status. First, this status was legally codified in the agreement and the memorandum signed in January 1998. Second, Russia as a full partner had government standing behind its partnership. This is very important to NASA, as opposed to just buying goods and services from Russian enterprises. Finally, NASA would be "very surprised if the Russian government or Russian people would want to stand by and say: 'Oh, yes. We will just take that 30–40 years of hard-won experience and just sell it.'"[143]

Schumacher's opinion was paralleled to some extent by the opinions of the RSA leaders. RSA Deputy General Director Alexander Medvedchikov saw no reasons for modifying the agreement and the memorandum that clearly established Russia's full-member status in the ISS program. According to Medvedchikov, the decision to rent some of the space and research time inside the Russian segment to the Americans had a "mutually beneficial character." Russia gets money to continue work on its contribution to the ISS and America gets "much needed space and time to conduct its scientific activities."[155] According to the RSA Manned Space Programs Department Director Mikhail Sinelshchikov,

NASA has a lot of equipment which it would like to accommodate in the International Space Station. However, the Americans do not have enough space for this [equipment] in their segment during the ISS assembly. This is why they offered us to rent them out a space inside the Russian Service Module, which will be left unoccupied after all the Russian equipment is put into this module. We carefully evaluated and estimated everything and came to the conclusion that there is a possibility [of accepting such an offer] without any damage to the Russian interests....

According to the agreement [with NASA], we lease to the NASA specialists from 25 to 75 percent of our time at certain stages of the ISS assembly. Besides, the Russian Space Agency has a right to buy this time back at any time it wishes to do so....

Part of the $60 million which we will get in the near future will be spent on the design and development of scientific equipment [for the Russian segment of the ISS].

Summarizing the above, one can say that the recent agreements with NASA is currently the only chance for our space science and technology to move forward and [for Russia] not to lose its status of a great space power.[156]

RSA Deputy Director of the Department of International Cooperation Alexei Krasnov agreed with Medvedchikov and Sinelshchikov. In his view, "Russia is not ready yet to discuss a possibility of becoming a contractor. What Russia could accept now is the reasonable combination of contracting and partnership relations which would help the country to go through its difficult economic period and maintain its participation in the ISS program at the same time."[157]

Although NASA's decision to buy goods and services from Russia saved Russia's participation in the ISS program and in the long-term perspective rescued the Russian manned space program overall, they were not unanimously welcomed by the Russian space community. Although the RSA considered the deal to be beneficial, some Russian experts did not approve of it and viewed it as "the selling of the standing Russian part" of the ISS.* In their opinion, as soon as the station was launched, the project participants would be able to raise the question of the revision of everyone's contribution.[159] However, the real lack of consensus inside the Russian space

---

*Overall, the Russian scientific community did not express any major concern regarding the sale of the Russian research time and space to the Americans. According to the Institute of Biomedical Problems (IBMP) Department Director Inessa Kozlovskaya, Russian researchers realize that most of the scientific activities onboard the station will be conducted by joint efforts anyway, which will involve the joint use of research facilities and time.[158] (See also Ref. 154.)

Kozlovskaya's opinion parallels that of Schumacher's, who also believes that the integrated character of the U.S.–Russian crews will naturally result in some mutual scientific activities in the ISS. Such cooperation will not extend, however, to commercially important experiments.[143]

community regarding particular aspects of U.S.–Russian cooperation in space was demonstrated by its attitude toward the Mir station.

DEORBITING THE MIR SPACE STATION

In May 1998, Yuri Koptev advanced a few arguments in favor of further investments into the Mir station that would bring big revenues through international contracts ($800 million in 1998 and $1 billion in 2000). They were, however, defeated by the Minister of Finance Mikhail Zadornov.[115] Some experts give smaller numbers than Koptev for the potential annual revenues that the commercial use of the scientific equipment of the orbital complex could potentially bring, but they still believe that it could be around $200–300 million.[160]

Currently Mir operational costs amount to $250 million annually, including the costs related to four Progress cargo flights (about $35 million each), two Soyuz missions ($40 million each), and ground support (about $30 million).[161]

The July 1998 decisions made at the meetings in the office of the Deputy Prime Minister Boris Nemtsov were aimed particularly at solving the financial problems of not only the Russian contribution to the ISS, but also of the Mir space station. The outlined measures did not bring any noticeable effect, and in October Koptev said that Russia must choose between continuing to operate its Mir orbital station or taking part in the ISS because it did not have the funds to do both. Koptev said that he made his choice and believed that Russia should concentrate its resources on developing the Russian part of the ISS.* He admitted, however, that Mir could be used for two more years.[165] RSA officials informally expressed their regret regarding Mir's deorbiting, but formally all of them condemned any

---

*The decision to deorbit Mir was officially codified in the protocol BN250-57, signed by Yuri Koptev and Boris Nemtsov on 16 July 1998.

According to Russian aerospace experts, Nemtsov was happy to get rid of the station that caused him a lot of headache. These experts also believed that this decision had a purely political character and was not based on any serious evaluation of Mir's technical health. This was confirmed by Marshal Evgeny Shaposhnikov, aerospace assistant to President Yeltsin, who stated on 30 September 1998 that the decision to terminate the Mir program would be a political one. However, even if members of the Russian government apparatus wanted to conduct an evaluation of the station's technical condition, they could not do it, because among them "there were no more specialists who would have any clue in technical issues."[162,163]

The Russian aerospace writer Vladimir Gubarev did not have a high opinion about Russian lawmakers' competence in aerospace issues either. He complimented the U.S. congressmen by noting that during the ISS hearing they "demonstrated so much knowledge about [the Russian space program] as if the hearing took place in Moscow, in Duma, not in Washington…. Although I gave Duma too much credit. Unfortunately, our lawmakers, unlike their overseas colleagues, have little clue in science issues, space ones in particular."[165]

talks concerning Mir's rescue. They were instructed not to discuss the issue because the federal government had already stated its position regarding the fate of the station.[162]

Objections from the Russian Space Community

Not everybody in the Russian space community shared the RSA's approach. By the end of the summer of 1998, it became clear that "systemic opposition" to the idea of Mir deorbiting in the summer of 1999 was taking shape. The campaign to save Mir called "We Will Not Allow Mir to Be Buried!" was launched under the patriotic flag. RKK Energia Deputy General Designer and leader of the Russian section of the Mir–Shuttle program Valery Ryumin (who flew STS-91 in June 1998) said, in particular:

> We have a station and the Americans do not. The Americans will be the leaders of the new International Space Station and we will be just partners. The Mir station is like a thorn in the flesh of the Americans. That is why they are so keen for us to get out of there.[166]

A similar opinion was expressed by cosmonaut Igor Volk. According to him, "if we deorbit Mir, our cosmonautics will go down the tubes the same way our aviation went. If we destroy Mir, nobody will take us seriously into consideration, neither in space, nor on Earth."[162]

Those who support the continuation of Mir's operation used more than patriotic slogans but also legal and technical arguments. The first kind of argument was presented by a former Yeltsin aide turned cosmonaut, Yuri Baturin:

> Over the last few years, Mir has been financially supported by RKK Energia. Actually, the station has already become the Energia's property. What kind of right does the government have to determine the future of somebody else's property? This is a completely Russian national program, not an international one in which Russia will play a minor role. If the government believes that our space program belongs to the state, it should be supported by the money from the federal budget.[167]

Technical arguments were advanced by the cosmonauts who flew on Mir or were involved in its operation. Cosmonaut Talgat Musabayev, who returned from Mir in August 1998, said that the station was in better shape than during his first flight in 1994. This opinion of the Mir's technical health was confirmed at the end of October by cosmonaut Vladimir Dezhurov, who said that the station's controllers at the Mission Control Center (MCC) were "half sleepy" because Mir was not causing them any problems at all. "It is almost like Mir listened to the talks between the MCC and the crews, realized that space officials want to deorbit it because of the problems it developed, and decided to show everybody that it is still a good station which could productively work in orbit for quite a while," said Dezhurov.[168] Baturin, who returned from Mir with Musabayev, also supported the idea of saving Mir. He believed that before Mir is deorbited, the ISS should prove that it is fully operational. According to Baturin and other aerospace experts, the overall cost of Mir's design, development, manufacturing, on-orbit assembly, and equipment (from 1980 through 1998) was about $3–3.5 billion. So far, Mir had spent only half of its recently estimated lifetime. Deorbiting it in 1999 would mean throwing away $1.5 billion. RSA Deputy General Director Boris Ostroumov agreed with Baturin by saying that Mir could fly for another 10 years. Currently Mir carries onboard more than 240 scientific instruments with a total mass of 11 metric tons, including 2 tons of foreign equipment.

A number of Russian aerospace experts and engineers believed that Mir should continue its operation as a testbed for the ISS. They argued that the ISS projected life span is 15 years. Thus, Mir should fly at least through 2000, until it is 15 years old, to test the hardware similar to that of the ISS. The last Mir crew could even simulate an emergency bailout procedure. Vladimir Nikitsky, RKK Energia deputy general designer, blamed the Americans for the decision not to use Mir's testing potential:

> It was stated in the shameful [NASA–Mir] contract (just $400 million) that Mir should serve as a testbed for verification of all technical solutions regarding ISS safety. Such contract a was a lie from the very beginning because now more than ever it is necessary to continue work in this direction. However, U.S. "kids" who were listening very carefully to our specialists, became over a five-year period specialists themselves. They realized that we had no money and they also realized that they could not build the Station without our help. NASA–Mir contract was a U.S. technical espionage. We are no longer competitors to them [the Americans], and after we passed to them all our experience, they don't really need us as partners. Who will we be in their space station? We should probably agree with them about what sides of their bodies our cosmonauts will be sleeping on.[162]

One solution for the Mir financial problems, according to its defenders, could be commercial activities onboard the station. They might not only support the operation of Mir but actually bring a profit.[162] According to Nikitsky, Russia officially offered 10 countries the chance to "donate" $24 million each in exchange for receiving "open access to the Mir program on a shared basis."[166]

All of the aforementioned pro-Mir arguments ultimately found a positive response at the state level. In October 1998, the State Duma asked
President Yeltsin "to take the personal part in the settlement of strategic
issues of the Russian piloted space complex till the complete deployment
of the International Space Station." The Parliament members called on the
president to instruct the government to prepare a decree on the extended
service life of Mir. The appeal said, in particular, that "the termination of
works with the Mir complex will result in the expulsion of the Russian
Federation from the international manned space programs and other large-
scale scientific projects.... [The Mir] issue has a special state importance,
and the decision [to deorbit the station] may have negative international
consequences in the financial and political spheres." The Duma deputies
who prepared this appeal estimated that destroying the station would result
in eliminating 100,000 jobs for highly skilled scientists and engineers and
might lead to the increase of social tensions and the liquidation of
advanced industries.[169]

The Duma's attitude toward the Mir space station as an issue of great
national significance was illustrated by its attitude to the Hollywood film
Armageddon. The film depicted a dilapidated Russian space station that
was blown apart because of a leaky pipe. In addition, a Russian cosmonaut
was portrayed as a psychopath wearing a fur hat in space and fixing the station by hitting machinery with wrenches. Alexei Mitrofanov, chairman of
the Committee on Geopolitics of the State Duma, said that the movie
"mocked the achievements of Soviet and Russian technology." Mitrofanov
expressed his extreme disappointment with the Americans who "for five
years...used our Mir space station and now...send us this insult." The
Duma voted to require Armen Medvedev, chairman of the Russian State
Cinema Committee, to appear before the chamber with an explanation on
23 October.[170]

At the end of October 1998 support for the preservation of the Mir
space station came from the Russian scientific community. The Space
Council of the Russian Academy of Sciences came to the conclusion that
the scientific equipment of the station preserved its viability and still had
a considerable service life. Creation of similar equipment was impossible
in Russia in the near future. Moreover, Russian researchers expressed
their serious doubts about the scientific value of the ISS. Although Russia
possesses one-third of the overall volume of the future station (providing
that part of this volume will be temporarily leased to the Americans), the

ISS, according to Russian scientists, would not be able to provide Russia with research opportunities that are equivalent to those offered by Mir because of the absence of Mir-type equipment on the ISS. The Space Council supported the idea of keeping Mir in orbit for another three to five years.[171]

Overall, according to RSA Deputy General Director Alexander Medvedchikov, the issue of Mir is a very complex one. An overwhelming majority of designers and manufacturers involved in the creation of the Mir space station believe that the orbital outpost could operate at least for another few years. However, the decision to keep Mir in orbit beyond 1999 is not a purely technical one. It is not only about the station's operation, but about the future fate of the Russian national manned space program, and this is why the decision to deorbit Mir should be made at the highest governmental level. Currently, there is no documented official agreement between the RSA and NASA envisaging Mir's deorbiting in the summer of 1999.[155]*

The approaching date—20 November 1998—of the ISS first element launch (FGB or Zarya), and the need to make a decision regarding Mir's deorbiting for technical reasons at least three to four months before the deorbiting itself,[†] spurred interest in space activities in Russia, both at public and governmental levels. In mid-November, space-related issues were discussed at a closed meeting in Samara, a region where Soyuz boosters are manufactured. Chairman of the meeting was First Deputy Prime Minister Yuri Maslyukov. Among the participants were RSA General Director Yuri Koptev, Governor of the Samara region Yuri Titov, First Deputy Minister of Defense Nikolai Mikhailov, and directors of the leading Russian space industry's enterprises. Koptev confirmed at the meeting that the Russian space program received only 32% of its funds from the Russian federal budget. Koptev also noted that the Russian Ministry of Finances planned to allocate to the whole federal space program only 2.8 billion rubles, whereas RSA

---

*The decision-making process in the Russian space program regarding the most important space activities usually has the following structure: the decision is made at the General Designers Review (GDR), then it is discussed and approved at the RSA board, and only after that the decision must be approved at the top governmental level.

†Every four to six months, depending on how much fuel remained in the Progress spacecraft (which determines the duration of the thrusting impulse of the spacecraft), Mir needs to be reboosted to maintain normal orbital altitude. The station loses altitude because of a natural aerobraking that occurs when it "scratches" the upper layers of the atmosphere. If at a certain point the station does not get a reboost, it will descend below the critical altitude of 320 km, after which Progress will not be able to lift it to its normal orbital operational altitude (320–400 km). Should this series of circumstances occur, the station's deorbiting will become only a matter of time.

Overall, the reboosting of the station requires so much fuel that RKK Energia is currently considering the development of a pure space tanker (to carry nothing but fuel) for ISS, a design based on the Progress spacecraft.

needs at least 3 billion rubles just to keep Mir operating and to fulfill its ISS obligations.

This meeting was significant not only because of the attendance by top governmental officials and leaders of the Russian space industry but also because Yuri Koptev unexpectedly changed his attitude toward the issue of Mir's deorbiting in mid-1999. He stated that Mir's preservation as a unique orbital outpost is both possible and necessary. Russian aerospace officials decided to evaluate Mir's technical health to see whether it would be safe to continue its operation beyond mid-1999. According to the last evaluation, the station's operational lifetime expires in early 1999. After that date the safety of cosmonauts onboard the station may be jeopardized.[172]

In November 1998, Alexander Serebrov, a veteran of four space missions and an advisor on space issues to President Yeltsin's administration, outlined his concept for a joint project to reap scientific benefit from Mir by cannibalizing it prior to its deorbit, a concept that would keep Mir in orbit ideally for another two years while integrating it into the ISS program. Serebrov proposed using two Shuttle missions to transport the large amount of samples and equipment from Mir to ISS or back to Earth.

What the Shuttle is supposed to do during its first proposed mission— flying cargo from one spaceship to another—was done successfully for the first time when equipment was transferred from Salyut-7 to Mir, argued Serebrov. During the second Shuttle mission, critical Mir elements (both interior and exterior) would be dismantled for further laboratory investigation on Earth. To accomplish this, the crew would be evacuated, and an American astronaut and Russian cosmonaut team would dismantle critical parts, perhaps as much as 2 tons of material (based on Shuttle capacity), to include shielding elements, heat tubes, and so forth. After the work is completed, Mir could be given the necessary commands to begin its descent and deorbit.

For the extended Mir program, Serebrov proposed using three Progress spacecraft and two Soyuz vehicles with two crewmembers onboard. Thus, the operational cost of the extended program would be about $215 million, $35 million less than the current annual cost of operation resulting from the saved Progress mission. Private funds and donations could cover the first half of that cost, and the second half could be covered by Russia, Japan, Europe, the United States, and all participants in the ISS program who, as Serebrov put it, "care about the safety of their future crews."[161]

On 12 November, after the meeting with the Russian government, Koptev said that the government would decide Mir's fate at the end of 1998. According to the RSA general director, everything would depend on the 1999 fiscal year budget. If the government finds enough money to continue

Mir's operation, the station will fly for another two years.[173]

Attempts by the Russian space community and politicians to save Mir sent a mixed message to the world and Russia's ISS partners. On one hand, such attempts showed Russia's continued interest in space exploration, particularly its human aspect. On the other hand, the question remained, What would Russia ultimately prioritize—its national manned space program or participation in the International Space Station. This question held special importance in light of the continuous delays of the Russian hardware for the ISS.

While Mir still has scientific, technological, and symbolic value for Russia, the key Russian space decision makers realize that without international cooperation Mir could not have maintained its place in orbit for so long and without Russia's participation in the ISS program the Russian manned space program will cease to exist. Vladimir Utkin, a TsNIIMash director, believes that the Americans deserve part of the credit for Mir's success. "I would not be objective if I would not share the glory for [Mir's] achievements with our U.S. colleagues. It is to a significant extent because of Mir–Shuttle program we managed to operate the station for so long."[165]

Even such an outspoken space nationalist as Valery Ryumin had to admit that

> if it had not been because of [U.S.–Russian] cooperation in space, [the Russian government] would have already forgotten [about the Russian space program]. However, [the government] has no choice but to support the Russian cosmonautics because Russian and U.S. presidents meet once a year and vice president and prime minister meet twice a year. [The government] has to discuss the state of the space program and to give it some money at least by the time of these summits and top governmental meetings. The cosmonautics would have died already without such support.[139]

On 12 November 1998, the Russian government confirmed its support of Russia's participation in the ISS program. One of the reasons for the support was presented by the RSA general director who stressed again that participation in the ISS project keeps 70,000–80,000 highly qualified professionals employed and "gives the Russian cosmonautics a chance to survive in general." According to the RSA general director, if Russia bails out of the project, the country will "hit the wall" in the international space market.[173]

The RSA Manned Space Programs Department Director Mikhail Sinelshchikov believes that

[i]f the ISS would not be built, if it would not contain the Russian scientific equipment, the Russian science will not be present in space. Everybody should have a clear understanding of this....

Mir space station, according to the specialists' evaluation, can normally operate for another three to four years. What kind of perspective will Russian cosmonautics have after this? There is no real alternative to Russia's participation in the ISS program.[156]

Unless the extension of Mir's operation beyond mid-1999 further affects the fulfillment of Russia's obligations to the ISS program, NASA would unlikely bring this issue up in its relationship with the Russians. Daniel Goldin, while saying that NASA will continue to press Russia to deorbit Mir by the summer of 1999, emphasized that Russia "is a sovereign nation with its own Mir space station."*

Lynn Cline confirmed that

NASA's preference is to have Mir de-orbited next year. But it is largely because Russia has told us that it cannot support two space stations If the Russians had said that they could support everything and there would be no difficulty in doing ISS in spite of Mir, we would have no opinion on Mir. But given that the Russians have made very clear that they are fully committed to ISS and they have limited funding and production capability for Progress and Soyuz, and that they could only support one station, obviously our priority is going to be to move to the new station.

Cline also said that if the Russian government finds enough funds to meet ISS commitments and also continue operating Mir, "that would be up to the

---

ISS Problems and Options

The NASA–RSA relationship around the Mir space station continues to be uneasy one. RSA wanted to integrate Mir into the initial stage of ISS operation and for this reason was planning to delay the FGB (Zarya) launch by 10 h. A delay would have allowed the placement of the FGB in the same orbital inclination as Mir, thus facilitating transfer of equipment and crews (in case of emergency) between the two outposts. NASA objected to this decision, and Zarya was launched according to the initial schedule. However, the reason for the rejection was not NASA's principal opposition to any Mir integration into the ISS program. According to Cline,

Over the years every once in a while, Russians have mentioned to us that they were studying some ideas about taking advantage of Mir modules or equipment for the new Space Station. But they have not made a formal proposal to us so that all the partners could study it and make a collective engineering decision, whether it is the right thing to do. So, we were rather surprised by the RSA's decision so close to the launch, because we could have studied it many months earlier and made a decision as partners on how to approach it. And our concern was that we really did not understand this decision, that it had not been thoroughly explored, and we have all already agreed on another schedule. So, we felt that it needed to be worked in a little bit more detail before we could say yes or no.

On the other hand, in November–December 1998, NASA agreed in principle to consider using the tenth Shuttle mission to Mir to retrieve hardware that could be used on the ISS. In particular the Russian-built docking module that was delivered to Mir by STS-74 in November 1995 could be retrieved; this module served as a permanent docking location for the Shuttle–Mir program.

Russian government, although NASA still has some residual concern about whether it really could be done."[174]

On 20 November 1998, many concerns regarding the future of ISS and U.S.–Russian cooperation in the framework of this program were left behind. The first ISS element, Zarya, was successfully launched into orbit. As Dmitry Paison, a Russian aerospace writer put it,

> [W]hatever financial difficulties the Russian space program will have to face in the future, whatever software-related problems will complicate NASA's life, and whatever doubts political leaders on both sides of the ocean would have regarding the necessity of the unprecedented ISS program, the job is done.... This unequivocally successful completion of the first step of the ISS assembly, will help the Russian space program to feel the firm ground under its feet.[175]

Moreover, according to RSA Deputy General Director Medvedchikov, the successful launch of Zarya imposed additional obligations on Russia in terms of meeting its ISS commitments: "With the first module in orbit, we will all now be following a tough schedule. Most of all for Russia, there is not time for delays, and I hope that funding will no longer be a problem."[88]

The Russians were not the only ones encouraged by the successful launch of the FGB module. In Lynn Cline's words, it definitely made the Americans feel better about working with the Russians in the ISS program because

> [It] is very important for the whole partnership to reach the milestone of getting hardware in orbit and to begin that assembly.... We are by no means done with our concerns since we know that we need Progress and Soyuz and many other things, but this is a good first step. Very good first step.[174]

In January 1999, the Russian government finally found a way to continue operating Mir without jeopardizing Russia's commitments to ISS. On 22 January, Prime Minister Primakov signed decree No. 76 authorizing the extension of Mir's operation until the year 2002. Starting from mid-1999, Mir's flight, including its controlled deorbiting, will be financed from an off-budget source. This source agreed to provide Mir with $750 million. (The name of the sponsor will be made public only after it makes an advance payment of $150 million.)

The decree gave RKK Energia exclusive rights as the main operator of the orbital outpost (while the government remains its only owner) and to realize commercial, technical, and scientific projects in the interests of Russian and foreign customers. RKK Energia was also ordered 1) to conduct a peer review of proposals from Russian and foreign customers and investors regarding Mir's utilization, 2) to ensure safety of work onboard the station, and 3) to ensure Mir's controlled deorbiting after termination of its operation or in case of emergency.

The decree also assigned RSA and RKK Energia to conduct negotiations with Russian and foreign customers and investors, and based on the results of the negotiations, to develop within the next three months the program of Mir's flight until 2002. Following these developments, appropriate suggestions are to made to the government of the Russian Federation.[183, 184]

WHY CONTINUE COOPERATING?

For the Russian manned space program, cooperation with the United States in building, operating, and utilizing the ISS is pretty much a matter of survival. But why should the United States continue cooperating with a country that bears a great deal of responsibility for the delay of ISS assembly? The U.S. manned space activities are in much better shape than Russia's, and they will not cease if Russia bails out of the ISS program.

Whether the Mir issue affects the U.S.–Russian relationship in the ISS program remains to be seen. Goldin believes that there are too many benefits to be realized for the U.S. and for Russia to let this new, effective partnership disintegrate.[120]

The first and probably the most important benefit to maintaining the partnership is related to a continuous U.S. dependence on the Russian contribution to the ISS. Although NASA started realizing contingency plans, it will take the U.S. space agency three or more years to develop a U.S. propulsion module without which (or using similar Russian hardware) the ISS cannot be assembled in orbit. According to Chabrow, the ICM that NASA has funded and developed

> cannot bridge the gap until a permanent propulsion module can be delivered. This almost dictates continued Russian involvement.... In the near term, maintaining Russian participation until the U.S. develops its own independent capabilities is significantly cheaper than the cost of developing the ISS without Russia.[120]

Another benefit is that the Americans, in Goldin's view, are learning from the Russians at the initial stage of ISS assembly and operation more than the latter are learning from the former. The Russians, according to the NASA administrator, "have an incredible capability that when we first entered this relationship, we had no idea of."[89] Among the major ISS applicable areas of knowledge that the Russians are bringing to the program are their expertise in how to build the space station; how to operate and use it; and how to take care of the astronauts/cosmonauts health, safety, and performance during long-term flights.

The third benefit is a political one. According to Deputy Chairman of Duma Defense Committee Alexei Arbatov,

> The euphoria [dominating U.S.–Russian relations] of the early 1990s, is over.... As paradoxical as it may seem, ten years after the end of the Cold War, Moscow and Washington understand each others' concerns, goals and motivations regarding international security, much less than before the hard confrontation and arms race were over.... Russia and the United States are drifting from each other further and further in the issues regarding arms reduction, non-proliferation of weapons of mass destruction and missile technologies, NATO extension and the foundation of the future European security system, regional peacekeeping missions and policy towards former Soviet republics.[175]

Cooperation in the space station program is one of the "fastening bolts" of U.S.–Russian relations that should prevent Russia and the United States from drifting further apart. Goldin believes that "this Station is a symbol, not of the end of the Cold War, but this station is a symbol of what nations could do not to build weapons, but to do things on a peaceful basis. And for that reason [we are]...not prepared to give up."[89]

There are a number of secondary factors that convince the United States to continue cooperating with Russia despite all of the drawbacks that Russia is causing the ISS program. One of them is that the Russian critical element of the station—the SM—is 95–98% complete according to U.S. and Russian aerospace experts. Chabrow believes "that we are so allied to that Service Module right now,...we are so close, that I would think that it would be wrong, and our whole team would think it would be wrong if we were to disassociate ourselves from that now."[89]

Another factor, according to Goldin, is a U.S.–Russian "effective team that can, and will, provide valuable insight and capability in the assembly, operations and maintenance of the ISS."[120] Goldin's opinion is paralleled by Chabrow's who stated that "the space agency people work very well with each other. The technical people work very well with each other. The subcontractors work very well with each other...."[89]

The third factor is that the Americans could clearly see the nature of the Russian problems and the reasons for their emergence. NASA's "confidence in Russian technical capability remains unshaken. The issue is the lack of commitment by the Russian Government."[89] The U.S. space agency's view is shared by Congress. House Science Committee Chairman Sensenbrenner confirmed in June 1998 that "the RSA personnel led by Yuri Koptev and the contractor, when they are financed adequately, do a superb job."[127]

Also, the Americans see how much the Russian space professionals are committed to the ISS program. Duncan Moore found it "truly amazing" that unlike the Americans, the Russians are working "in many cases without getting paid. There is a huge amount of pride on the Russians' side to maintain their continuity in space" (Ref. 89).*

Moore's words could be confirmed by two examples. Khrunichev enterprise is building at its own initiative and expense FGB 2, which will be put into orbit if the U.S.-paid-for FGB (Zarya) is destroyed as a result of a launch failure.[177] After Russia's contributions to the ISS were restructured in January 1998, FGB 2 was baselined to be used as a substitution for two smaller DSMs and partly for two LSMs previously planned to be launched by Soyuz launch vehicles. However, if anything goes wrong with Zarya before FGB 2 is put into orbit in its new role, the latter will be used to replace the former.[178]

Another example is the strong commitment of the Russian scientific community to human space flight and particularly to the ISS program. The Institute of Biomedical Problems (IBMP)—the leading Russian scientific research institute involved in studies of space biology and medicine—like many other Russian research institutions found itself in a very difficult economic situation. However, despite all of the economic hardships, in 1999 IBMP is preparing to conduct a long-lasting psycho-physiological experiment in confinement SFINCSS-99 (Simulation of Flight of International Crews in the Space Station). The goal of the experiment is to study crew performance during ISS assembly, when astronauts/cosmonauts will have to cope with an especially heavy workload. SFINCSS-99 will last 240 days. One group of the participants in the experiment will live 240 days in IBMP's two Moscow-located isolated chambers that roughly simulate ISS modules. The chambers will be connected so that the crew will have an opportunity to do joint studies. There will be also two other groups that will live with the first group for 110 days each, and visiting "crews" that will stay with the main groups for 7–14 days.

The benefits of the U.S.–Russian ISS partnership in the Space Station alliance were reinforced by the successful FGB launch. "Now it is more than just paper," said the U.S. independent observers who attended the launch.[179] According to Daniel Goldin, "Russia has fully demonstrated its capacity during the creation of the Zarya module. Fifty more launches with other ISS elements, 1500 hours of EVA and countless difficulties are ahead of us, and no one would be able to cope with this alone."[175] However, with all of the benefits and achievements of U.S–Russian cooperation in space, is there any point at which this cooperation will find itself in real jeopardy, or even fall apart, if the Russians do not improve their performance in meeting the schedule? At what point will the Americans decide that the actual and potential

---

*Evidence of the high requirements that the Russian hardware must meet is a low leak rate. Major General Ralph Jacobson, USAF (Ret.), member of the NASA Advisory Council Task Force on The International Space Station Operational Readiness, stated that "the leak rate specifications for the ISS pressurized elements vary among the ISS partners, with Russia having the lowest leak rate specification.... Russia would like the leak rate specifications to be standardized at the most stringent levels."[176]

advantages of keeping Russia in the ISS program do not outweigh problems that this country is causing to it? According to John Schumacher,

> A lot of people think that it is either yes, or no. But there is so much involved in this relationship and there are so many pieces in it. And this is a very important thing to understand for people on both sides of the ocean. There is no threshold in U.S.–Russian cooperation in the ISS program where the Americans will tell their Russian colleagues: "Do it right now, or we will throw you out of the program." That is not the way it works. We will say: "We are all cooperative partners. We really expect you to meet your commitments. If you can't meet your commitments, you need to talk with us, we need to talk together." It is a cooperative program. Every partner signed up based on their desire to come in, be part of the program and meet certain commitments. If we are faced with the situation, where any partner, let's say Russia, continuously does not meet its commitments, then we will look at contingency and backup that capability and say: "Is it essential for the Space Station? If it is, then what do we have to do to ensure that we have that capability." Being late is one thing. Being so late that you are really throwing concern over the whole program, is another thing.[108]

Schumacher believes that even if Russia's role is rescoped to SM, Soyuz, and Progress, Russia will still play a very important role in the space station program. RMs are designed to be used mostly by Russians, therefore it is really Russia's decision whether to bring these modules. RMs do not affect other partners of the ISS unless these partners start joint experiments with the Russians. Currently the U.S., Japanese, Canadian, and European research plans are interrelated, and the Russians coordinate on them, but the RMs are Russia's to use.*

---

*In 1995, the ISS international partners (excluding Russia) formed the International Microgravity Strategic Planning Group (IMSPG). The goal of the group is to coordinate research activities of the partners onboard the ISS to prevent the unnecessary duplication of costly experiments. Aaccording to the U.S. journal, *Microgravity News*, IMSPG "believed that the participation of the Russians was critical" and offered them membership in the group. The Russians, however, for quite a while were reluctant to accept IMSPG's invitation. NASA's senior representative to IMSPG and Microgravity Research Division Director Robert Rhome explained Russian reluctance as an "unfamiliarity with…'the mores and forays' of other national agencies." Communicating to RSA officials "that consensus, not majority, rules the IMSPG has been difficult."[153]

Russian researchers have, however, a slightly different view of the situation. According to the IBMP Department Director Inessa Kozlovskaya, the Russian scientists have three major concerns regarding IMSPG. First, each country-member of the group has only one representative to IMSPG, whereas the United States has two. Second, each year IMSPG selects a new chairman who is a representative of one of the country-members. A U.S. representative, however, is a permanent head of the group. This is why Russian scientists consider IMSPG as a U.S.-dominated international body, or as Kozlovskaya put it, "all members are equal but some are more equal than the others." Finally, the resources of the station will be distributed among IMSPG members according to their financial contribution to the ISS. But Russia's financial contribution is far smaller than its real contribution to the station in terms of hardware and experience. This will put Russia in an unfairly disadvantageous position compared to other more "cash-rich" IMSPG members.

With all of these concerns in view, Russian scientists do not reject cooperation with IMSPG. They accepted the group's offer to join it as observers in 1998. Unfortunately, the invitation was sent to the RSA too late and for this reason Russian researchers could not join the group's annual meeting.[158]

However, there is good reason to assume that the Russians will deliver the RMs and the science power platform because that is pretty much the course the Mir followed. Moreover, after RSA sold part of its research time and space to NASA inside the Russian segment, "Russia's involvement in the ISS program without its own research modules makes sense for Russia only in terms of fulfillment of international commitments." Without these modules Russia will become "just a 'warden' of the international scientific complex."[179]

One of the most important distinguishing features of U.S.–Russian cooperation in the ISS program is that it has an open-ended character. This was clearly stated by Daniel Goldin:

> if the Russians don't deliver the Service Module, if they don't provide the propulsion activities over a significant period of time, they run the risk of falling out of being a real member. However, the Space Station is designed so that at some point in the future, if they did want to come back, they could then deliver all the things they said they would, and restructure the arrangement.[127]

As long as Russia is willing to participate in the space station program, it will always have a open door in the ISS, and even if for some reason Russia does not walk through the door on time, it does not mean that the door will be closed for Russia to step through it in the future.

CONCLUSION

One of the factors contributing to U.S.–Russian cooperation in the ISS program was the clear interest in it from major parts of the U.S. and Russian space communities working together on the project. Such interest not only contributed to the broadening and deepening of U.S.–Russian cooperation in space, but also facilitated it by helping to avoid bureaucratic obstacles, which were especially significant in Russia.

The obstacles created by bureaucracy, however, were not the major problem standing in the way of U.S.–Russian cooperation in the ISS program after Russia joined it. The greatest difficulty was Russia's inability to live up to its commitments to the program, caused by the deepening economic crisis in Russia that was deteriorating the economic health of the Russian space industry. Delays in the production and delivery of Russian hardware delayed the construction of the ISS by more than a year.

A continuous postponement of the delivery of Russia's elements for the station became one of the major factors contributing to the ISS cost growth. This began eroding the political support of Congress to U.S.–Russian cooperation in the ISS program. Although the White House did not stop backing this cooperation, a number of legislators, including House Science

Committee Chairman James Sensenbrenner and Subcommittee on Space and Aeronautics Chairman Dana Rochrabacher increased their criticism of NASA's attitude toward cooperation with the RSA. The congressmen required that NASA remove Russia from the critical path by replacing Russian elements with U.S. ones. Congress only approved NASA's $60 million bailout for Russia to continue work on the ISS elements contingent on the satisfaction of this requirement.

To avoid further delay of ISS assembly, NASA came up with the contingency plan aimed at removing Russia from the critical path of the station. However, it did not change the nature of U.S.–Russian cooperation in the ISS program. The U.S. space agency has a number of reasons for maintaining Russia as a full partner.

1) Cooperation in space with Russia has already proved its soundness and potential. The ISS Phase I Shuttle–Mir program ended in success and the first U.S.-paid-for Russian-manufactured ISS module FGB (Zarya) was launched on 20 November 1998.
2) A considerable period of time will be required to build and launch the U.S. replacing elements, whereas Russian "critical" elements are practically completed and, with some financial help, could be put in orbit much earlier than their American substitutes.
3) Russia continues to bring a lot of experience to the ISS program.
4) U.S.–Russian joint work on the space station remains one of the "fastening bolts" of the overall relationship between the two countries that has experienced growing problems.
5) The Russian space industry and scientific community remain committed to the ISS program and do a superb job when adequately financed.
6) Russian and American space professionals involved in the ISS program formed a very effective team with a strong potential for boosting realization of the program.

For all these reasons NASA chose to have open-ended cooperation with Russia in the ISS, i.e., even if at some point Russia could not fulfill its commitments and withdraws from the program, it could still join it at any time in the future.

At the end of 1998, one more issue arose that could potentially create some problems for U.S.–Russian cooperation in space. NASA and the RSA tentatively agreed that Mir should be deorbited in the summer of 1999, so that Russia could concentrate all of its efforts on the ISS. However, the improved technical health of the station, which was plagued by multiple malfunctions in 1997, a belief that Mir can still productively work in orbit for a number of years, doubts that the ISS could become operational early enough to provide continuity between the Mir and ISS science programs, a concern that Russia will not have enough autonomy onboard the new station, and overall growing nationalism in Russian foreign policy increased support in Russia for Mir's operation beyond 1999. In January 1999, Prime Minister Primakov officially approved the extension of Mir's operation through 2002.

Although the Mir issue is important, it is doubtful that Russia would ultimately prefer Mir to the ISS. The Russian space station is running out of its useful lifetime (i.e., when the benefits of its operation outweigh the cost of its maintenance in orbit) and currently Russia cannot afford to build its own space station. Taking into consideration the great amount of pride on Russia's side to maintain its continuity in space, there is a good reason to believe that Russia will ultimately concentrate all of its efforts on the ISS, and U.S.–Russian cooperation in the joint program will continue through the period of construction and actual operation of the station.

REFERENCES

[1]Congress, House, Committee on Science, Space, and Technology, *Oversight Visit. Baikonur Cosmodrome.* Chairman's Rept, 103d Cong., 2d sess., 23 March 1994, p. 85.

[2]"Russia Integration into Space Station Partnership Discussed. Joint Statement on Negotiations Related to the Integration of Russia into the Space Station Partnership," *NASA News* N94-23, 18 March 1994. NASA Historical Reference Collection, NASA History Office, Washington, DC.

[3]"Joint Statement—Space Station Heads of Agencies Meeting," *NASA News* N94-28, 5 April 1994. NASA Historical Reference Collection, NASA History Office, Washington, DC.

[4]"NASA and Russian Space Agency Sign Space Station Interim Agreement and $400 Million Contract," *NASA News* N94-101, 23 June 1994. NASA Historical Reference Collection, NASA History Office, Washington, DC.

[5]Congress, Senate, Committee on Commerce, Science, and Transportation, Subcommittee on Science, Technology and Space, *NASA Space Station Program: Hearing Before the Subcommittee on Science, Technology and Space*, 104th Cong., 1st sess., 23 May 1995, p. 23.

[6]Semyonov, Y., (ed.), *Raketno-Kosmicheskaya Korporatsiya "Energia" Imeny S.P. Korolyova* [Korolev Rocket-Space Corporation], NPO Energia, Moscow, 1996.

[7]Alexander, B. S. F., *1996 Year in Review. A Look at the Year in the Russian Space Program*, ANSER Center for International Aerospace Cooperation, Arlington, VA, Jan. 1997, p. 5.

[8]Tarasenko, M. V., "Current Status of Russian Space Program and Its Implications to Global Cooperation and Competition," 46th International Astronautical Congress, Oslo, Norway, 2–6 Oct. 1995.

[9]Shuryguin, V., "Russkii Kosmos ne Zakryt!" [The Russian Space Program Will Not Be Shut Down!], *Zaytra*, No. 46, Nov. 1998, p. 259.

[10]Covault, C., "Russian Service Module Fabrication Advances," *Aviation Week and Space Technology*, 19 May 1997, p. 62.

[11]Krikalev, S., Soviet/Russian cosmonaut, interview by author, Moscow, 10 May 1996.

[12]Congress, Senate, Committee on Commerce, Science, and Transportation, Subcommittee on Science, Technology and Space, *NASA Fiscal Year 1997 Budget: Hearing Before the Subcommittee on Science, Technology and Space*, 104th Cong., 2d sess., 26 March 1996, p. 16.

[13]Congress, House, Committee on Science, Subcommittee on Space and Aeronautics of the Committee on Science, F. James Sensenbrenner, Opening Statement at the *FY 1998 NASA Authorization: The International Space Station: Hearing Before the Subcommittee on Space and Aeronautics*, 105th Cong., 1st sess., 9 April 1997.

[14]Decree of the Government of the Russian Federation no. 791, Duma Library, Moscow, 7 Aug. 1995.

[15]Decree of the Government of the Russian Federation no. 1163, Duma Library, Moscow, 26 Aug. 1995.

[16]Decree of the Government of the Russian Federation no. 422, Duma Library, Moscow, 12 April 1996.

[17]"Concept of the National Space Policy of the Russian Federation," Duma Library, Moscow, 1 May 1996.

[18]Reuters, Moscow, 13 May 1997.

[19]*Aerospace Daily*, 28 Jan. 1997.

[20]*Aerospace Daily*, 7 Feb. 1997.

[21]"Joint Statement on Human Space Flight and Science Cooperation," 7 Feb. 1997, NASA Historical Reference Collection. NASA History Office, Washington, DC.

[22]Congress, House, Committee on Science, *The Status of Russian Participation in the International Space Station Program: Hearing Before the Committee on Science*, 105th Cong., 1st. sess., 12 Feb. 1997.

[23]Congress, House, Committee on Science, Subcommittee on Space and Aeronautics, *FY 1998 Authorization: The International Space Station: Hearing Before the Subcommittee on Space and Aeronautics*, Statement by Wilbur C. Trafton, NASA associate administrator for Space Flight, 105th Cong., 1st sess., 9 April 1997.

[24]"Concerning Measure Aimed at Realization of the Russian Federal Space Program and Fulfillment of the International Space Treaties," Decree no. 153, Duma Library, Moscow, 10 Feb. 1997.

[25]"Concerning Measures Aimed at Fulfillment of International Space Treaties," Decree no. 391, Duma Library, Moscow, 2 April 1997.

[26]Congress, House, Committee on Science, Subcommittee on Space and Aeronautics, *FY 1998 Authorization: The International Space Station: Hearing Before the Subcommittee on Space and Aeronautics*, Statement by James Sensenbrenner, 105th Cong., 1st sess., 9 April 1997.

[27]Verda, A., "Prezident Rossii Boris Yeltsin Zayavil o Bezuslovnoi Podderzhke Rossiiskoi Kosmonavtiki" [Russian President Boris Yeltsin Expressed His Full Support to the Russian Space Program], *Nezavisimaya Gazeta* [Independent newspaper], 12 April 1997.

[28]Congress, Senate, Committee on Commerce, Science and Transportation, Subcommittee on Science, Technology, and Space, *National Aeronautics and Space Administration FY '98 Budget: Hearing Before the Subcommittee on Science, Technology, and Space*, 105th Cong., 1st sess., 24 April 1997, Alderson Reporting Co., pp. 15–16.

[29]Miura, S., executive director of the Japanese National Aerospace Development Agency, interview by author, Washington, DC, 9 April 1997.

[30]Dulin, J., "Mir and Supply Ship Collide; Crew Safe But Power Is Down," *Washington Times*, 26 June 1997.

[31]Nise, J., "Space Station and the International Human Exploration and Development of Space," 1998 AAS National Conference and 45th Annual Meeting, 17 Nov. 1998.

[32]Clark, R., former NASA associate administrator for International Relations and Policy, interview by author, Rosslyn, VA, 10 June 1997.

[33]Interview with professor Vladimir Kovalyonok, Soviet cosmonaut and chief of the Zhukovski Air Force Engineering Academy, *Pravda Pyat* [Pravda five], 12 April 1997.

[34]Nygren, R. "Space Station and the International Human Exploration and Development of Space," 1998 AAS National Conference and 45th Annual Meeting, 17 Nov. 1998.

[35]"Interview with F. James Sensenbrenner," *Space News*, 26 May–1 June, 1997, p. 22.

[36]Goldin, D., NASA administrator, Press Conference About NASA FY 1998 Budget, NASA Headquarters, Washington, DC, 6 Feb. 1997.

[37]"GAO: Russia Will Not Save Much Money on Space Station," *Washington Times*, 25 June 1994. NASA Historical Reference Collection, NASA History Office, NASA Headquarters, Washington, DC.

[38]A synopsis and commentary given by NASA Administrator Daniel Goldin during *NASA Posture Hearing* before the Subcommittee on Space and Aeronautics, House Committee on Science, 105th Cong., 1st sess., 4 March 1997.

[39]"NASA Announces Revised Plan for Mir Staffing," *NASA News*, N87-163, 30 July 1997. NASA Historical Reference Collection, NASA History Office, NASA Headquarters, Washington, DC.

[40]Congress, House, *Civilian Space Authorization Act, Fiscal Years 1998 and 1999, A Report*, 105th Cong., 1st sess., 21 April 1997, p. 33.

[41]Congress, Senate, *Department of Veteran Affairs and Housing and Urban Development, and Independent Agencies Appropriation Bill, 1995, A Report*, 103d Cong., 2d sess., 14 July 1994, p. 123.

[42]Congress, Senate, Committee on Commerce, Science, and Transportation, Subcommittee on Science, Technology and Space, *NASA Space Station Program: Hearing Before the Subcommittee on Science, Technology and Space*, 104th Cong., 1st sess., 23 May 1995, p. 5.

[43]Congress, Senate, Committee on Commerce, Science, and Transportation, Subcommittee on Science, Technology and Space, *Fiscal Year 1997 NASA Budget: Hearing Before the Subcommittee on Science, Technology and Space*, 104th Cong., 2d sess., 26 March 1996.

[44]Congress, Senate, Committee on Commerce, Science, and Transportation, Subcommittee on Science, Technology and Space, *Space Station and Space Shuttle Programs: Hearing Before Subcommittee on Science, Technology and Space*, 104th Cong., 2d sess., 24 July 1996, p. 2.

[45]Congress, House, Committee on Science, Subcommittee on Space and Aeronautics, statement of Representative Roemer before the Subcommittee on Space and Aeronautics at *FY 1998 NASA Authorization: The International Space Station: Hearing Before the Subcommittee on Space and Aeronautics*, 105th Cong., 1st sess., 9 April 1997.

[46]*Aerospace Daily*, 7 Feb. 1997.

[47]*Aerospace Daily*, 26 Feb. 1997.

[48]*Aerospace Daily*, 17 April 1997.

[49]*Aerospace Daily*, 13 Feb. 1997.

[50]"Interview with Dana Rohrabacher," *Aerospace America*, Vol. 35, April 1997, pp. 12–13.

[51]Sensenbrenner, J., "Russia Blocks the Station's Critical Path," *Aerospace America*, Vol. 35, April 1997, p. 3.

[52]Sensenbrenner, J., to the House of Representatives, "Stop Payment on NASA's Blank Check," 11 July 1997, NASA Headquarters, Washington, DC.

[53]*Aerospace Daily*, 17 July 1997.

[54]Congress, Senate, Committee on Commerce, Science and Transportation, Subcommittee on Science, Technology and Space, A statement by Kenneth F. Galloway, Dean of the School of Engineering at Vanderbilt Univ. before the Subcommittee on Science, Technology and Space during *NASA FY'98 Budget: Hearing Before the Subcommittee on Science, Technology and Space*, 105th Cong., 1st sess., 24 April 1997.

[55]Congress, Senate, Committee on Commerce, Science and Transportation, Subcommittee on Science, Technology and Space, A statement of Louis Friedman, executive director of the Planetary Society before the

Subcommittee on Science, Technology and Space during *NASA FY'98 Budget: Hearing Before the Subcommittee on Science, Technology and Space*, 105th Cong., 1st sess., 24 April 1997.

[56]Logsdon, J. M., "Russian Roulette," *Launchspace*, June/July 1997, p. 8.

[57]Sietzen, F., Jr., "Are We Better Together? In Defense of Shuttle–Mir and Alpha," *Space Times*, Vol. 36, No. 4, 1997, p. 27.

[58]Congress, House, Committee on Science, Subcommittee on Space and Aeronautics, *FY 1998 Authorization: The International Space Station: Hearing Before the Subcommittee on Space and Aeronautics*, Statement by Robert L. Park, 105th Cong., 1st sess, 9 April 1997.

[59]Zimmerman, R., "Just Say *Nyet!*" *Sciences*, July/Aug. 1997, p. 19.

[60]"The Buck Stops with Al Gore: Veep-Approved Rip-Off by Russia of U.S. Taxpayer, Technology Now Threatens an American Life," *A Decision Brief* No. 97-D 89, Center for Security Policy, Washington, DC, 27 June 1997.

[61]"A Rescue Assignment for NASA," *New York Times*, 17 July 1997.

[62]Camhi, E., "Is the Flame Worth the Candle?" *Aerospace America*, Vol. 35, Aug. 1997, p. 3.

[63]Russian Space Budget for FY 1997, Ministry of Economics, Moscow.

[64]*Aerospace Daily*, 31 Jan. 1997.

[65]*Kosmicheskaya Deyatelnost Rossii* [Russia's Space Activities], Russian Space Agency, Moscow, 1996.

[66]"Interview with RSA General Director Yuri Koptev," *Rossiiskaya Gazeta*, 25 Feb. 1997.

[67]Congress, House, Committee on Science, Space, and Technology, Subcommittee on Space, *United States-Russian Cooperation in the Space Station Program: Parts I and II: Hearing Before the Subcommittee on Space*, 103d Cong., 1st sess., 14 Oct. 1993, pp. 9–10.

[68]Koretski, A., "O Problemakh Kosmonavtiki President Uznayot Po Telephonu" [The President Learns About Cosmonautics' Problems by Phone], *Segodnya* [Today], 12 April 1997.

[69]Pyrieva, G., "Na Puti K Alphe," [On the way to Alpha], *Narodnaya Gazeta* [People's Newspaper], 18 April 1996.

[70]Semyonov, Y., Speech at the celebration of the 35th anniversary of Yuri Gagarin's flight, Moscow Region, Star City, 12 April 1996.

[71]Utkin, V., *O Programme Nauchnykh e Prikladnykh Issledovanii Na Pilotiruemykh Kosmicheskikh Kompleksakh Rossii* [Concerning the Program of Scientific and Applied Research Conducted in the Russian Space Complexes], TsNIIMash Archive, Moscow, 1996.

[72]"Interview with Vladimir Utkin," *Rabochaya Tribuna* [Workers' Tribune], 12 April 1997.

[73]Titov, G., Soviet cosmonaut and Duma Deputy, interview by author,

Moscow, 18 Feb. 1996.

[74]Krikalev, S., Soviet/Russian cosmonaut, interview by author, Star City, Moscow, April 1996.

[75]Dezhurov, V., Soviet/Russian cosmonaut, interview by author, Star City, Moscow, April 1996.

[76]Titov, V., Soviet/Russian cosmonaut, interview by author, Star City, Moscow, April 1996.

[77]Strekalov, G., Soviet/Russian cosmonaut, interview by author, Star City, Moscow, April 1996.

[78]Glazkov, Y., deputy chief of Star City, interview by author, Star City, 29 March 1996.

[79]Krikalev, S., Soviet/Russian cosmonaut, phone interview by author, 15 Sept. 1996.

[80]Harwood, W., "New Agreement Reached on Station Crew Selection," *Space News*, 19–25 May 1997.

[81]Raushenbakh, B., "V Moei Knigue Sobrany Raboty Raznykh Lyet" [My Book Consists of My Works Written in Different Years], *Samolyot* [Airplane], no. 2, 1997, p. 58.

[82]Sevastyanov, V., Soviet cosmonaut and chairman of the Duma Mandate Commission, interview by author, 20 Feb. 1996.

[83]Savynykh, V., Soviet cosmonaut, president of the Moscow University of Geodesy, Aerial Photography and Cartography, interview, *Sovetskaya Rossiya* [Soviet Russia], 21 May 1996.

[84]Umarov, M., "Cherez Ternii k Bolshim Kosmicheskim Dengam" [Through the Thorns to the Big Space Money], *Kommersant*, daily edition, 1 Feb. 1996.

[85]Kiselyov, A., speech at the celebration of the 35th anniversary of Yuri Gagarin's flight, Moscow Region, Star City, 12 April 1996.

[86]Semyonov, Y., RKK Energia chief designer, interview by author, Star City, 12 April 1996.

[87]Vlasov, P., "Amerikantsy Uletely Is Rossii" [The Americans Flew Away from Russia], *Ekspert* [The expert], no. 16, 22 April 1996.

[88]Medvedchikov, A., interview by ANSER representatives, Moscow, 2 Dec. 1998.

[89]Congress, House, Committee on Science, *Hearing on ISS Problems and Options*, 105th Cong., 2d sess., 6 May 1998.

[90]Breed, T., communication to David Portree, received from Moscow via Internet, 12 Feb. 1997, ANSER/CIAC Library, Arlington, VA.

[91]Zyuganov, G., leader of Russian Communists and 1996 Russian presidential candidate, interview by author, Moscow, 25 April 1996.

[92]Lebed, A., 1996 Russian presidential candidate, interview by author, Moscow, 11 May 1996.

[93]Zorkin, V., justice of the Constitutional Court of the Russian Federation, interview by author, Moscow, 24 April 1998.

[94]Yachmennikova, N., "Vsyo Nizhe E Blizhe" [It Comes Lower and Closer], *Rossiiskaya Gazeta* [The Russian Newspaper], 3 Sept. 1996.

[95]Kirillov, V., and Mikheev, P., "Rossiya Skatyvaetsya Na Rol Subpodryadchika" [Russia Is About to Be Downgraded to the Subcontractor Level], *Segodnya* [Today], 29 Nov. 1996.

[96]Golovanov, Y., "Gagarinskaya Vesna Smenilas Oseniyu" [Gagarin's Spring Has Been Replaced by the Autumn], *Komsomolskaya Pravda*, 12 April 1997.

[97]Popova, N., "Kuda Letit Alpha" [Where Alpha Flies], *Rossiiskiye Vesti*, 29 April 1997.

[98]"Rossiya Mozhet Lishitsya Sovmestnogo Kosmosa" [Russia May Lose Joint Space], *Moskovskiy Komsomolets*, 29 April 1997.

[99]Yurova, Y., "Alpha Otbrasyvayet Stupeni" [Alpha Jettisons Its Stages], *Moskovskiy Komsomolets*, 12 April 1997.

[100]Rakukin, N., and Umarov, M., "Alpha Zaderzhitsa Na Zemle Po Vine Rossii" [Russian Needs to Be Blamed for the Alpha's Launch Delay], *Kommersant-Daily*, 28 Feb. 1997.

[101]Lovyaguin, D., "Alpha – Proyekt S Letalnym Iskhodom" [Alpha is the Project Which Will End Up Flying], *Kommersant-Daily*, 15 April 1997.

[102]Otto, G., "Nekotoriye Osobennosti Kosmicheskoi Kooperatsii Rossii E Se.She.A." [Some Aspects of U.S.–Russian Cooperation in Space], *Novosti Kosmonavtiki*, 1–12 Jan. 1997, p. 58.

[103]Dmitrenko, O., and Svistunov, S., "Bez Rossii Stroitelstvo Alphy Obrecheno Na Bolshiye Izderzhki," [Without Russian Participation, ISS Construction Will Significantly Overrun Its Cost], *Finansoviye Izvestiya* [Financial news], 15 April 1997.

[104]Popov, E., "Podnozhki Na Orbite" [Obstacles in Orbit], *Sovetskaya Rossiya* [Soviet Russia], 30 Jan. 1997.

[105]Cline, L., NASA deputy associate administrator for External Relations, interview by author, Washington, DC, 14 Aug. 1997.

[106]Decree no. 467 "On the Termination of Guarantees and Securities at the Expense of the Federal Budget," Duma Library, Moscow, 12 May 1997.

[107]Decree of the President of the Russian Federation no. 848 "On Measures for the Fulfillment of International Agreements on Space," Duma Library, Moscow, 8 Aug. 1997.

[108]Schumacher, J., NASA associate administrator for External Relations, interview by author, 1 Sept. 1998, Washington, DC.

[109]*Agreement Among the Government of Canada, Governments of Member States of the European Space Agency, the Government of Japan, the Government of the Russian Federation, and the Government of the United*

States of America Concerning Cooperation on the Civil Space Station, Article 1, 29 Jan. 1998.

[110]*Memorandum of Understanding Between the National Aeronautics and Space Administration of the United States of America and the Russian Space Agency Concerning Cooperation on the Civil International Space Station,* Article 2, 29 Jan. 1998.

[111]*Moscow Office Report* no. 289, ANSER/CIAC, Arlington, VA, 15 May 1998.

[112]*Aerospace Daily,* 15 April 1998.

[113]UPI, Moscow, 28 April 1998.

[114]Itar-Tass, Moscow, 18 May 1998.

[115]Safronov, I., "Rossia Teriayet Mir e Uchastiye v Mezhdunarodnoi Kosmicheskoi Stantsii" [Russia is Losing Mir and Its Participation in the ISS], *Commersant-Daily,* 29 May 1998.

[116]AP World News, Moscow, 9 June 1998.

[117]FBIS-SOV-98-299, 26 Oct. 1998.

[118]Protocol of the Meetings in the Office of the Deputy Chairman of the Government of the Russian Federation Boris Nemtsov "Regarding the Use of Additional Sources of Financing for Programs Within the Framework of the Federal Space Program...," 16, 22 July, 1998, Moscow, Library of the Government of the Russian Federation.

[119]AP, Moscow, 5 Aug. 1998.

[120]Congress, House, Committee on Science, Subcommittee on Space and Aeronautics, *Statement of NASA Administrator Daniel Goldin at NASA's Hearing: The White House Perspective on the International Space Station's Problems and Solutions,* 105th Cong., 2d sess., 5 Aug. 1998.

[121]*Moscow Office Report* no. 289, ANSER/CIAC, Arlington, VA, 21 Aug. 1998.

[122]*Moscow Office Report* no. 296, ANSER/CIAC, Arlington, VA, 9 Oct. 1998.

[123]Goldin, D., letter in response to Cost Assessment and Validation Task Force, 16 June 1998.

[124]*Moscow Office Report* no. 297, ANSER/CIAC, Arlington, VA, 16 Oct. 1998.

[125]Sensenbrenner, J., communication to Sergey Kiriyenko, 21 July 1998.

[126]Sensenbrenner, J., communication to Al Gore, 21 July 1998.

[127]Congress, House, Committee on Science, *Hearing on ISS: "Houston, We Have a Problem,"* 105th Cong., 2d sess., 24 June 1998.

[128]Botvinko, A., deputy chairman of RSA Manned Space Flights Department, and Karnaukhov, M., RSA senior specialist, interview by the author, Moscow, 30 July 1998.

[129]Lantratov, K., Khrunichev enterprise Public Relations Department

senior expert, fax interviews by author, 23 Oct. and 3 Nov. 1998.

[130]*Report of the Cost Assessment and Validation Task Force on the International Space Station*, prepared under the leadership of Jay Chabrow, NASA Advisory Council, Washington, DC, April 1998.

[131]Congress, House, Committee on Science, *CAV Leader Jay Chabrow's Statement at NASA Hearing: The Administration's Proposed Bail-Out for Russia*, 105th Cong., 2d sess., 7 Oct. 1998.

[132]Wheeler, L., "Goldin Rips Russians, Promises New Station Schedule by June 15," *Florida Today*, 7 May 1998.

[133]Interfax, 12 Oct. 1998.

[134]*NASA Release* No. 98-93, 1 June 1998.

[135]*NASA Release* No. 98-94, 1 June 1998.

[136]Culbertson, F., Speech, Science and Technology Policy Seminar, Washington, DC, 10 Oct. 1997.

[137]Congress, House, Committee on Science, Subcommittee on Space and Aeronautics, *Statement of Associate Administrator Joseph Rothenberg at NASA Hearing: FY 1999 Budget Request*, 105th Cong., 2d sess., 19 March 1998.

[138]*Mission: Possible. Culmination of Shuttle–Mir Docking Program Provides Measure of Confidence for International Space Station*, Shuttle–Mir Web site at http://Shuttle–Mir.nasa.gov/ and Mission Control Center Status Rept no. 20.

[139]Marinin, I., "Valerii Ryumino Polyote na Shuttle" [Valery Ryumin Is Telling About Shuttle Flight Experience], *Novosti Kosmonavtiki*, no. 15/16, Aug. 1998.

[140]*Aerospace Daily*, 6 Aug. 1998.

[141]*Unofficial Minutes of the General Designers Review*, Moscow, 28 Sept. 1998.

[142]Goldin, D., communication to Dana Rohrabacher, chairman of the House Subcommittee on Space and Aeronautics of the Committee on Science, 29 Sept. 1998.

[143]Schumacher, J., NASA associate administrator for External Relations, interview by author, Washington, DC, 19 Oct. 1998.

[144]AP, 5 Oct. 1998, Moscow.

[145]*The House Science Committee Press-Release*, 2 Oct. 1998.

[146]Sensenbrenner, J., communication to Daniel Goldin, NASA administrator, 9 Oct. 1998.

[147]McCain, J., and Frist, B., communication to Daniel Goldin, NASA administrator, 13 Oct. 1998.

[148]Bond, C., Mikulski, B., Lewis, J., and Stokes, L., communication to Daniel Goldin, NASA administrator, 13 Oct. 1998.

[149]*House Science Committee Press-Release*, 14 Oct. 1998.

[150]*U.S. Contingency Plans to Address Russian Shortfalls,* Enclosure 1 to NASA Administrator Daniel Goldin's Letter to House Science Committee

James Sensenbrenner, 15 Oct. 1998.

[151]Davis, B., "NASA Sends Millions to Russia," *Newhouse News Service*, 16 Oct. 1998.

[152]Cline, L., deputy NASA associate administrator for External Relations, interview by author, Washington, DC, 23 Nov. 1998.

[153]"The International Microgravity Strategic Planning Group," *Microgravity News*, Vol. 5, No. 2, Summer 1999, p. 12.

[154]*NASA/RSA Protocol Establishing a Basis for a Modification to NAS15-10110*, NASA Headquarters, Washington, DC, 30 Sept. 1998.

[155]Medvedchikov, A., RSA deputy general director, phone interview by author, 2 Nov. 1998.

[156]Yachmennikova, N., "Spasatelniy Krug Dlya Nashego Kosmosa" [Lifebelt for Our Space], *Rossiiskaya Gazeta*, 4 Oct. 1998.

[157]Krasnov, A., RSA deputy director of Department of International Cooperation, phone interview by author, 21 Oct. 1998.

[158]Kozlovskaya, I., IBMP department director, phone interview by author, 31 Oct. 1998.

[159]Itar-TASS, 16 Oct. 1998.

[160]*Moscow Office Report* no. 296, ANSER/CIAC, Arlington, VA, 9 Oct. 1998.

[161]*Moscow Office Report* no. 302, ANSER/CIAC, Arlington, VA, 20 Nov. 1998.

[162]Leskov, S., "Mir mozhet e dolzhen rabotat dalshe" [Mir Can and Must Continue Its Operation], *Izvestia*, 20 Oct. 1998.

[163]*Moscow Office Report* no. 296, ANSER/CIAC, Arlington, VA, 9 Oct. 1998.

[164]AP, 14 Oct. 1998, Moscow.

[165]Gubarev, V., "Odnin 'Mirom' Mazany" [All of Us Have Only One Mir], *Novaya Gazeta*, 21–27 Sept. 1998.

[166]Payson, D., "No Unanimity on Mir. Energia Space Rocket Corporation Proposing to Extend Service Life of Russian Orbital Complex," *Nezavisimaya Gazeta*, 29 Sept. 1998.

[167]Goltz, A., "Ne ot 'Mira' Sego" [He Is a Stranger Among Us], *Itogi*, 1 Sept. 1998, p. 33.

[168]Dezhurov, V., Russian cosmonaut, phone interview by author, 29 Oct. 1998.

[169]Itar-TASS, Moscow, 9 Oct. 1998.

[170]"Russian Lawmakers Slam 'Armageddon' film," *Reuters News Service*, 9 Oct. 1998.

[171]Itar-TASS, Moscow, 22 Oct. 1998.

[172]Miryazeva, E., "'Mir' Mozhet Izbezhat Zatopleniya" [Mir Could Avoid the Sinking], *Nezavisimaya Gazeta*, 17 Nov. 1998.

[173]Bronzova, M., "Pravitelstvo Zanimalos Problemami Kosmonavtiki" [The Government Was Solving the Cosmonautics' Problems], *Nezavisimaya Gazeta*, 13 Nov. 1998.

[174]Berger, B., "NASA Weighs Another Shuttle–Mir Mission," *Space News*, 7–13 Dec. 1998.

[175]Paison, D., "Rossia Vyvela na Orbitu Klyuchevoi Modul Mezhdunarodnoi Kosmicheskoi Stantsyi" [Russia Has Put into Orbit a Key Module of the International Space Station], *Hezavisimaya Gazeta*, 21 Nov. 1998.

[176]Arbatov, A., "Pora nachat 'perenaladku' otnosheniy" [Time Is Ripe to Start the "Readjastment" of Relations], *Nezavisimaya Gazeta*, 21 July 1998.

[177]*NASA Advisory Council Task Force on the International Space Station Operational Readiness*, unofficial minutes, 16 June 1998.

[178]Kiselyov, A., director of Khrunichev Space Center, interview by author, Moscow, 22 April 1998.

[179]Zhiltsov, S., Khrunichev Enterprise Public Relations Department director, phone interview by author, 28 Oct. 1998.

[180]Paison, D., "Russkiy den Mirovoi Kosmonavtiki" [Russian Day of the World Cosmonautics], *Nezavisimaya Gazeta*, 25 Nov. 1998.

[181]*Russian Issues Briefing to the Office of the Vice President, the Office of Management and Budget, and the Office of Science and Technology Policy*, 30 July 1998.

[182]*Aerospace Daily*, 25 Jan. 1999.

[183]Informatsionnyi byulleten posolstva Rossiiskoi Federatsii, Vypusk II, 22 Jan. 1999 [Information Bulletin of the Embassy of the Russian Federation, Issue II].

[184]Leskov, S. "Stantsii 'Mir' Prikazano Vyzhit" [Mir Station Has Been Ordered to Survive], *Izvestia*, 23 Jan. 1999.

Chapter 7

COOPERATION IN SPACE: LESSONS AND PERSPECTIVES

Analyzing the three periods of Soviet/Russian–American cooperation in space—from the late 1950s through the mid-1960s, from the late 1960s through 1985, and from 1985 until the end of 1998—enables us to identify a number of political, economic, and technical factors that influenced this cooperation in the past and may influence it in the future. Some of these factors played a centrifugal role by creating obstacles to U.S.–Soviet/Russian space cooperation; some, a centripetal role by promoting joint activities in space; and others, a centrifugal/centripetal role, depending on the overall state of bilateral relations and the technical–economic status of the two space programs. Soviet/Russian–American cooperation in space has never been influenced by only one of these factors. It has always been a balance between centrifugal and centripetal forces that made the possibility of cooperation a reality or that prevented potential cooperation from being realized.

STRUCTURE OF THE INTERNATIONAL SYSTEM

The first factor that played a centrifugal role, at least until the mid-1980s, was the structure of the international system. Such structure could help explain why different states behave similarly and, "despite their variations, produce outcomes that fall within expected ranges."[1] The structural approach will help explain why countries as different as the Soviet Union and the United States attached such great importance to their respective space programs and ultimately to victory in the space race. This approach also explains why there was no real bridging of the two space programs until the early 1990s.

During the first two periods of Soviet/Russian–American cooperation in space, the international system could be characterized as bipolar: two powerful states controlled and regulated interactions between and within their respective spheres of influence and were concerned with the extension of these spheres. The mid-1950s to mid-1980s was an era of struggle between the Soviet Union and the United States for governance of the international system.

One of the most important components of this governance is the hierarchy of prestige among states. According to E. H. Carr, a prominent scholar in international relations, prestige is "enormously important" because, "if your strength is recognized, you can generally achieve your aims without having to use it."[2] Prestige in the modern era is based on economic, technological, and military power. Space exploration is still one of the types of human activity that more than anything else reflects these components of prestige, and this was especially true from the 1950s through the early 1980s. Global competition/confrontation between the two countries, which continued with varying strength during this period of time, emphasized a competitive element in their space relationship.

However, if the Soviet Union's and the United States' behavior were predetermined by the structure of the international system, why did these two nations have different attitudes toward space cooperation in the late 1950s–early 1960s, even though both were superpowers and were equally concerned about their prestige? Why did the United States want to involve the Soviet Union in space cooperation, and why was the Soviet Union so reluctant to join in? To answer such questions, it is necessary to use unit level factors analysis, which helps explain "why different units [states] behave differently despite their similar placement in a system."[1] Such types of analyses focus on the differences between the driving mechanisms behind U.S. and Soviet politics.

SPACE COOPERATION AND INTERNATIONAL FUNCTIONALISM

The first ever serious attempt to involve the Soviets in space cooperation with the United States was made by President John F. Kennedy. Kennedy was concerned about damage to U.S. prestige and reputation from Soviet space triumphs. However, at the beginning of his presidency he was unsure of victory in a space race. Thus, his initial idea was to get involved in space cooperation with the Soviet Union so that the two countries could share the glory as space pioneers. This was the first reason for Kennedy's pro-cooperation attitude.

The second reason was that Kennedy recognized there was a new balance of world forces and that the Soviet military might was equal to that of the United States. This made Kennedy revise the Eisenhower–Dulles "position of strength" policy toward the Soviet Union. The revised policy included Kennedy's approach toward developing areas of common interest with the Soviet Union, which, if extended, would have an overall healing impact on bilateral relations and would diminish the possibility of a nuclear holocaust. Space, in the president's view, could become one of these areas.

Such an approach to pooling efforts with the Soviets in space indicates that Kennedy was thinking about Soviet–American space cooperation in terms of international functionalism. One of the major principles of functionalism is that "individuals and groups could begin to learn the benefits of cooperation and would be increasingly involved in an international cooperative ethos, creating interdependencies, pushing for further integration...."[3]

SOVIET UNION'S REBUFF OF EARLY COOPERATIVE EFFORTS

Kennedy's cooperative overtures did not meet with an adequate response from the Soviet leadership. First, U.S. attempts to build bridges between the two countries were perceived by the Kremlin as a new effort to undermine socialism. Second, the Soviets saw peaceful coexistence not only as a form of struggle with the West but also as a strategy of struggle, aimed at achieving Soviet global objectives. Third, the Soviet leadership refused to consider any form of space cooperation outside the overall Cold War context of Soviet–American relations. Fourth, the Kremlin was overly complacent about its space achievements and did not want to erode its "Sputnik diplomacy" by sharing the glory of space pioneers with the United States. Finally, the Soviet space program was an extension of the Soviet ballistic missile program and was concealed with a special secretiveness. This secrecy was not consistent with cooperation with a potential adversary.

Overall, the Soviet Union, as it was characterized by Henry Kissinger, was at that time a "revolutionary" state, pretending to revolutionize, i.e., to change the world according to the views and ideas of the Soviet leadership.[4] Whereas the Soviet Union was the "attacker," the United States during this period was more of an "active defender" than a "counter attacker," containing rather than pushing back communism. Although the U.S. goal was to incorporate the Soviet Union into the system by promoting any kind of cooperation, the revolutionary and totalitarian character of the Soviet Union subordinated all elements of Soviet policy, the space program, in particular, to the major Soviet goal, which was squeezing the capitalist order out of the international system.

DISINTEREST OF COMMUNITIES IN COOPERATION

The first period in the history of Soviet–American space cooperation identified another factor that played a centrifugal role in Soviet/Russian–American space cooperation in the beginning of the space era. Besides the top U.S. and Soviet leaders, there were other major players in U.S.–Soviet/Russian cooperation in space: the space communities of the two countries, especially the parts that related to the manned space programs. Neither of them was actually interested in cooperation with the other because financial injections into the programs depended to a considerable degree on a space race between the Soviet Union and the United States. The Soviet and, to a lesser extent, the U.S. space communities had influence on top policymakers in their countries and could be considered interest groups that "the state must accommodate…, and they must have a share in public decision-making."[5]

DÉTENTE AND ASTP

The second period in Soviet/Russian–American cooperation in space, which started in the late 1960s and lasted through 1985, climaxed in the Apollo–Soyuz Test Project (ASTP) in 1975. This project was possible above all because of the changes in Soviet–American relations. Although the Soviet Union and the United States were still involved in global competition, confrontation in their relationship was replaced by détente. Leaders in the two countries viewed a "joint manned mission in space—an area in which 15 years ago [the Soviet Union and the United States] saw [themselves] in almost mortal rivalry"—as a symbol of the distance they had traveled.[6] Decreased tensions in U.S.–Soviet relations alleviated concern on both sides about the transfer of dual-use technology as a result of joint work on a space project.

Another factor that made ASTP possible was the changed attitude of the U.S. and Soviet space communities toward cooperation with each other after the successful mission of Apollo 11, which ended the moon race. On the Soviet side, the weakening of the competitive drive behind space activities resulted in an attempt to compensate for the defeat in the moon race by seeking cooperation with the Americans in projects in which both could work as equal partners. Also, the Soviets hoped to gain docking experience from their American colleagues, because the latter had advanced further than the former in this area. The shift in the leadership of the Soviet space program, from those who actively supported competition with the United States in space to those who at least did not mind cooperating with them, also contributed to a space rapprochement between the two countries.

Representatives of the U.S. space community were also interested in cooperating with their Soviet colleagues. The end of the moon race could have had a more destructive effect on the U.S. program than on the Soviet one, since the existence of the former was considerably dependent on public support. Flight with the Soviets could have sparked the interest of the American people in space ventures again. A joint mission also had reasonably high chances of being supported by the U.S. Congress, which might view the project as an area of common interest with the Soviet Union. Also, a Soviet–American mission could have utilized the remaining Apollo hardware; filled a gap in U.S. manned space flights; provided NASA with time to define more fully its requirements to Shuttle-era subsystems; built mutual confidence and trust to ensure future joint operations in space, including the creation of permanently manned space stations in Earth orbit and a possible journey to Mars; and shared the costs of the projects, in the context of the shrinking NASA budget.

ASTP involved satisfaction and disappointment on both sides. On one hand, it was an example of unprecedented cooperation in the field of high and dual-use technologies between the United States and the Soviet Union, reflecting a high degree of trust and a new relationship between the two countries. On the other hand, the implementation phase of ASTP, which involved joint work on the realization of a concrete project, inevitably exposed some of the previously hidden weaknesses of the Soviet space program to U.S. space officials and specialists. Among these weaknesses was the relative backwardness and not very high safety record of Soviet space hardware. Contrary to U.S. expectations, the Soviets did not reciprocate the degree of openness of the U.S. space program to Soviet specialists.

On the Soviet side, there was also some disappointment about cooperation with the Americans. First, public comparison of U.S. and Soviet space technologies stressed the superiority of the former, undermining further the prestige of one of "the greatest Soviet achievements." Second, the Soviets expected that ASTP would be followed by a joint flight of Soyuz with space station Skylab, thus giving them more access to U.S. space technology and experience. However, the American refusal to implement this kind of plan diminished Soviet interest in the coming Apollo–Soyuz mission.

Overall, although the project helped the Soviets and the Americans become better acquainted, ASTP did not involve any significant joint development of space hardware or increased technological and economic dependence between the two countries. Moreover, although the Soviets and Americans were generally interested in cooperation with each other, the U.S. and Soviet space programs were still pretty much driven by competitive rationales, and pooling efforts in space was not of immediate interest to the space officials of the two countries. These factors, coupled with the lack of will of the White House and the Kremlin to make space an area of common interest, predetermined a decline of U.S.–Soviet cooperation in the most complicated and demanding area of space exploration—manned space flights—as well as a decline of détente in the late 1970s.

After ASTP (late 1970s–1985) and through the end of the second period of Soviet/Russian–American cooperation in space, another factor began playing a centripetal role in this cooperation. This factor was a cooperative regime that manifested itself at two levels of U.S.–Soviet space-related interactions: at the level of the two space communities and at the top political level. This regime can explain the existence of cooperation "even when ...conditions that initially gave rise to [it] change."[7]

Although Apollo–Soyuz-type projects were out of the question in the late 1970s and early 1980s, the limited U.S.–Soviet space cooperation that was the ASTP byproduct continued, mostly in the area of space science. This cooperation fit one of the definitions of a regime's origin: "regimes and cooperation in one issue-area may arise as an unintended consequence of cooperation in some other area" (Ref. 7, pp. 506–507). The regime factor, coupled with the intention of the Kremlin and White House to use cooperation in space as a means of warming U.S.–Soviet relations as they did in the early 1970s, created preconditions for the intensification of bilateral cooperation in space as soon as the overall context of U.S.–Soviet relations improved.

CHANGES IN THE INTERNATIONAL SYSTEM

Such improvement started taking shape in the second half of the 1980s, after *perestroika* was launched. Unlike détente, it not only reflected the intention of Soviet and U.S. leaderships to move from "an era of confrontation to an era of negotiation," but also reflected the beginning of deep structural changes in the international system. The government of Mikhail Gorbachev started disengaging the Soviet Union from global competition with the United States, predetermining the end of a bipolar system. This disengagement contributed to Soviet/Russian–American cooperation in space in two ways. It improved bilateral relations and eased fears in the two countries about the possible transfer of dual-use technology to a potential adversary. However, it also undermined one of the major rationales for the vigorous space effort—to serve as a tool in the struggle for global hegemony—and became one of the preconditions for the blossoming of Soviet/Russian–American cooperation in space.

Although political devaluation had an impact on both space programs, the Soviet program was affected especially hard by the loss of a competitive rationale. Many Soviets saw no other reason for the space program except to beat the Americans in space. When this goal ceased to exist, a significant portion of Soviet society started perceiving the program as a burden on the country's troubled economy. The populist leadership of Gorbachev gave a twofold response to the people's attitude toward space exploration. It started decreasing the Soviet space budget, while opening its space program to the West to become engaged in space cooperation with Western countries. Such cooperation would have enabled the Soviet leadership to achieve a number of goals. First, it would have allowed the Kremlin to share the costs of major space projects. Second, it would have enabled the Soviet space industry to earn money through commercial contracts. Third, after cancellation of the Buran program, Soviet space professionals were very interested in cooperating with the Americans who designed and operated the Space Shuttle. The Mir space station and U.S. reusable spacecraft would make a perfect match. Finally, like ASTP in 1975, renewed cooperation in space would have become a symbol of a new relationship between the United States and the Soviet Union.

NASA was also interested in sharing the costs of space projects with the Soviet Union. Space cooperation with the Soviets was supported by U.S. politicians who viewed space as one of the means of promoting friendly relations with the Soviet Union. However, U.S. officials did not initially subject U.S.–Soviet relations to such dramatic reconsideration as did their Soviet colleagues. This is why some NASA managers were still using space race arguments to convince the White House and Congress not to cut the agency's budget. Also, in the 1980s NASA flew a number of classified Department of Defense missions. For these reasons the agency, although generally welcoming cooperation with the Russians, preferred to limit it to small or medium-sized projects.

The end of the bipolar system caused by the disintegration of the Soviet Union introduced new elements into U.S.–Russian cooperation in space. One of the centrifugal factors associated with the bipolar system, which was standing in the way of this cooperation, was gone—space was no longer a tool in the struggle for global hegemony in Russia and the United States, and these countries were no longer attempting to race each other into space.

JOINT SPACE EXPLORATION

Space exploration, particularly its manned component, kept its political value for both countries. The U.S. and Soviet leaderships clearly indicated their willingness to make space cooperation an area of common interest for the two countries. Also, continuing manned missions into space had a special importance for the Kremlin: It was evidence of Russia's ability to preserve its cutting-edge technological potential despite the unfavorable economic and political situation in this country. For the White House, human space flight retained its significance as an important element of the U.S. claim for global leadership. However, in the changed geopolitical environment of the world, the goals of both countries were better served by cooperation, not competition.

The economic survival of the Russian manned space program and consequently of a number of Russian space enterprises became impossible without cooperation with the West, particularly with the United States in the ISS project. Understanding this factor, the Russians even considered giving priority to the ISS project over the greatest symbol of their national space program—the Mir space station. Russian space officials made a tentative decision to terminate the operations of Mir a year earlier than scheduled, in late 1998–mid 1999, when the first crew was scheduled to arrive on the ISS. The reason for the decision was Russia's inability to support two programs—Mir and the ISS—simultaneously.

The Mir issue, however, demonstrated that space exploration still has a dual national/international value and that opposite sides do not always live in harmony. Mir's operation beyond 1999 was overwhelmingly supported by Russian space professionals, scientists, and ultimately the State Duma. On 22 January 1999, Prime Minister Primakov signed a decree authorizing the extension of Mir's operation until the year 2002. This decision is not likely to undermine U.S.–Russian cooperation in the ISS program. First, starting from mid-1999, Mir's operation will be financed from an off-budget source, meaning that the Russian space station will not drain federal funding allocated for the ISS. Second, in three years Mir will run out of its useful lifetime anyway, and currently Russia cannot afford to build its own space station. Considering the great amount of pride that Russia places on its continuity in space, there is good reason to believe that Russia will ultimately concentrate all of its efforts on the ISS.

The U.S. manned space program, although in much better economic shape than the Russian program, also became dependent to a considerable degree on cooperation with Russia, specifically in the ISS program. This cooperation has been aimed at achieving a number of U.S. space-, security-, and general policy-related goals.

1) It was to help NASA build the station faster, better, and cheaper.
2) The U.S. invitation for Russia to join the ISS program was partly a tradeoff for Russia's refusal to sell its cryogenic engines technology to India.
3) By creating job opportunities for Russian rocket and missile specialists at home, the United States attempted to prevent them from being hired by countries that practice state terrorism.
4) Cooperation with Russia in the ISS program will be used to highlight the U.S. ability to lead through cooperation, i.e., to play a leading role in organizing and completing large-scale, complex international projects.
5) The United States was concerned about its competitors strengthening their positions by gaining access to advanced and inexpensive Russian space technology; the United States wanted to get access to this technology itself to remain on the cutting edge.
6) All of the aforementioned benefits of cooperation with Russia ensured congressional support for the ISS project, which U.S. lawmakers were threatening to cancel. The reason for this decision was Russia's perceived inability to support both Mir and the ISS.

The joint U.S.–Russian work on the ISS program (1993–1998) was the first serious test of the vitality of broadscale bilateral space cooperation in the absence of global competition/confrontation between the two space super-powers. The development of this cooperation has been influenced by a number of factors that have weakened or strengthened it. The outcome of this cooperation will depend on the balance between these two types of factors.

FACTORS WEAKENING AND STRENGTHENING COOPERATION

The most serious factor weakening U.S.–Russian cooperation in the ISS project is Russia's inability to meet its commitments to the space station program on time. This inability delayed the beginning of ISS assembly by more than a year, increased the ISS costs, and consequently undermined one of the key rationales for Russian involvement in the project—to build the station faster, better, and cheaper. Bureaucratic obstacles caused by the residual Soviet-type mentality and business style of some of the Russian space officials also hindered cooperation.

There are a number of factors, however, that have strengthened cooperation despite all of the setbacks. The first factor is the continuous support that top political leaders of both countries provide to joint U.S.–Russian work on the ISS project.

The second factor is the interest of U.S. and Russian space communities in cooperating with each other. The major interest on the U.S. side is based on the enabling, not enhancing, role that Russia plays in the ISS project, i.e., Russia controls the critical path of the station. This critical path consists of the SM (propulsion and crew quarters unit) and Soyuz and Progress space-craft (rescue and resupply vehicles). Although NASA has ultimately come up with a plan to replace Russian critical elements, it will take the U.S. space agency three or more years to realize this plan. This makes Russia's involvement virtually indispensable. U.S. space experts came to the conclusion that maintaining Russian participation until the United States develops its own independent capabilities is significantly cheaper than the cost of developing the ISS without Russia.

Four more factors keep the ISS managers interested in maintaining their partnership with Russia. One of them is a number of assembly EVAs to be performed by Russian cosmonauts to assemble the space station. Second is the Russian space station's operational experience. Third is the potential of U.S.–Russian cooperation in space as proven by the sound success of the ISS Phase I Shuttle–Mir program. Fourth, in January 1998, the U.S.–Russian space station partnership was legally codified in an intergovernmental agreement among the ISS partners, including Russia and the United States, and in a memorandum of understanding between NASA and the RSA.

The aforementioned positive factors coupled with the support of the White House for the U.S.–Russian space station alliance, and overall support of Congress for the ISS, largely determined NASA's wish to maintain Russia's ISS full-member status and to keep the door open for the future participation of Russia in the program, if for some reason Russia withdraws temporarily from it. At the same time, the United States and Russia do not totally exclude a limited client–contractor relationship that will provide the United States with additional research opportunities onboard the Russian segment of the station in exchange for money that will help Russia fulfill its ISS obligations.

The interest of the Russian space professionals in cooperating with their American counterparts is also partly based on the possibility of learning from the American manned spacecraft operational experience. However, Russian specialists are more knowledgeable than their U.S. colleagues about building, using, and operating space station-type hardware—the main purpose of the ISS program. Thus, the major interest of the Russians in their joint work with the Americans is based not so much on what they could learn from the latter, but on the need to save the Russian manned space program from total disintegration because of a lack of money. Russia's manned space-flying capability is certainly the most significant achievement of the Russian aerospace sector that "largely determines the status of Russia as a great power."[8]

One more factor has special significance for the present and future of U.S.–Russian cooperation in the ISS and similar projects because it presents further evidence of the strengthening cooperative regime in the space area. It is a stable working relationship between the RSA and NASA made possible because Russian and U.S. space officials were able to understand and accept each other's rules, norms, procedures, and business culture.

Growing disagreements between Russia and the United States concerning a number of regional and global security issues did not affect cooperation between the two countries in the ISS program. One of the most serious tests of the vitality of U.S.–Russian cooperation occurred at the time of NATO expansion. U.S. officials, remembering the negative impact that cooling bilateral relations had on space cooperation in the past, were concerned about Russia reconsidering its participation in the ISS. However, Russia did not make "any linkage stated or implied between fulfilling Space Station commitments and NATO expansion."[9] Among the responses of Russian officials to NATO expansion was the refusal of the Duma to ratify the START II treaty. However, this action of the lower house of the Russian Parliament, which further strained U.S.–Russian relations, did not make the United States disengage from cooperation with Russia in the ISS program.

U.S.–Russian partnership in the ISS program was subjected to an equally serious test in December 1998 when, as a sign of a protest against the U.S. bombing of Iraq, Russia recalled its ambassador to the United States "for consulations." However, even this, one of the deepest rifts in the bilateral relationship over the last 60 years, did not have any impact on Russian–American cooperation in building the new space station—proof that Russia was breaking with its traditional "policy continuum" approach toward cooperation with the United States in space and that the two countries are really determined to make space an area of common interest.

An assessment of the current period of U.S.–Russian cooperation in space indicates that interaction between the two countries could also be analyzed in terms of international functionalism. A number of statements made by top U.S. and Russian state officials and politicians signal a growing functional role played by space cooperation in a broader context of U.S.–Russian contemporary relations. Vice President Albert Gore stressed that nowhere will a new partnership between the United States and Russia "be so keenly felt as in the area of high-technology cooperation."[10] NASA Administrator Daniel Goldin said that U.S.–Russian cooperation in space also has a "critical underpinning of the success of the U.S.–Russian Joint Commission on Economic and Technological Cooperation" (Ref. 9, p. 22). Former Prime Minister Victor Chernomyrdin told Vice President Gore about "the tremendous benefits of the Shuttle–Mir program in providing a common thread between the two nations" (Ref. 9, p. 22). According to Chernomyrdin, "millions of Russian citizens have seen how well Americans and Russians can work together in space," and he indicated that the two countries have understood each other's problems and achieved mutually acceptable solutions (Ref. 9, p. 22).

CONCLUSION

The concluding observations, based on the analysis of a 40-year-long space relationship between the two countries, reveal the conditions required for the emergence and maintenance of meaningful U.S.–Russian space cooperation.

The first and the most important condition is the absence of global competition between Russia and the United States, which would make these countries again use space as one of the primary means for strengthening their prestige against each other, greatly diminishing the possibility of space cooperation. Another condition is the absence of confrontation, which hinders any cooperative activity involving dual-use technology.

Third, top political leaders in the two countries should be clearly determined to engage in space cooperation and to make it an area of common interest. This is especially important because joint work on space projects encounters numerous political, technical, and economic problems that can be solved only with the help and support of top public officials.

The fourth condition is the Russian and U.S. space communities' interest in cooperation. This interest should be go beyond the organization of a joint "space show" meant to preserve the existing level of budget funding and should involve meaningful cooperation in a concrete project with a "shadow of the future." A cooperative regime developed by the space communities of the two countries would greatly facilitate such cooperation.

Fifth, the U.S. and Russian space sectors should be technologically and economically dependent on each other in their joint work on a particular project.

Finally, cooperation should bring tangible results and adequate compensation for the difficulties encountered during the realization of projects as complex as joint space ventures. One such compensation was the successful launch of the ISS first element (Zarya or FGB), which considerably secured the future of the Russian–American partnership in the ISS program.

REFERENCES

[1]Waltz, K. N., *Theory of International Politics*, Random House, New York, 1979, p. 72.

[2]Gilpin, R., *War and Change in World Politics*, Cambridge Univ. Press, Cambridge, MA, 1979, p. 31.

[3]Mitrany, D., *The Functional Theory of Politics*, St. Martin's Press, New York, 1975, p. x.

[4]Graubard, S. R., *Kissinger: Portrait of a Mind*, W. W. Norton, New York, 1973, p. 273.

[5]Fleron, F. J., Jr., and Hoffmann, E. P., (eds.), *Post-Communist Studies and Political Science. Methodology and Empirical Theory in Sovietology*, Westview Press, Boulder, CO, p. 46.

[6]Kissinger, H. A., *American Foreign Policy*, W. W. Norton, New York, 1977, p. 208.

[7]Haggard, S., and Simmons, B. A., "Theories of International Regimes," *International Organizations*, Vol. 41, Summer 1987, p. 506.

[8]Yeltsin, B., Russian president's radio address on Russian space issues, ITAR-TASS, 22 Aug. 1997.

[9]Congress, House, Committee on Science, *The Status of Russian Participation in the International Space Station Program: Hearing Before the Committee on Science*, 105th Cong., 1st sess., 12 Feb. 1997, p. 54.

[10]Gore, A. J., Remarks by vice president during a signing ceremony with Prime Minister Chernomyrdin of Russia, Washington, DC, NASA History Office, 2 Sept. 1993.

CHRONOLOGY OF MAJOR EVENTS

EISENHOWER–KHRUSHCHEV PERIOD

mid–1950s	The Soviet Union and the United States begin developing satellites.
1956	Soviet space General Designer Sergey Korolev convinces Soviet Premier Nikita Khrushchev to start racing Americans in space.
12 January 1957	The United States suggests to the United Nations international cooperation in the control of space-based weapons.
4 October 1957	The Soviet Union launches the first satellite (Sputnik) in space.
31 January 1958	The U.S. satellite Explorer I is successfully launched into space.
15 February 1958	President Eisenhower refers to the use of outer space for military purposes as a "terrible menace" in a letter to Soviet Premier Khrushchev.
1 October 1958	NASA is formed. International cooperation in space is stated as one of the goals of the new agency.
17 November 1958	Senator Lyndon B. Johnson addresses the United Nations, urging adoption of the resolution to establish the Ad Hoc Committee on the Peaceful Uses of Outer Space.
November 1959	Hugh Dryden approaches Soviet Academicians Leonid Sedov, Alexander Blagonravov, and V. Krassovski regarding U.S.–Soviet cooperation in the use of satellites.
12 December 1959	The United Nations Committee on the Peaceful Uses of Outer Space is formed as a result of the U.S. initiative. The goal of the Committee is to promote international cooperation in space,

particularly between the United States and the Soviet Union.

KHRUSHCHEV–KENNEDY PERIOD

1960	Presidential candidate John F. Kennedy considers developing "areas of common interest" with the Soviet Union.
30 January 1961	In his first State of the Union address, Kennedy invites the Soviet Union to join the United States in a number of space projects, including communication satellites and deep space probes.
12 February 1961	The Soviet Union successfully launches the un manned Venera probe to explore planet Venus.
March–April 1961	The joint NASA–President's Science Advisory Committee–Department of State panel prepares a report on international cooperation in space and presents draft proposals for U.S.–USSR cooperation.
12 April 1961	The Soviet Union launches the first man in space —Yuri Gagarin.
Winter–Spring 1961	Kennedy makes two more offers to the Soviets— on the occasions of the Venera and Gagarin launches—to cooperate in space. These offers are ignored.
5 May 1961	Astronaut Alan Shepard makes the first U.S. suborbital space flight.
25 May 1961	Kennedy addresses the U.S. Congress with a special message, "Urgent National Needs," in which he introduces the U.S. moon manned program.
June 1961	During the Vienna summit, Kennedy proposes to Khrushchev a joint moon mission. Khrushchev initially reacts positively but then links Soviet–American cooperation in space to disarmament.
25 September 1961	In his address to the U.N. General Assembly, Kennedy advances the idea of international cooperation in the control of space-based weapons.

20 February 1962	Astronaut John Glenn makes the first U.S. orbital space flight.
21 February 1962	Khrushchev sends a congratulatory telegram to Kennedy on the occasion of John Glenn's flight in which he acknowledges potential benefits to cooperation in space.
7 March 1962	Kennedy responds to Khrushchev's telegram by proposing five space-related areas where the Soviet Union and the United States can begin immediate cooperation. Khrushchev responds positively.
1962–1963	Dryden–Blagonravov talks regarding U.S.–USSR space cooperation occur. Four meetings are held but little progress is made.
Summer 1963	Khrushchev reportedly considers cooperation with the Americans in a moon landing.
20 September 1963	In his address to the U.N. General Assembly, Kennedy again proposes a joint Soviet–American expedition to the moon.

KHRUSHCHEV–JOHNSON AND BREZHNEV–JOHNSON PERIODS

2 December 1963	U.S. Representative to the U.N., Adlai Stevenson, following Johnson's instructions, reaffirms Kennedy's proposal to the Soviets to go to the moon together.
January 1964	Soviets provide performance data for the U.S. ECHO II satellite to the Americans.
1964–1967	Johnson approaches the Soviet Union six times regarding cooperation in space.
May 1968	A joint manned lunar surface laboratory is pro posed by the Johnson Administration.

APOLLO–SOYUZ TEST PROJECT AND RETURN TO THE COLD WAR

1969	U.S. President Richard Nixon expresses his intention to move from "a period of confrontation" to "an era of negotiation."
Spring–Winter 1969	President of the USSR Academy of Sciences Mstislav Keldysh and NASA Administrator

	Thomas Paine exchange letters regarding possible areas of cooperation in space between the two countries.
July 1969	Apollo 11 successfully carries the first men to the moon and back to Earth.
October 1969	Soviet cosmonauts Georgi Beregovoy and Konstantin Feoktistov visit the United States.
April 1970	Vice President of the Soviet Academy of Sciences Mikhail Millionshchikov announces that conditions are favorable for the Soviet Union and the United States to negotiate an agreement on space exploration.
10 July 1970	Nixon publicly confirms his interest in pursuing discussions on space cooperation with the Soviet Union.
October 1970	A group of 8 senators and 39 congressmen call on Nixon to seek talks with the Soviet Union concerning its participation in the Skylab experiments and other future U.S. space projects.
October 1970	Two more Soviet cosmonauts, Nikolai Andrianov and Vitaly Sevastyanov, visit the United States.
26–27 October 1970	U.S. and Soviet space hardware designers and engineers meet in Moscow to discuss possible joint space projects. (This meeting marks the beginning of a productive five-year period in the history of Soviet–American space cooperation, culminating in the July 1975 Apollo–Soyuz space flight.)
January 1971	Soviets agree to exchange moon samples with the Americans.
25 February 1971	Nixon officially incorporates cooperation with the Soviet Union into U.S. space policy.
24 May 1972	U.S.–Soviet (Nixon–Kosygin) space agreement. The two countries officially commit themselves to the Apollo–Soyuz Test Project (ASTP).
15 July 1975	Apollo and Soyuz spacecraft dock in orbit.
11 May 1977	NASA–USSR Academy of Sciences agreement is signed. Its intent is to maintain the momentum generated by ASTP.

18 May 1977	Soviet Foreign Minister Andrei Gromyko and Secretary of State Cyrus Vance renew the bilateral U.S.–Soviet agreement on space cooperation.
December 1979	Soviets invade Afghanistan.
May 1982	President Ronald Reagan retaliates against the declaration of martial law in Poland by refusing to renew the 10-year old agreement on Soviet–American space cooperation.
January 1984	The United States privately proposes to the Soviet Union the idea of a simulated space rescue mission involving the U.S. Space Shuttle and the Salyut 7 space station.
February–October 1984	The U.S. Congress launches several initiatives aimed at resuming cooperative efforts with the Soviet Union.

RENEWED COOPERATION

May 1985	Mikhail Gorbachev is elected General Secretary of the Communist Party of the Soviet Union. *Perestroika* is launched.
July 1985	NASA Administrator James Beggs confirms the readiness of the United States to resume cooperation with the Soviet Union on civilian space missions.
Fall 1985	Congressman Bill Nelson proposes a "political show in space"—A U.S.–Soviet communication via a Space Shuttle–Salyut signal exchange.
Spring 1986	The National Commission on Space recommends to the White House colonization of Mars by 2007 through joint U.S. and Soviet efforts.
June 1986	Senator Albert Gore proposes a manned mission to Mars in concert with the Soviet Union.
September 1986	A 10-member U.S. space team, headed by Jet Propulsion Laboratory Director Lew Allen, discusses possible joint Mars studies with their Soviet counterparts.
15 April 1987	U.S. Secretary of State George Shultz and Soviet Foreign Minister Eduard Shevardnadze

	sign the "Agreement Between the United States of America and the Union of Soviet Socialist Republics Concerning Cooperation in the Exploration and Use of Outer Space."
1987	Association of Space Explorers (ASE) encourages the Soviet Union and the United States to cooperate in space.
1989	ASE proposes that Soviet cosmonauts fly on the Space Shuttle and that American astronauts fly on the Mir space station.
January 1990	The first official visit by U.S. astronauts to the Soviet Union since 1975.
June 1990	Mikhail Gorbachev and U.S. Vice President Dan Quayle meet in Washington and tentatively agree to send an American astronaut to Mir and a Soviet cosmonaut on a Space Shuttle mission.
October 1990	A NASA delegation headed by Administrator Richard Truly visits the Soviet Union to discuss possible joint projects.
November 1990	Washington learns about the Soviet's intention to sell cryogenic engine hardware and technology to India. (This precipitates a three-year-long controversy that ultimately results in Russia securing a place in the International Space Station project in exchange for not transferring the engine technology to India.)
1991	A report titled *America's Space Exploration Initiative* is prepared under the leadership of former Apollo–Soyuz astronaut Tom Stafford. The report encourages the United States to pursue cooperative opportunities presented by the Mir space station.
June 1991	An official delegation from the Ministry of General Machine-Building (*Minobshechemash*) visits the Johnson Space Center for the first time since ASTP.
15 August 1991	The Soviet Meteor 3 satellite carries NASA's Total Ozone Mapping Spectrometer. It is the first Soviet satellite to carry American instruments.

Fall 1991	Senator Barbara Mikulski asks RKK Energia General Designer Yuri Semyonov and NASA Administrator Richard Truly to consider the possibility of using the Soyuz-TM spacecraft as a rescue vehicle.
October 1991	Department of Defense and NASA officials travel to the Soviet Union to consider the use of Soviet launch vehicles.
December 1991	The Soviet Union disintegrates. Russia takes responsibility for the majority of the Soviet space program.
January 1992	The report *A Post Cold War Assessment of U.S. Space Policy* encourages the United States to actively seek cooperation with the former Soviet Union in space.
February 1992	Russia creates the Russian Space Agency.
1 April 1992	Daniel S. Goldin becomes administrator of NASA. He introduces a new management style, "Faster, Better, Cheaper." Cooperation with Russia is seen as an integral part of this ideology.
June 1992	NASA Administrator Goldin and RSA General Director Yuri Koptev meet for the first time and agree that there are many opportunities for enhanced cooperation, particularly in the area of human space flight.
June 1992	U.S. President George Bush and Russian President Boris Yeltsin announce their intention to boost space cooperation between the two countries.
August 1992–December 1993	RKK Energia signs contracts with Lockheed, Rockwell, and NASA regarding the transfer of Soyuz-TM technology and its use as a "lifeboat" for Space Station Freedom.
5 October 1992	Koptev and Goldin sign an agreement concerning the flight of Russian cosmonauts on the U.S. Space Shuttle, the flight of U.S. astronauts on Mir, and a joint mission involving the rendezvous and docking of the Space Shuttle with Mir.

10 November 1992   The Russian Parliament issues a statement emphasizing the need to deepen international cooperation on a mutually advantageous basis and to ensure the fulfillment of Russia's obligations under international agreements.

PRELUDE TO THE INTERNATIONAL SPACE STATION

9 March 1993   President Clinton orders the redesign of Space Station Freedom.

3–4 April 1993   During the Vancouver summit, Clinton indicates to Yeltsin that if Russia agrees to follow the guidelines of the Missile Technology Control Regime the United States will guarantee substantial Russian participation in the Space Station project and permit greatly increased Russian participation in the commercial space-launch market.

March–July, October–November 1993   Representatives from the Russian space program negotiate with their U.S. colleagues regarding the creation of the new International Space Station (ISS) and also regarding joint U.S.–Russian missions to Mir.

April 1993   The Advisory Committee on the Redesign of the Space Station (Vest Committee) is established and promotes broad-scale involvement of Russia in the Space Station program.

July 1993   The Russian–Indian cryogenic engines controversy is resolved.

2 September 1993   Prime Minister Victor Chernomyrdin and Vice President Albert Gore sign the "Joint Statement on Cooperation in Space," which covers future U.S.–Russian cooperation in the area of orbital space stations.

6 October 1993   Russia adopts the "Law of the Russian Federation On Space Activities," which includes international cooperation as a goal.

16 October 1993   NASA and international partners (Canada, Europe, and Japan) agree that an intense process

	is needed at all levels leading to Russia becoming a full partner in the ISS program.
1 November 1993	Goldin and Koptev sign the "Addendum to Program Implementation Plan" containing a detailed description of the three phases in U.S.– Russian cooperation in the ISS program.
6 December 1993	Russia is formally invited to join the ISS program.
11 December 1993	The Russian Federation adopts the "Federal Space Program up to the Year 2000" covering Russian participation in a number of projects, including the ISS program.
February 1994	Sergey Krikalev flies on STS-60, the first joint U.S.–Russian Space Shuttle mission.

IMPLEMENTING THE ISS PARTNERSHIP

18 March 1994	The ISS partners discuss practical steps toward bringing Russia into the program.
5 April 1994	The first joint meeting of all ISS partners, including Russia.
23 June 1994	The "Interim Agreement for the Conduct of Activities Leading to Russian Partnership in Permanently Manned Civil Space Station" and a separate contract for Russian space hardware, services, and data relevant to the Shuttle–Mir program are signed.
February 1995	Second Russian cosmonaut Vladimir Titov flies on the Space Shuttle (STS-63). The Shuttle makes a rendezvous with Mir as a general rehearsal for the Shuttle–Mir docking.
14 March 1995	U.S. astronaut Norm Thagard blasts off in Soyuz-TM-21 with Vladimir Dezhurov and Gennady Strekalov to the Mir space station.
July 1995	Space Shuttle STS-71 docks with Mir, initiating the flight stage of the Shuttle–Mir program.
August 1995	The Russian government issues a number of decrees aimed at fulfilling its commitments to the ISS.

December 1995	Russia proposes using the existing Mir Core Module (CM) for the ISS instead of a Service Module (SM).
9 January 1996	The United States responds skeptically to the Russian idea of replacing the SM with the Mir CM.
April–May 1996	The Russian government issues several decrees aimed at improving the economic situation in the Russian space program so that commitments to the United States can be met.
16 July 1996	Gore, Goldin, Chernomyrdin, and Koptev sign "The Schedule for the Development and Deployment of the Elements of the International Space Station."
7 February 1997	Gore and Chernomyrdin issue a statement con firming the commitment of both sides to the ISS schedule from the beginning of on-orbit assembly through the completion of construction in 2002.
1997	NASA begins working on an Interim Control Module as a potential replacement for the Service Module and reallocates $200 million from the Space Shuttle program to the U.S.–Russian Cooperation budget.
10 February 1997	The Russian government issues Decree No. 153 approving the launch schedule of the initial elements of the ISS.
2 April 1997	The Russian government issues Decree No. 391 authorizing RSA to get funding from commercial banks to fulfill its commitments to the ISS program.
April 1997	Goldin proposes three decision-making points for continued Russian participation in the ISS program.
May 1997	The FGB launch is postponed from November/December 1997 to June 1998; the SM launch is postponed from April 1998 to December 1998.
May 1997	Russian and U.S. space officials agree on the selection of Space Station commanders.
29 January 1998	The *Agreement Among the Government of Canada, Governments of Member States of the*

	European Space Agency, the Government of Japan, the Government of the Russian Federation, and the Government of the United States of America Concerning Cooperation on the Civil Space Station (IGA) and the *Memorandum of Understanding Between the National Aeronautics and Space Administration of the United States of America* and the *Russian Space Agency Concerning Cooperation on the Civil International Space Station* (MOU) are signed, finally codifying Russia's full-member status in the ISS program.
June 1998	NASA opens the Human Space Flight program office in Moscow.
June 1998	Andy Thomas, the last U.S. astronaut to fly on Mir, returns to Earth, completing the Shuttle–Mir program.
July 1998	The Russian government tries to secure broad-based support from within to solve the economic problems of the Russian space program.
July 1998	NASA suggests three options for restructuring cooperation with Russia in the event that delayed delivery of ISS hardware continues.
August 1998	Economic crisis hits Russia and greatly affects Russia's ability to fulfill its ISS commitments.
September 1998	Russia postpones the SM launch to July 1999 while keeping the FGB launch on schedule.
September 1998	NASA proposes reallocating $60 million in ISS resources to Russian Program Assurance in exchange for space in the Russian segment and crew time so that Russia can fulfill its ISS commitments. NASA also considers spending $600 million over the next four fiscal years to buy goods and services from Russia to ensure timely delivery of critical Russian ISS components.
15 October 1998	NASA develops a large-scale contingency plan to address Russian shortfalls.
12 November 1998	The Russian government confirms its support of the ISS program.

20 November 1998   The first ISS element Zarya (FGB) is successfully put into orbit.

5 December 1998   Space Shuttle *Discovery* successfully delivers U.S. element Node 1 to the FGB and docks to the module.

22 January 1999   Prime Minister Eugueni Primakov signs a decree extending Mir's operation through 2002.

INDEX